AF352892

Editors
Javier Menéndez
Mining and Civil Department
SADIM, S.A.
Oviedo
Spain

Jorge Loredo
Mining Exploitation and
Prospecting Department
University of Oviedo
Oviedo
Spain

Editorial Office
MDPI
St. Alban-Anlage 66
4052 Basel, Switzerland

This is a reprint of articles from the Special Issue published online in the open access journal *Applied Sciences* (ISSN 2076-3417) (available at: www.mdpi.com/journal/applsci/special_issues/ Underground_Energy).

For citation purposes, cite each article independently as indicated on the article page online and as indicated below:

LastName, A.A.; LastName, B.B.; LastName, C.C. Article Title. *Journal Name* **Year**, *Volume Number*, Page Range.

ISBN 978-3-0365-1880-0 (Hbk)
ISBN 978-3-0365-1879-4 (PDF)

Contents

About the Editors

Javier Menéndez

Javier Menendez received his M.Sc. degree in Mining Engineering from the School of Mines, Energy and Materials of the University of Oviedo in 2005 and the Ph.D. with outstanding "Cum Laude" in Energy and Mining from the University of Oviedo in 2018. He worked for 15 years at mining and energy companies. He currently works as Engineering Maganer at SADIM, S.A., a public engineering company specialized in mining and civil projects. He has published 12 articles in JCR journals and has participated in 20 international congresses as a speaker, session chair and as a member of the techical committe. His current research interests include renewable energies, underground space and energy storage systems in closed mines. He also currently works as a Professor of the Department of Mining Technology, Topography and Structures at the University of Leon and as an External Professor in the Master of Renewable Energies at the University of Cantabria.

Jorge Loredo

Jorge Loredo received his Degree in Mining Engineering from the University of Oviedo (Spain) in 1976, and Master's in Enviromental Engineering by Minister of Industry of Spain in 1977-1978. He obtained his Ph.D. in Mining Engineering from the University of Oviedo (Spain) in 1981, and Master's in Mining Geology by Paris School of Mines (France) in 1981-1982. He has been Senior Lecturer of Mining and Environmental Engineering at the Oviedo School of Mines of the University of Oviedo (Spain) from 1985 to 2008 and Professor since 2008. He has served as Head of the Department of Mining Engineering of Oviedo University (Spain) between 2004 and 2012. Visiting Researcher in the Department of Geology and Geophysics of the University of California at Berkeley in 1987-1988. He has been the author and/or editor of 13 books and author or co-author of more than 100 book chapters and more than 200 research papers that have appeared in international indexed reference journals and international refereed conference proceedings.

applied sciences

Editorial

Advances in Underground Energy Storage for Renewable Energy Sources

Javier Menéndez [1,*] and Jorge Loredo [2]

[1] Mining Department, SADIM Engineering, 33005 Oviedo, Spain
[2] Mining Exploitation and Prospecting Department, School of Mines, Energy and Materials, University of Oviedo, 33004 Oviedo, Spain; jloredo@uniovi.es
* Correspondence: javier.menendez@sadim.es

Citation: Menéndez, J.; Loredo, J. Advances in Underground Energy Storage for Renewable Energy Sources. *Appl. Sci.* **2021**, *11*, 5142. https://doi.org/10.3390/app11115142

Received: 28 May 2021
Accepted: 31 May 2021
Published: 1 June 2021

Publisher's Note: MDPI stays neutral with regard to jurisdictional claims in published maps and institutional affiliations.

The use of fossil fuels (coal, fuel, and natural gas) to generate electricity has been reduced in the European Union during the last few years, involving a significant decrease in greenhouse gas emissions. The global climate goal would be to reach zero emissions in 2050, and a reduction in the last portion of the CO_2 emissions could come from renewables, green hydrogen, and renewable-based electrification. In the current energy transition towards a sustainable economy, large-scale energy storage systems are required to increase the integration of intermittent renewable energies, such as wind and solar photovoltaics. Underground energy storage systems with low environmental impacts using disused subsurface space may be an alternative to provide ancillary services in the European electricity grids. In this Special Issue, advances in underground pumped storage hydropower, compressed air energy storage, and hydrogen energy storage systems are presented as promising solutions to solve the intermittency problems caused by variable renewable energy sources.

Nowadays, pumped storage hydropower (PSH) is the most mature large-scale form of storage technology. PHS systems are the primary technology used to provide electricity storage services to the grid, accounting for 161 GW of installed global storage capacity. PHS would need to double, reaching 325 GW in 2050. PSH systems consist of two water reservoirs at different heights. The stored energy depends on the mass of water moved and the net hydraulic head between both upper and lower reservoirs. The round trip energy efficiency is between 0.7–0.8. Topographic limitations in flat areas and environmental impacts currently hinder the development of these systems around the world. Conversely, disused underground space could facilitate the installation of underground pumped storage hydropower (UPSH) systems, where at least one water reservoir is underground. Menendez et al. [1] analyzed the economic feasibility of UPSH plants in closed mines providing ancillary services in the Iberian electricity market. Two different options of lower reservoirs were considered: (i) to make use of current mining infrastructure, and (ii) to excavate a new network of tunnels. Secondary regulations, deviation management, and tertiary regulation services considering daily turbine cycle times at full load between 4–10 h were employed to optimize the economic results. Investment costs of 366 M€ were obtained when the existing underground infrastructure was used as lower reservoir. Finally, an internal rate of return of 7.10% was estimated to participate in the Iberian ancillary services markets, considering turbine cycle times at a full load of 8 h. Due to the high investment costs, the profitability is reduced whenever a new reservoir has to be drilled.

The feasibility study of UPSH plants must also include geomechanical and hydrogeological aspects. Menendez et al. [2] studied the geomechanical performance of an underground water reservoir in a closed coal mine. Sandstone and shale rock masses were considered as rock masses to excavate the tunnel networks with a cross-section 30 m^2 and 200 m long. Three-dimensional numerical models were conducted to analyze the deformations and thickness of the plastic zones around the excavations. Systematic grouted rock bolts and reinforced shotcrete were applied as support systems. The results obtained showed that the excavation of the underground reservoir is technically feasible. Pujades et al. [3] carried out a study to determine the impact of hydrogeological features on the performance of an underground pumped storage hydropower plant in Belgium. The subsurface water exchange between the surrounding medium and the lower reservoir of UPSH plants was investigated. They developed a numerical study to evaluate the influence of groundwater exchanges of UPSH plants using abandoned mines as lower reservoirs. The hydraulic conductivity and the elevation of the piezometric head were analyzed. They concluded that water quality can deteriorate under the influence of UPSH systems when abandoned coal mines are used as lower water reservoirs. Dianellou et al. [4] carried out research considering large-scale wind and photovoltaic power plants and the potential contribution of PSH plants in the Greek power system. They concluded that the increase of PSH systems is required to integrate large-scale wind and photovoltaic power plants in non-interconnected grids.

Compressed air energy storage (CAES) systems consist of one underground reservoir where the compressed air is stored at high pressures. The pressurized air is released and expanded in the turbines in on-peak periods to produce electricity. Currently, there are two commercial diabatic compressed air energy storage (D-CAES) plants using abandoned salt caverns as subsurface reservoirs. The round trip efficiency of D-CAES systems is lower than PSH, reaching typical values of 0.4–0.5. Unlike D-CAES systems, adiabatic compressed air energy storage (A-CAES) systems include a thermal energy storage system, and therefore, fossil fuels are not required to heat the compressed air before the expansion in the gas turbines. Some researchers have determined that the global efficiency of A-CAES systems can reach 0.7–0.8. Prado et al. [5] investigated the thermodynamic performance of A-CAES plants in abandoned mines. An underground reservoir in lined mining tunnels at operating pressures from 5 to 8 MPa and two different sealing layers was considered in the simulations. Analytical and CFD numerical models were conducted for 100 charge (consumption) and discharge (generation) processes. They concluded that the air temperature and the heat transfer through the sealing layer depends on the sealing layer's thermal conductivity. Evans et al. [6] developed a study about the exergy storage capacity potential in United Kingdom's massively bedded halites. Massively bedded halite deposits existing in the UK were considered as CAES reservoirs. They concluded that the exergy storage capacity in salt caverns could provide important support to the electricity grid.

Hydrogen energy storage is a form of chemical energy storage in which the electrical power of renewable energies is converted into hydrogen. High pressures (35–70 MPa) are required to store hydrogen as a gas. Gajda and Lutyński [7] carried out an experimental study to compare the hydrogen permeability considering different materials, such as concrete, polymer concrete, epoxy resin, salt rock, and mudstone. The results obtained showed that epoxy resin can be a promising sealing liner for hydrogen storage. Hydrogeological concerns are also very important to determine the feasibility of hydrogen energy storage. Lafortune et al. [8] simulated a sudden hydrogen leak into an aquifer in France. They carried out an injection test of organic and ionic tracers and helium-saturated water to design the future protocol related to hydrogen storage.

Funding: This research received no external funding.

Conflicts of Interest: The authors declare no conflict of interest.

References

1. Menéndez, J.; Fernández-Oro, J.M.; Loredo, J. Economic Feasibility of Underground Pumped Storage Hydropower Plants Providing Ancillary Services. *Appl. Sci.* **2020**, *10*, 3947. [CrossRef]
2. Menéndez, J.; Schmidt, F.; Konietzky, H.; Bernardo Sánchez, A.; Loredo, J. Empirical Analysis and Geomechanical Modelling of an Underground Water Reservoir for Hydroelectric Power Plants. *Appl. Sci.* **2020**, *10*, 5853. [CrossRef]
3. Pujades, E.; Poulain, A.; Orban, P.; Goderniaux, P.; Dassargues, A. The Impact of Hydrogeological Features on the Performance of Underground Pumped-Storage Hydropower (UPSH). *Appl. Sci.* **2021**, *11*, 1760. [CrossRef]

4. Dianellou, A.; Christakopoulos, T.; Caralis, G.; Kotroni, V.; Lagouvardos, K.; Zervos, A. Is the Large-Scale Development of Wind-PV with Hydro-Pumped Storage Economically Feasible in Greece? *Appl. Sci.* **2021**, *11*, 2368. [CrossRef]
5. Prado, L.; Menéndez, J.; Bernardo-Sánchez, A.; Galdo, M.; Loredo, J.; Fernández-Oro, J. Thermodynamic Analysis of Compressed Air Energy Storage (CAES) Reservoirs in Abandoned Mines Using Different Sealing Layers. *Appl. Sci.* **2021**, *11*, 2573. [CrossRef]
6. Evans, D.; Parkes, D.; Dooner, M.; Williamson, P.; Williams, J.; Busby, J.; He, W.; Wang, J.; Garvey, S. Salt Cavern Exergy Storage Capacity Potential of UK Massively Bedded Halites, Using Compressed Air Energy Storage (CAES). *Appl. Sci.* **2021**, *11*, 4728. [CrossRef]
7. Gajda, D.; Lutyński, M. Hydrogen Permeability of Epoxy Composites as Liners in Lined Rock Caverns—Experimental Study. *Appl. Sci.* **2021**, *11*, 3885. [CrossRef]
8. Lafortune, S.; Gombert, P.; Pokryszka, Z.; Lacroix, E.; Donato, P.; Jozja, N. Monitoring Scheme for the Detection of Hydrogen Leakage from a Deep Underground Storage. Part 1: On-Site Validation of an Experimental Protocol via the Combined Injection of Helium and Tracers into an Aquifer. *Appl. Sci.* **2020**, *10*, 6058. [CrossRef]

 *applied sciences*

Article

Economic Feasibility of Underground Pumped Storage Hydropower Plants Providing Ancillary Services

Javier Menéndez [1],[*] , Jesús Manuel Fernández-Oro [2] and Jorge Loredo [3]

1   Hunaser Energy, Avda. Galicia 44, 33005 Oviedo, Spain
2   Energy Department, University of Oviedo, 33271 Gijón, Spain; jesusfo@uniovi.es
3   Mining Exploitation Department, University of Oviedo, 33004 Oviedo, Spain; jloredo@uniovi.es
*   Correspondence: jmenendezr@hunaser-energia.es; Tel.: +34-98-510-7300

Received: 14 May 2020; Accepted: 5 June 2020; Published: 6 June 2020

Abstract: The electricity generated by some renewable energy sources (RESs) is difficult to forecast; therefore, large-scale energy storage systems (ESSs) are required for balancing supply and demand. Unlike conventional pumped storage hydropower (PSH) systems, underground pumped storage hydropower (UPSH) plants are not limited by topography and produce low environmental impacts. In this paper, a deterministic model has been conducted for three UPSH plants in order to evaluate the economic feasibility when considering daily turbine cycle times at full load (DTCs) between 4 and 10 h. In the model, the day-ahead and the ancillary services markets have been compared to maximize the price spread between the electricity generated and consumed. Secondary regulation, deviation management and tertiary regulation services have been analyzed to maximize the income and minimize the cost for purchasing energy. The capital costs of an open-loop UPSH plant have been estimated for the case of using the existing infrastructure and for the case of excavating new tunnels as lower reservoirs. The net present value (NPV), internal rate of return (IRR) and payback period (PB) have been obtained in all scenarios. The results obtained show that the energy generation and the annual generation cycles decrease when the DTC increases from 4 to 10 h, while the NPV and the IRR increase due to investment costs. The investment cost of a 219 MW UPSH plant using the existing infrastructure reaches 366.96 M€, while the NPV, IRR and PB reached 185 M€, 7.10% and 15 years, respectively, participating in the ancillary services markets and considering a DTC of 8 h.

Keywords: energy storage; underground pumped storage; economic feasibility; ancillary services; day-ahead market; underground space

---

1. Introduction

The rapid growth of intermittent renewable energy sources (RESs) for electricity generation requires flexible large-scale energy storage systems (ESSs). Electricity generated by some forms of RESs, such as wind or solar photovoltaic (PV), is difficult to forecast; therefore, ESSs are required for balancing electricity supply and demand [1]. Pumped storage hydroelectricity (PSH) is the most mature and efficient storage technology and accounts for 98% of storage capacity worldwide [2]. However, the development of new PSH projects is limited by topographic and environmental restrictions. Thus, disused underground space may be used as reservoir for large-scale storage systems such as underground pumped storage hydropower (UPSH) or adiabatic compressed air energy storage (A-CAES) [3,4] where the typical round trip energy efficiencies exceed 0.7–0.8 [5,6]. Unlike conventional PSH plants, which consist of two water reservoirs located at the surface, both upper and lower reservoirs may be underground in the case of UPSH plants. For that purpose, two different options may be considered: (1) to make use of existing infrastructure; or (2) to dig new tunnels. Winde et al. [7,8] explored the use of deep-level gold mines in South Africa for UPSH schemes. Pujades et al. [9,10] carried out a study considering the use of a closed slate mine located at Belgium with a capacity of 550,000 m^3 as a lower reservoir for UPSH. Bodeux et al. [11] analyzed the interactions between groundwater and the slate chambers used as a subsurface water reservoir. Closed coal mines in Spain and Germany have also been proposed as underground reservoirs for UPSH [12–15]. Wong [16] proposed digging new tunnels or shafts as a lower reservoir for UPSH in the Bukit Timah granite of Singapore. The economic feasibility of UPSH plants depends on the capital costs and the price spread between the electricity generated and consumed in turbine and pumping modes, respectively. In the Iberian electricity system, RES generation is granted priority during the dispatch and receives a fixed feed-in tariff. The day-ahead prices (spot prices), are set around noon on the day preceding the delivery. The day-ahead markets are complemented by intraday markets and ancillary services in the case of unforeseen events and changing weather conditions, which mainly could affect wind and solar PV generation. Finally, the balance of the electricity demand and supply is achieved through the ancillary services, which are managed by the system operator, taking the form of auctions in the Iberian electricity system.

Traditionally, a PSH plant has been operated by the price-arbitrage strategy. PSH plants have participated in the day-ahead market, selling the electricity generated at peak periods (peak price hours) and purchasing the electricity at off-peak periods [17,18]. However, the electricity price spread between the on-peak periods and off-peak periods has been reduced significantly, and the economic feasibility of a PSH might not be guaranteed participating just in the day-ahead electricity market. Lobato et al. [19] carried out an overview of ancillary services in Spain, including a technical description and the management of the ancillary services markets. The study highlights the paramount importance of the ancillary services in a power system. Pérez-Díaz et al. [20] reviewed the trends and challenges of PSH plants with the aim of optimizing their operation in the balancing markets. The study showed that ancillary services markets, particularly those related to balancing supply and demand, emerge as a valuable source of income for PSH plants.

Krishnan and Das [21] studied the feasibility of CAES plants participating in the day-ahead and balancing markets of PJM and Midcontinent Independent System Operator (MISO), concluding that the profit may be increased 10-fold by providing ancillary services. Berrada et al. [22] estimated the income of different ESSs (PSH, CAES and gravity energy storage) that participate in the day-ahead market, the real-time energy market and the regulation market of the New York Independent System Operator (NYISO). The results obtained showed that PSH and CAES may be economically feasible when operating in the regulation market. Chazarra et al. [23] studied the economic viability of twelve PSH plants participating in the secondary regulation of the Iberian electricity system. The PSH plants were equipped with different fixed-speed and variable-speed units and with and without considering hydraulic short-circuit operation. They concluded that PSH plants equipped with variable speed technology, along with full converters with and without the possibility to operate in hydraulic short-circuit mode, and the PSH plants with ternary units obtain the lowest payback periods.

Recently, Maciejowska et al. [24] developed a model that is able to predict the price spread between the day-ahead prices and the corresponding volume-weighted average intraday markets in the German electricity system. The research concluded that the sign of the price spread can be successfully predicted with econometric models, such as ARX and probit. Ekman and Jensen [25] conducted a study of a generic ESS with a global energy efficiency of 0.7 participating in the day-ahead market and some balancing markets in Denmark. They concluded that only UPSH might be profitable as long as it participates both in the day-ahead and ancillary services markets. The contribution of a variable speed PSH to increasing the revenue has been assessed for participation in the day-ahead and secondary regulation reserve markets in the Iberian electricity system [26]. Chazarra et al. [27] developed a stochastic model for the weekly scheduling of a hydropower system to optimize the revenue in the Spanish electricity system. The obtained solution protects a multireservoir system against risk of water and storage unavailability. The effect of the complementarity between the variable renewable energy sources and the load on the flexibility of the power system was examined in the Korean electricity system [28]. They examined an optimal mix ratio between the wind and solar PV and concluded that the ratios of the wind and solar PV to the total variable generation resource were 1.3% and 93.4%, respectively. Lago et al. [29] proposed a deep neural network by using Bayesian optimization and functional analysis of variance to improve the predictive accuracy in a day-ahead energy market. Chazarra et al. [30] estimated the maximum theoretical income of a PSH plant participating as a price-taker in the day-ahead and the secondary regulation reserve markets while considering different configurations of power plants in the Iberian electricity system. The results obtained demonstrate that the operation with the variable speed technology could be of considerable help in enlarging the income of the hydropower plant.

In this work, the economic feasibility of three UPSH plants is analyzed considering DTCs between 4 and 10 h. The generation and consumption of electricity and the number of annual generation cycles have been estimated assuming a round trip energy efficiency of 0.77. The day-ahead and ancillary services markets in the Iberian electricity system are analyzed for optimizing the profitability of investment. Secondary regulation, deviation management and tertiary regulation services have been considered to maximize the income from selling energy and minimize the cost for purchasing energy. The capital costs of UPSH plants have also been calculated for the first time, considering the case of using the existing infrastructure and the case of excavating new tunnels or caverns as a lower water reservoir. Finally, a profitability analysis has been carried out using the net present value (NPV), internal rate of return (IRR) and payback period (PB).

2. Methodology

A three-step scheme of the methodology proposed to study the economic feasibility and profitability of UPSH plants is shown in Figure 1.

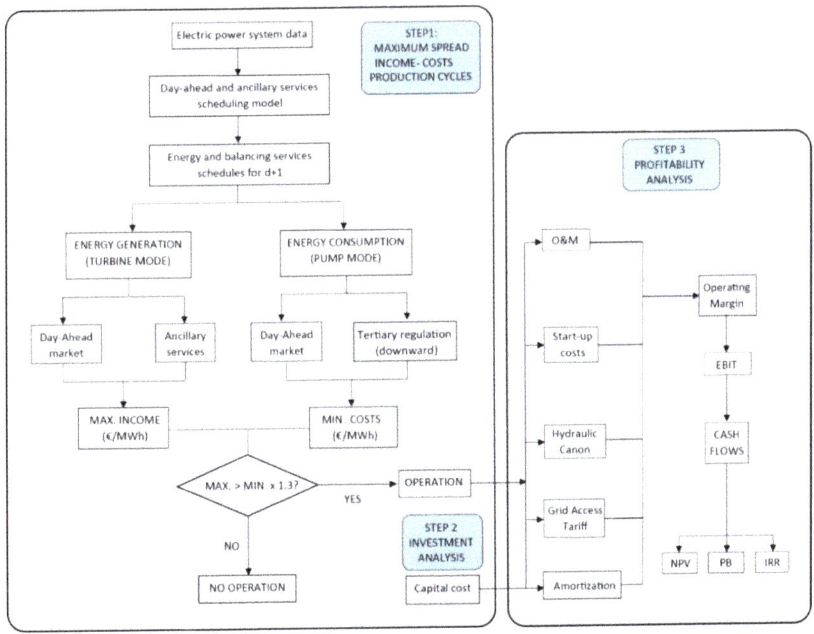

Figure 1. Methodology scheme to analyze the economic feasibility of underground pumped storage hydropower (UPSH) plants.

The model considers the electricity prices in the day-ahead and ancillary services markets hour by hour for a time period of three years (2016–2018). In step 1, the maximum and minimum prices are analyzed for the energy generated and consumed in the Iberian electricity system, considering DTCs between 4 and 10 h. The number of annual generation cycles and the amount of energy generated are obtained for all scenarios. In step 2, the investment costs of a UPSH plant are estimated for the case of using the existing infrastructure and the case of excavating new tunnels or caverns as a lower reservoir. Amortization costs are calculated considering a typical operation period of 35 years. Finally, in step 3, the operating margin, which is defined as the income from selling energy minus the costs for purchasing energy and the operation costs (O&M, start-up costs, hydraulic cannon and grid access tariffs), is estimated for the three hydropower stations (HPSs) considered in this study.

The earnings before interest and taxes (EBIT) and the cash flows have been calculated to analyze the economic feasibility and the profitability. The NPV, i.e., the difference between the present value of cash inflows and the present value of cash outflows over a period of time; IRR, i.e., a discount rate that makes the NPV of all cash flows from a project equal to zero; and PB, i.e., the time in which the initial outlay of an investment is expected to be recovered, have been calculated in all scenarios.

2.1. Technical Data of UPSH Plants

Figure 2 shows two different schemes of UPSH plants in a closed mine. Figure 2a shows a UPSH scheme with an upper surface reservoir and an underground lower reservoir. Conversely, Figure 2b depicts a shallow upper reservoir and an underground lower reservoir. Note that the gross head is reduced when a shallow upper reservoir is considered. This reduction in gross head implies a decrease in storable amount of energy.

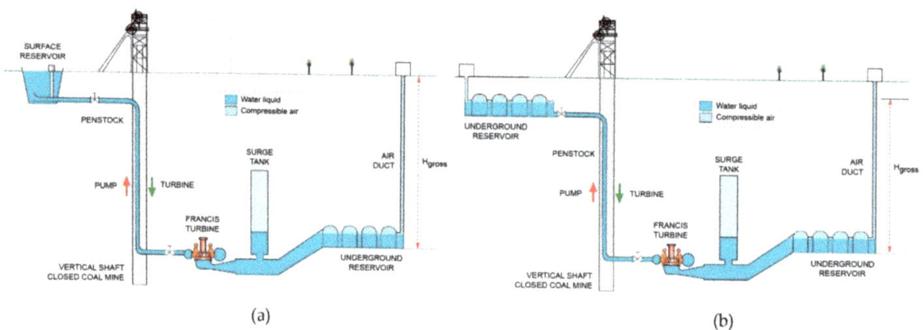

Figure 2. UPSH scheme in a closed mine. (**a**) Surface upper reservoir and underground lower reservoir; (**b**) Shallow upper reservoir and underground lower reservoir.

The technical data of the three hydropower stations (HPSs) are shown in Figure 3, considering DTCs between 4–10 h day^{-1}. The water flow rate is lower in pumping mode and therefore the pumping cycle time at full load increases to 5.40 and 13.30 h day^{-1}. The storage capacities are 0.46, 0.8 and 1.6 Mm^3 for HPS 1, HPS 2 and HPS 3, respectively, and remain constant in each DTC (4–10 h). Although the amount of storable energy does not vary, the output powers decrease from 125, 219 and 440 MW to 50, 87 and 176 MW for HPS 1, HPS 2 and HPS 3, respectively, when the turbine cycle time increases from 4 to 10 h day^{-1}. A maximum gross pressure of 4.41 MPa has been considered. The water flow rates in turbine mode are 12.78, 22.23 and 44.45 m^3 s^{-1} for HPS 1, HPS 2 and HPS 3, respectively, considering a DTC of 10 h, while the efficiency values are 0.91 and 0.90 in turbine and pump mode, respectively. The water flow rates in pumping mode are 9.61, 16.71 and 33.42 m^3 s^{-1} for HPS 1, HPS 2 and HPS 3, respectively, when the pumping cycle time at full load is 13.30 h day^{-1}. In addition, the round trip energy efficiency is assumed to be 0.77 [6].

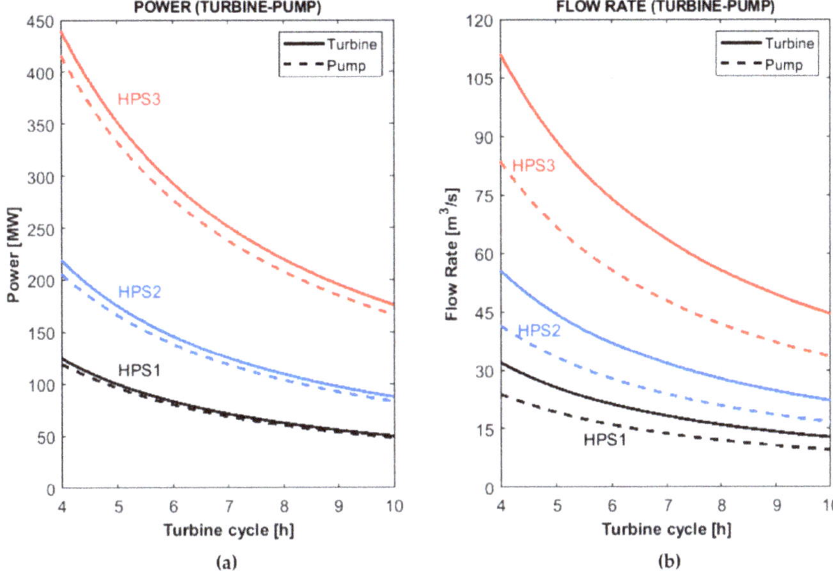

Figure 3. Technical data of HPS 1, HPS 2 and HPS 3, considering daily turbine cycle times at full load (DTCs) between 4–10 h day^{-1}. (**a**) Power in turbine and pump modes; (**b**) water flow rate in turbine and pump modes.

2.2. Electric Power System Data

The economic feasibility of UPSH plants is obtained considering the historical hourly values of the prices in the day-ahead and ancillary services markets in the Iberian electricity system during a period of three years (2016–2018). The main objective of this work is to estimate the maximum income from selling energy (turbine mode) and the minimum cost for purchasing energy (pumping mode) while participating in the day-ahead and balancing markets. The following markets and services for the mentioned years have been studied: (i) the day-ahead market; (ii) the upward secondary regulation; (iii) the upward deviation management; and (iv) the tertiary regulation (upward and downward).

The overall amounts of energy managed in secondary regulation, deviation management and tertiary regulation in 2018 were 2592, 2358 and 3301 GWh, respectively. The purpose of the secondary regulation service is to maintain the generation-demand balance, automatically correcting deviations with respect to the anticipated power exchange schedule and the system frequency deviations. The objective of the deviation management service is to resolve the deviations between generation and demand which could appear in the period between the end of one intraday market and the beginning of the next intraday market horizon. Finally, the purpose of the tertiary regulation service is to resolve the deviations between generation and consumption and the restoration of the secondary control band reserve used. These ancillary services are managed and remunerated by market mechanisms in the Iberian electricity system.

3. Results and Discussion

3.1. Investment Cost of UPSH Plants

In this section, the investment costs of UPSH plants are estimated for the case of using existing underground infrastructure and the case of digging new tunnels or caverns as a lower reservoir. Currently, there are no research works where the investment costs of UPSH plants are analyzed. However, several authors have studied the investment cost of conventional PSH plants. Connolly et al. [18] proposed a range between 0.47 and 2.17 M€ MW^{-1} from projects in some countries in the European Union. Steffen [31] proposes a range between 0.77 and 1.28 M€ MW^{-1} from projects of PSH in Germany and Luxembourg. An increase between 7 and 15% is produced when the PSH plant is equipped with variable speed units. The investment cost of a UPSH plant depends strongly on the facility's location and type of underground infrastructure.

The investment costs of a UPSH plant in a closed mine considering the case of excavation of new tunnels and the case of using existing infrastructure are shown in Table 1. A Francis pump-turbine (FT) with a maximum output power of 219 MW and a maximum input power of 208 MW, a water flow rate in turbine mode of 55.56 m^3 s^{-1} (turbine cycle of 8 h at full load) and a water reservoir capacity of 1.6 Mm^3 have been considered. In addition, in both scenarios, due to the dimensions and weight of the hydropower equipment (FT and motor-alternator), the excavation of a new access tunnel from the surface to the powerhouse is required. This tunnel would be used during the construction and operation phases. The investment cost reaches 687.34 M€, which represents a unit cost per MW of installed power of 3138 € kW^{-1}. The main cost corresponds to civil works, reaching 525.69 M€ and representing 76.48% of the total investment cost. The cost of excavating the new tunnels as a lower reservoir reaches 453.33 M€, which is 86.23% of the total civil works cost. Conversely, the construction and waterproofing of the upper reservoir reach 30.73 M€. More details can be found in Appendix A, where the capital cost of a UPSH plant are detailed.

When the existing infrastructure is used as a lower reservoir, the investment costs are reduced to 366.96 M€, representing 46.66% less than the previous investment cost. The existing underground infrastructure would be secured with a reinforced shotcrete layer and waterproofed. Consequently, the unit cost per MW of installed capacity is also reduced to 1675 € kW^{-1}. The cost of civil works decreases down to 218.08 M€, which is 59.40% of the total cost. The cost of conditioning the lower reservoir reaches 145.73 M€, representing 66.82% of the total civil works cost. Electrical grid connection cost includes the electric substation located in the UPSH plant and the electric power line. The excavation materials of the upper and lower reservoirs could be used for restoring the open pit mines existing in the study area. As indicated in Figure 2b, both upper and lower reservoirs could be underground. In this scenario, where the environmental impact is reduced, the investment cost reaches 1076 M€.

Table 1. UPSH investment costs analysis for construction of new tunnels or caverns and for making use of existing infrastructure.

UPSH Investment Costs (k€)	New Tunnels	Existing Infrastructure
Civil works	525,692.80	218,083.58
Hydromechanical equipment and penstock	16,986.52	16,986.52
Hydropower equipment	99,709.12	99,709.12
Electrical grid connection	4510.50	4510.50
Commissioning	4140.85	3767.38
Project management	36,301.30	23,907.56
Total (k€)	**687,341.10**	**366,964.67**

3.2. Estimation of Income and Expenses

The storable amount of energy in UPSH plants depends on the net head and the water mass moved. The maximum income from selling energy has been studied considering the amount of electricity generated in turbine mode considering the maximum prices in the day-ahead and ancillary services markets. Likewise, the downward tertiary regulation service and the day-ahead market have been analyzed to obtain the minimum price for purchasing energy. Figure 4a shows the maximum prices in the day-ahead and ancillary services and the minimum prices in the downward tertiary regulation service, considering DTCs between 4 and 10 h. The price spread between the day-ahead and the ancillary services markets for generation and consumption modes is shown in Figure 4b.

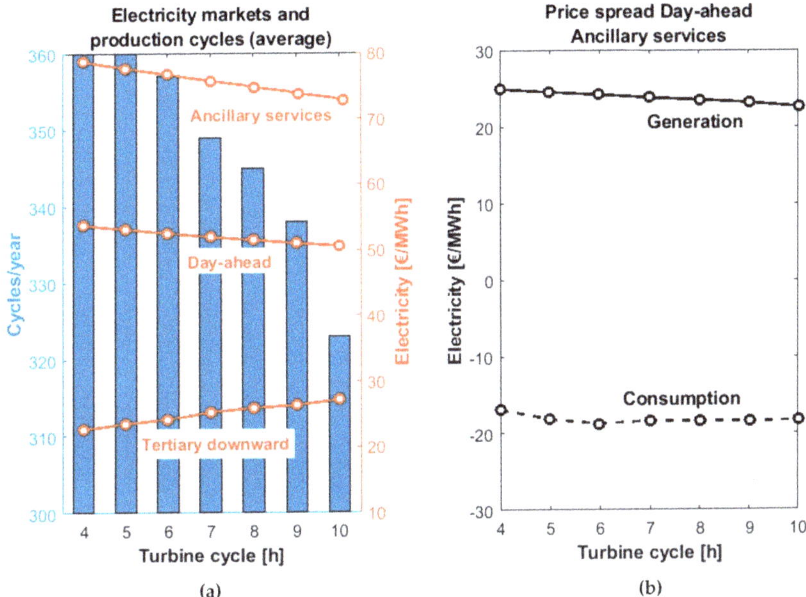

Figure 4. Analysis of electricity markets. (**a**) Maximum electricity prices in the day-ahead and ancillary services markets; minimum costs in the downward tertiary regulation service and annual generation cycles. (**b**) Price spread between the day-ahead and the ancillary services markets.

Concerning Figure 4a, the number of annual generation cycles is calculated to obtain the amount of electricity generated in turbine mode. The electricity consumed in pumping mode is estimated assuming a round trip energy efficiency of about 0.77. The annual generation cycles decrease from 360 to 323 when the DTC increases from 4 to 10 h. The maximum price of electricity is 78.74 € MWh^{-1} and is reached when participating in the ancillary services markets. That value, which corresponds to a DTC of 4 h, is progressively reduced by 7% when the DTC increases to 10 h. The minimum price in consumption mode is 22.13 € MWh^{-1}. This cost for purchasing energy is increased by 22.81%, reaching 27.18 € MWh^{-1} when the DTC increases to 10 h. In Figure 4b, it is shown that the maximum price spread between the day-ahead and the ancillary service markets in generation mode is 24.92 € MWh^{-1}. In consumption mode, the maximum price spread reaches −16.88 € MWh^{-1}. The price spread decreases by 2.36 € MWh^{-1} in generation mode and 1.42 € MWh^{-1} in consumption mode when the DCT increases to 10 h.

For the purpose of analyzing the economic feasibility of the three HPSs considered in this study, the amount of energy generated and consumed and the maximum income from selling energy in generation mode (turbine) and minimum costs for purchasing energy in consumption mode (pumping) are shown in Figure 5 for the three HPSs. The maximum income decreases while the minimum expenses increase as the DTC increases from 4 to 10 h (see Figure 5d).

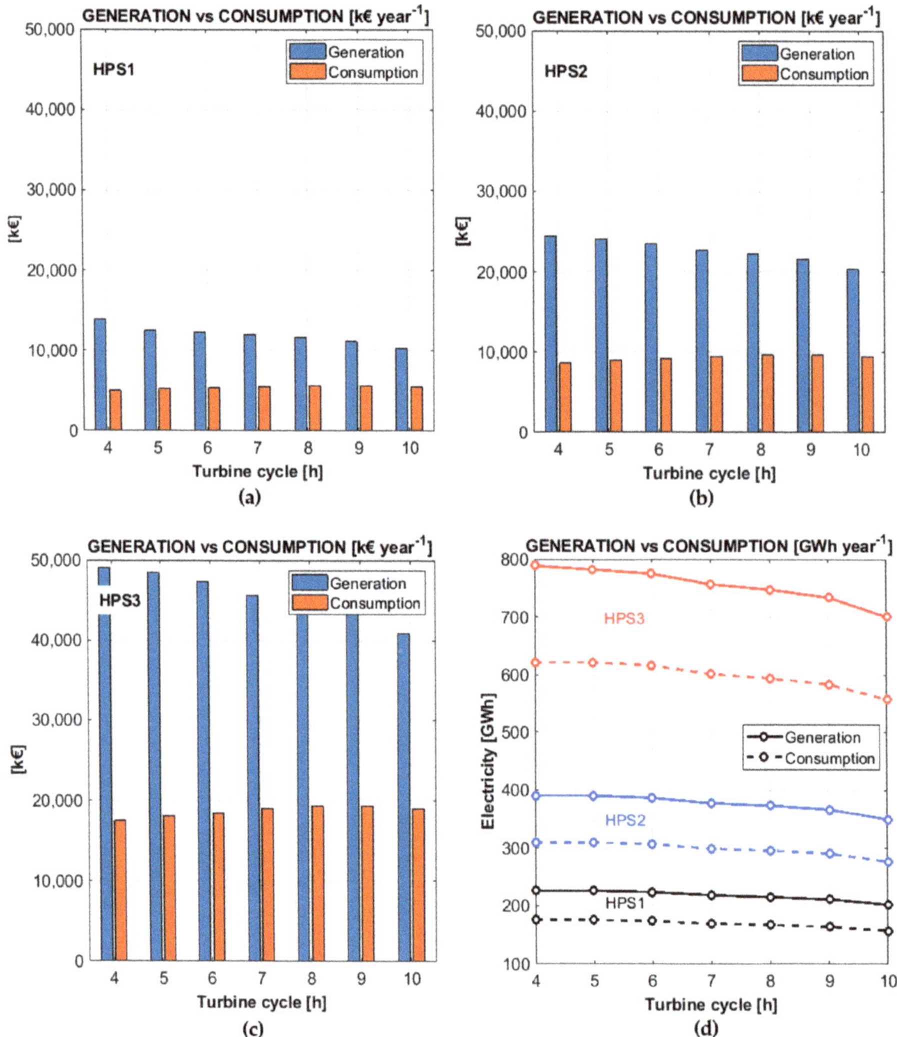

Figure 5. Maximum income from selling energy and minimum cost for purchasing energy, considering DTCs between 4 and 10 h: (**a**) HPS 1; (**b**) HPS 2; (**c**) HPS 3. (**d**) Amount of energy generated and consumed for HPS 1, HPS 2 and HPS 3.

The design of the DTC influences the annual number of production cycles and therefore the amount of electricity generated. The electricity generated is reduced from 226.46 to 203.00 GWh year^{-1} in HPS 1, from 391.19 to 350.16 GWh year^{-1} in HPS 2 and from 789.73 to 701.32 GWh year^{-1} in HPS 3 when the DTC increases from 4 to 10 h. The maximum theoretical income reaches 49.06 M€ year^{-1} and is obtained in HPS 3 when the DTC is 4 h. The maximum income is reduced by 16.70% and the minimum costs are increased by 8.56% in HPS 3 when the DTC is increased to 10 h. Consequently, the spread between the income from selling electricity and the cost for purchasing electricity is also reduced from 31.59 to 21.91 M€ year^{-1} in HPS 3 when the DTC increases to 10 h. The maximum incomes reach 13.91 and 24.41 M€ year^{-1} in HPS 1 and HPS 2 when the DTC is 4 h. In addition, the spreads between the income and expenses are reduced by 45.86% and 31.12% in HPS 1 and HPS 2, respectively, when the DTC increases from 4 to 10 h day^{-1}.

As a presumable guideline, Table 2 shows the maximum theoretical income and the expenses of the three HPSs when considering a DTC of 8 h. In this scenario, the maximum output powers (turbine mode) for HPS 1, HPS 2 and HPS 3 are 62, 109 and 219 MW, respectively, while the maximum input powers (pumping mode) are 53, 92 and 185 MW for HPS 1, HPS 2 and HPS 3, respectively. The maximum income and the minimum cost for purchasing energy are obtained participating in the day-ahead and the ancillary services markets. The costs for purchasing energy represent 84%, 85% and 87% of the total costs in HPS 1, HPS 2 and HPS 3, respectively. In addition, O&M costs, start-up costs, grid access tariff and hydraulic cannon have been considered. As established in Spanish electrical regulation, a cost of 0.5 € MWh^{-1} has been considered as grid access tariff. O&M costs include personnel, insurance, spare parts and external service costs. Finally, the operating margins (income from selling electricity minus the cost for purchasing electricity and operation costs) have been estimated, reaching 6.07, 10.99 and 22.71 M€ year^{-1} in HPS 1, HPS 2 and HPS 3, respectively. Repeating all these considerations when a DTC of 4 h is designed, the operating margins increase to 7.83, 14.11 and 28.81 M€ year^{-1} in HPS 1, HPS 2 and HPS 3, respectively.

Table 2. Income and operation costs of HPS 1, HPS 2 and HPS 3, considering a DTC of 8 h.

UPSH Operation	HPS 1	HPS 2	HPS 3
Electricity generation income (k€ year^{-1})	11,602.55	22,222.67	44,655.73
Electricity consumption costs (k€ year^{-1})	5582.04	9642.45	19,284.89
O&M (k€ year^{-1})	464.10	651.75	961.79
Start-up costs (k€ year^{-1})	202.88	296.79	417.47
Grid access tariff (k€ year^{-1})	84.47	148.27	297.95
Hydraulic cannon (k€ year^{-1})	278.51	488.90	982.43
Operating margin (k€ year^{-1})	6047.60	10,994.52	22,711.22

3.3. Feasibility Analysis

The previous section has revealed that the maximum spreads between the income from selling electricity and the cost for purchasing electricity when participating in the ancillary services markets as well as the operating margins of UPSH plants are significantly reduced when the designed DTC increases from 4 to 10 h. However, when the DTC increases, the output power of the FT decreases and the investment costs of UPSH are reduced. This means that the economic feasibility of UPSH plants strongly depends on the investment costs. To introduce this variable in the analysis, the EBIT, PB, NPV and IRR are shown in Figure 6 for HPS 3, considering an investment cost between 1000 and 3000 € kW^{-1} for DTCs between 4 and 10 h. In this scenario, a UPSH plant with surface upper reservoir is considered (Figure 2a). Although the lifetime of UPSH plants could be between 60 and 100 years, a typical operation time of 35 years has been considered here in order to calculate the amortization costs [32,33]. Due to amortization costs, the EBIT decrease in all DTCs when the investment costs increase from 1000 to 3000 € kW^{-1} (see Figure 6a). The red line represents the limits where the EBIT turn into negative numbers. Precisely, the minimum EBIT are −8.32 M€ year^{-1} when the investment cost is 3000 € kW^{-1} and the DTC is 4 h.

In Figure 6b, a minimum PB of 9 years is obtained when the investment cost is 1000 € kW^{-1} and the DTC is 10 h. The PB increases to 34 years when the investment cost is 3000 € kW^{-1} and the DTC is 4 h. PBs lower than or equal to 20 years are reached when the investment costs are lower than 2000 € kW^{-1} and the DTC is greater than 6 h. In Figure 6b, the NPV decreases sharply when the investment cost increases. Again, a red line represents the border between positive and negative values. An NPV of −581 M€ is reached when the investment cost is 3000 € kW^{-1}. The NPV increases to 294 M€ when the investment cost is reduced to 1000 € kW^{-1} and the designed DTC is 10 h. Finally, a maximum IRR of 12.92% is obtained when the investment costs are 1000 € kW^{-1} and the DTC is 10 h.

IRRs greater than or equal to 6% are reached when the investment costs are lower than 2000 € kW^{-1} and the DTC is greater than 5 h.

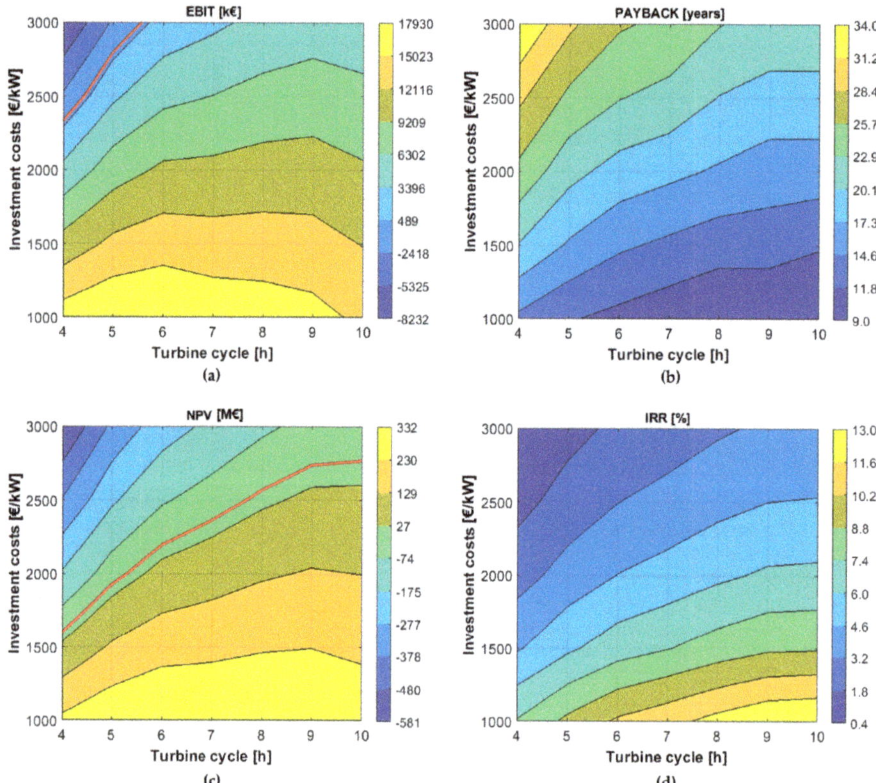

Figure 6. Feasibility analysis of HPS 3 with surface upper reservoir, considering DTCs between 4 and 10 h and investment costs between 1000 and 3000 € kW^{-1}. (**a**) Earnings before interest and taxes (EBIT), where the red line represents zero EBIT; (**b**) payback period; (**c**) net present value (NPV), where the red line represents zero NPV; (**d**) internal rate of return.

According to the calculated investment costs, the profitability of each HPS is finally presented in Table 3, considering a DTC of 8 h as well as the digging of new caverns or tunnels as a lower reservoir for UPSH plants, where an investment cost of 3138 € kW^{-1} was estimated in Section 3.1. A pumping cycle time of 10.64 h has been assumed. As indicated previously, maximum output powers of 62, 109 and 219 MW have been considered for HPS 1, HPS 2 and HPS 3, respectively. The EBIT increase from 554.85 k€ year^{-1} in HPS 1 to 3335.28 k€ year^{-1} in HPS 3. The NPVs are negative in all HPSs, while the IRRs obtained are lower than 2.75%. Finally, high PBs (greater than 24 years) have been reached. Therefore, the construction of new underground infrastructure as the lower reservoir of a UPSH plant is not economically feasible.

Table 3. Profitability analysis of HPS 1, HPS 2 and HPS 3, considering the construction of new tunnels or caverns, DTCs of 8 h and a lifetime of 35 years.

EBIT and Profitability	HPS 1	HPS 2	HPS 3
EBIT (k€ year^{-1})	554.85	1352.20	3335.28
NPV (k€)	−41,570.22	−65,194.66	−118,307.90
IRR (%)	2.43%	2.61%	2.75%
PB (years)	25	25	24

The profitability analysis of HPS 1, HPS 2 and HPS 3 is shown in Table 4, considering an investment cost of 1675 € kW^{-1} (estimated in Section 3.1) and a DTC of 8 h. The results obtained when the existing underground infrastructure is used are much better than those of the previous scenario. The maximum NPV is 185.70 M€ and has been obtained in HPS 3, while the IRR increases to 7.10%. The minimum IRR is reached in HPS 1 (6.6%). Finally, the minimum PBs decrease to 15 years in all HPSs. This demonstrates that a UPSH plant could become economically feasible using the existing infrastructure and participating in the ancillary services markets.

Table 4. Profitability analysis of HPS 1, HPS 2 and HPS 3, making use of existing infrastructure and considering DTCs of 8 h and a lifetime of 35 years.

EBIT and Profitability	HPS 1	HPS 2	HPS 3
EBIT (k€ year^{-1})	3115.57	5847.65	12,368.74
NPV (k€)	44,614.77	86,094.49	185,702.73
IRR (%)	6.66%	6.90%	7.10%
PB (years)	15	15	15

3.4. Design Considerations of UPSH Plants

The proper dimensioning of a UPSH plant is highly important for optimizing its economic feasibility. The net head and the amount of water moved must be maximized for increasing the storable amount of energy and the economic feasibility. Although the amount of energy remains constant, different dimensioning of UPSH could be carried out depending on the DTC. When the DTC increases, the output power of the FT decreases, consequently reducing the investment costs. DTCs greater than 7 h would be suitable for UPSH plants. The profitability also depends on the lifetime considered. Therefore, the profitability analysis was carried out considering lifetimes of 50 and 75 years. In addition, when the lifetime is increased, the amortization costs decrease and the EBIT increase.

Table 5 shows the NPVs, IRRs and PBs for HPS 1, HPS 2 and HPS 3 when considering the construction of new tunnels or caverns as a lower reservoir. The NPVs and IRRs of HPS 1, HPS 2 and HPS 3 increase with respect to the scenario that considers a lifetime of 35 years. However, IRRs lower than 5% and NPVs lower than 167.88 M€ are reached for 75 years, and therefore the projects are not economically feasible.

Table 5. Profitability analysis of HPS 1, HPS 2 and HPS 3, considering the construction of new tunnels, DTCs of 8 h and lifetimes of 50 and 75 years.

Lifetime	50 Years			75 Years		
HPS	HPS 1	HPS 2	HPS 3	HPS 1	HPS 2	HPS 3
NPV (k€)	−3017.07	4110.40	23,617.80	36,419.81	74,753.03	167,882.32
IRR (%)	3.92%	4.06%	4.17%	4.68%	4.79%	4.88%
PB (years)	25	25	25	25	25	25

Table 6 shows the profitability analysis for lifetimes of 50 and 75 years in UPSH plants where the existing infrastructure is used as a lower reservoir. The NPV and the IRR reach 471.89 M€ and 8.13% when the lifetime of the UPSH increases to 75 years. The NPVs of HPS 3 for a lifetime of 75 years are increased by 284% and 108% in comparison with the values obtained for HPS 1 and HPS 2, respectively. Finally, the PB remains constant in each HPS.

Table 6. Profitability analysis of HPS 1, HPS 2 and HPS 3, making use of existing infrastructure and considering DTCs of 8 h and lifetimes of 50 and 75 years.

Lifetime	50 Years			75 Years		
HPS	HPS 1	HPS 2	HPS 3	HPS 1	HPS 2	HPS 3
NPV (k€)	83,167.91	155,399.55	327,628.43	122,604.79	226,042.18	471,892.95
IRR (%)	7.47%	7.68%	7.86%	7.78%	7.97%	8.13%
PB (years)	15	15	15	15	15	15

3.5. Comparison with Other Storage Technologies

The installation cost of energy storage technologies (€ kWh^{-1}) has been compared with UPSH plants. Figure 7 shows the planned installation cost of a number of energy storage types for 2030 and highlights the low cost of conventional PSH (19 € kWh^{-1}), followed by CAES systems (38.26 € kWh^{-1}) [2,34]. Electrochemical storage like lithium-ion is still more expensive to install, but it is more efficient at storing and releasing energy, opening it up to a wider range of potential applications [34]. The installation cost of UPSH plants using the existing infrastructure reaches 20.90 € kWh^{-1}, while the cost of UPSH considering the excavation of new tunnels increases to 38 € kWh^{-1}.

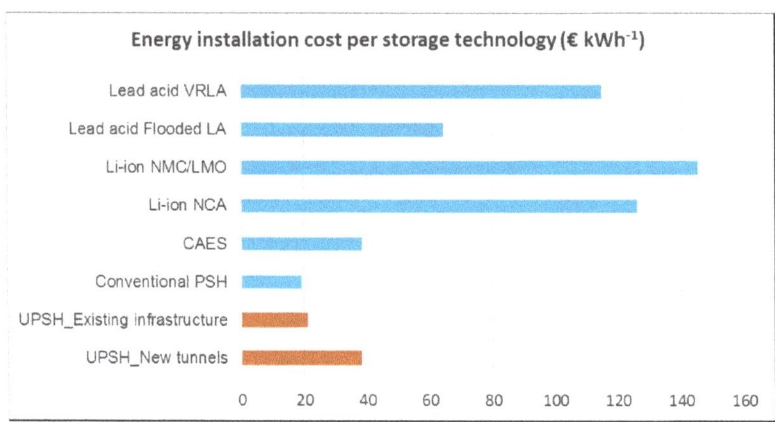

Figure 7. Energy installation cost of battery storage technologies vs. UPSH plants.

4. Conclusions

The economic feasibility of UPSH plants participating in the day-ahead and ancillary services markets in the Iberian electricity system is presented for three HPSs. A deterministic model has been applied in order to maximize the income and minimize the costs for purchasing electricity. Different DTCs between 4 and 10 h have been considered in order to evaluate the economic feasibility of UPSH plants. In addition, the investment costs when making use of existing underground infrastructure and when excavating new tunnels or caverns as a lower reservoir have been estimated in order to evaluate the profitability of the investment.

The results obtained show that the number of annual production cycles and the amount of electricity generated decrease when the DTC increases. The maximum number of annual production cycles is 360 when the DTC is 4 h and decreases to 323 when the DTC increases to 10 h. Although the spread between the income from selling electricity and the costs for purchasing energy decreases when the DTC increases, the operating margin increases due to investment costs. In the profitability model, IRRs greater than or equal to 6% are reached when the investment costs are lower than 2000 € kW^{-1} and the DTC is greater than 5 h. In general, it can be concluded that the IRR increases when the investment costs decrease and the DTC increases.

A UPSH plant is not economically feasible when new infrastructure has to be built. Maximum IRRs of 2.43%, 2.61% and 2.75% have been obtained for HPS 1, HPS 2 and HPS 3, respectively, with a minimum PB of around 24 years. On the contrary, the investment cost of a UPSH plant is reduced by 46.6% (1675 € kW^{-1}) when the existing underground infrastructure is used as a lower reservoir. Under these conditions, a UPSH plant could be economically feasible (IRRs greater than 6% and PBs lower than 15 years) when participating in the ancillary services markets, dimensioning DTCs greater than 6 h and using the existing underground infrastructure as a lower reservoir.

Author Contributions: Conceptualization, J.M.; investigation, J.M., J.M.F.-O. and J.L.; methodology, J.M.; software, J.M. and J.M.F.-O.; validation, J.M. and J.M.F.-O.; writing—original draft, J.M.; writing—review and editing, J.M. and J.M.F.-O.; supervision, J.L. All authors have read and agreed to the published version of the manuscript.

Funding: This research received no external funding.

Conflicts of Interest: The authors declare no conflict of interest.

Nomenclature

A-CAES	Adiabatic compressed air energy storage
CAES	Compressed air energy storage
DTC	Daily turbine cycle time at full load
EBIT	Earnings before interest and taxes
ESS	Energy storage system
FT	Francis pump-turbine
HPS	Hydropower station
IRR	Internal rate of return
LA	Lead-acid
Li-ion	Lithium ion
LMO	Lithium manganese oxide
MISO	Mid-continent Independent System Operator
NPV	Net present value
NCA	Nickel cobalt aluminum
NMC	Nickel manganese cobalt
NYISO	New York Independent System Operator
O&M	Operation and maintenance
PB	Payback period
PSH	Pumped storage hydropower
PV	Photovoltaic
RES	Renewable energy sources
UPSH	Underground pumped storage hydropower
VRLA	Valve-regulated lead-acid

Appendix A

Table A1. Assessment of the investment costs of a UPSH plant, considering an FT with 219 MW of output power and a DTC of 8 h. Civil works, hydromechanical equipment and penstock, hydropower equipment, electric substation and grid connection.

UPSH—Investment Cost (k€)	
Civil works	**525,692.80**
Surface works	**31,110.40**
Upper reservoir	30,730.00
Excavation and support	*23,680.00*
Waterproofing	*7050.00*
Electric substation	380.40
Underground works	**494,582.40**
Tunnel of access	24,923.30
Excavation and support system	*24,800.00*
Lighting system	*48.50*
Ventilation system	*74.80*
Powerhouse cavern	12,555.63
Excavation and support system	*8950.00*
Cavern equipment	*2485.00*
Vent shaft and electric cables	*975.30*
Drainage drift	*145.33*
Lower reservoir	453,333.33
Submergence tunnel	2550.00
Vent shaft	652.40
Surge tank	567.74
Hydromechanical equipment and penstock	**16,986.52**
Intake	3280.00
Penstock	4056.00
Auxiliary systems	9650.52
Hydropower equipment	**99,709.12**
Francis pump-turbine	37,640,85
Motor-generator	35,321.85
Electrical and control systems	25,895.53
Fire protection system	850.90
Electrical grid connection	**4510.50**
Electric substation	3250.30
Powerline	1260.20
Total Cost (k€)	**639,209.18**

Table A2. Assessment of the investment costs of a UPSH plant, considering an FT with 219 MW of output power and a DTC of 8 h. Commissioning and project management.

UPSH—Investment Cost (k€)	
Commissioning	**4140.85**
Spare parts and staff training	895.30
Civil and structure works and penstock	450.20
Mechanical review	790.25
Electrical review	675.80
Control system review	560.40
Commissioning testing	768.90
Project management	**36,301.30**
Building permits and others	16,950.00
Engineering	3250.00
Construction management	4230.90
Security and health	2420.40
Waste management	9450.00
Total Cost (k€)	**40,442.15**

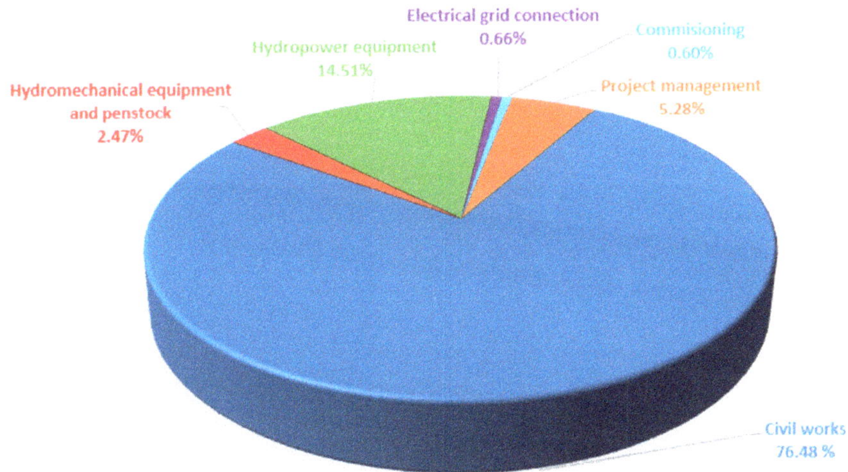

Figure A1. Capital cost breakdown for a UPSH plant with surface upper reservoir and underground lower reservoir.

References

1. Weron, R. Electricity price forecasting: A review of the state-of-the-art with a look into the future. *Int. J. Forecast* **2014**, *30*, 1030–1081. [CrossRef]
2. Mongird, K.; Fotedar, V.; Viswanathan, V.; Koritarov, V.; Balducci, P.; Hadjerioua, B.; Alam, J. *Energy Storage Technology and Cost Characterization Report*; U.S. Department of Energy's Water Power Technologies Office by Pacific Northwest National Laboratory: Washington, DC, USA, 2019.
3. Matos, C.R.; Carneiro, J.F.; Silva, P.P. Overview of large-scale underground energy storage technologies for integration of renewable energies and criteria for reservoir identification. *J. Energy Storage* **2019**, *21*, 241–258. [CrossRef]

4. Pummer, E.; Schüttrumpf, H. Reflection phenomena in underground pumped storage reservoirs. *Water* **2018**, *10*, 504. [CrossRef]

5. Tola, V.; Meloni, V.; Spadaccini, F.; Cau, G. Performance assessment of Adiabatic Compressed Air Energy Storage (A-CAES) power plants integrated with packed-bed thermocline storage systems. *Energy Convers. Manag.* **2017**, *151*, 343–356. [CrossRef]

6. Menendez, J.; Fernández-Oro, J.M.; Galdo, M.; Loredo, J. Efficiency analysis of underground pumped storage hydropower plants. *J. Energy Storage* **2020**, *28*, 101234. [CrossRef]

7. Winde, F.; Kaiser, F.; Erasmus, E. Exploring the use of deep level gold mines in South Africa for underground pumped hydroelectric energy storage schemes. *Renew. Sustain. Energy Rev.* **2016**, *78*, 668–682. [CrossRef]

8. Winde, F.; Stoch, E.J. Threats and opportunities for post-closure development in dolomitic gold mining areas of the West Rand and Far West Rand (South Africa)—A hydraulic view part 1: Mining legacy and future threats. *Water SA* **2010**, *36*, 69–74. [CrossRef]

9. Pujades, E.; Orban, P.; Archambeau, P.; Erpicum, S.; Dassargues, A. Numerical study of the Martelange mine to be used as underground reservoir for constructing an Underground Pumped Storage Hydropower plant. *Adv. Geosci.* **2018**, *45*, 51–56. [CrossRef]

10. Pujades, E.; Orban, P.; Archambeau, P.; Kitsikoudis, V.; Erpicum, S.; Dassargues, A. Underground Pumped-Storage Hydropower (UPSH) at the Martelange Mine (Belgium): Interactions with Groundwater Flow. *Energies* **2020**, *13*, 2353. [CrossRef]

11. Bodeux, S.; Pujades, E.; Orban, P.; Brouyère, S.; Dassargues, A. Interactions between groundwater and the cavity of an old slate mine used as lower reservoir of an UPSH (Underground Pumped Storage Hydroelectricity): A modelling approach. *Eng. Geol.* **2016**, *217*, 71–80. [CrossRef]

12. Menendez, J.; Loredo, J.; Galdo, M.; Fernandez-Oro, J.M. Energy storage in underground coal mines in NW Spain: Assessment of an underground lower water reservoir and preliminary energy balance. *Renew. Energy.* **2019**, *134*, 1381–1391. [CrossRef]

13. Menéndez, J.; Ordóñez, A.; Álvarez, R.; Loredo, J. Energy from closed mines: Underground energy storage and geothermal applications. *Renew. Sustain. Energy Rev.* **2019**, *108*, 498–512. [CrossRef]

14. Madlener, R.; Specht, J.M. *An Exploratory Economic Analysis of Underground Pumped-Storage Hydro Power Plants in Abandoned Coal Mines*; FCN Working Paper No. 2/2013; FCN: Aachen, Germany, 2013.

15. Menéndez, J.; Fernández-Oro, J.M.; Galdo, M.; Loredo, J. Transient Simulation of Underground Pumped Storage Hydropower Plants Operating in Pumping Mode. *Energies* **2020**, *13*, 1781. [CrossRef]

16. Wong, I.H. An underground pumped storage scheme in the Bukit Timah granite of Singapore. *Tunn. Undergr. Space Technol.* **1996**, *11*, 485–489. [CrossRef]

17. Graves, F.; Jenkin, T.; Murphy, D. Opportunities for electricity storage in deregulating markets. *Electr. J.* **1999**, *12*, 46–56. [CrossRef]

18. Connolly, D.; Lund, H.; Finn, P.; Mathiesen, B.V.; Leahy, M. Practical operation strategies for pumped hydroelectric energy storage (PHES) utilising electricity price arbitrage. *Energy Policy* **2011**, *39*, 4189–4196. [CrossRef]

19. Lobato, E.; Egido, I.; Rouco, L.; López, G. An overview of ancillary services in Spain. *Electr. Power Syst. Res.* **2008**, *78*, 515–523. [CrossRef]

20. Pérez-Díaz, J.I.; Chazarra, M.; García-González, J.; Cavazzini, G.; Stoppato, A. Trends and challenges in the operation of pumped-storage hydropower plants. *Renew. Sustain. Energy Rev.* **2015**, *44*, 767–784. [CrossRef]

21. Krishnan, V.; Das, T. Optimal allocation of energy storage in a co-optimized electricity market: Benefits assessment and deriving indicators for economic storage ventures. *Energy* **2015**, *81*, 175–188. [CrossRef]

22. Berrada, A.; Loudiyi, K.; Zorkani, I. Valuation of energy storage in energy and regulation markets. *Energy* **2016**, *115*, 1109–1118. [CrossRef]

23. Chazarra, M.; Pérez-Díaz, J.I.; García-González, J.; Praus, R. Economic viability of pumped-storage power plants participating in the secondary regulation service. *Appl. Energy* **2018**, *216*, 224–233. [CrossRef]

24. Maciejowska, K.; Nitka, W.; Weron, T. Day-Ahead vs. Intraday—Forecasting the Price Spread to Maximize Economic Benefits. *Energies* **2019**, *12*, 631. [CrossRef]

25. Ekman, C.K.; Jensen, S.H. Prospects for large scale electricity storage in Denmark. *Energy Convers. Manag.* **2010**, *51*, 1140–1147. [CrossRef]

26. Chazarra, M.; Pérez-Díaz, J.; García-González, J. Optimal Operation of Variable Speed Pumped Storage Hydropower Plants Participating in Secondary Regulation Reserve Markets. In Proceedings of the 2014 11th International Conference on the European Energy Market (EEM), Krakow, Poland, 28–30 May 2014.

27. Chazarra, M.; García-González, J.; Pérez-Díaz, J.I.; Arteseros, M. Stochastic optimization model for the weekly scheduling of a hydropower system in day-ahead and secondary regulation reserve markets. *Electr. Power Syst. Res.* **2016**, *130*, 67–77. [CrossRef]

28. Min, C.-G.; Kim, M.-K. Impact of the Complementarity between Variable Generation Resources and Load on the Flexibility of the Korean Power System. *Energies* **2017**, *10*, 1719. [CrossRef]

29. Lago, J.; Ridder, F.D.; Vrancx, P.; Schutter, B.D. Forecasting day-ahead electricity prices in Europe: The importance of considering market integration. *Appl. Energy* **2018**, *211*, 890–903. [CrossRef]

30. Chazarra, M.; Pérez-Díaz, J.; García-González, J. Optimal Joint Energy and Secondary Regulation Reserve Hourly Scheduling of Variable Speed Pumped Storage Hydropower Plants. *IEEE Trans. Power Syst.* **2018**, *33*, 103–115. [CrossRef]

31. Steffen, B. Prospects for pumped-hydro storage in Germany. *Energy Policy* **2012**, *45*, 420–429. [CrossRef]

32. Pearre Nathaniel, S.; Swan Lukas, G. Technoeconomic feasibility of grid storage: Mapping electrical services and energy storage technologies. *Appl. Energy* **2015**, *137*, 501–510. [CrossRef]

33. Ardizzon, G.; Cavazzini, G.; Pavesi, G. A new generation of small hydro and pumped hydro power plants: Advances and future challenges. *Renew. Sustain. Energy Rev.* **2014**, *31*, 746–761. [CrossRef]

34. IRENA. *Electricity Storage and Renewables: Costs and Markets to 2030*; International Renewable Energy Agency: Abu Dhabi, UAE, 2017.

applied
sciences

Article

Salt Cavern Exergy Storage Capacity Potential of UK Massively Bedded Halites, Using Compressed Air Energy Storage (CAES)

David Evans [1],*, Daniel Parkes [1], Mark Dooner [2], Paul Williamson [1], John Williams [1], Jonathan Busby [1], Wei He [2], Jihong Wang [2] and Seamus Garvey [3]

1   British Geological Survey, Keyworth, Nottingham NG12 5GG, UK; dparkes64@gmail.com (D.P.); jpw@bgs.ac.uk (P.W.); jdow@bgs.ac.uk (J.W.); jpbu@bgs.ac.uk (J.B.)
2   School of Engineering, University of Warwick, Coventry CV4 7AL, UK; m.dooner.1@warwick.ac.uk (M.D.); W.He.1@warwick.ac.uk (W.H.); jihong.wang@warwick.ac.uk (J.W.)
3   Faculty of Engineering, University of Nottingham, University Park, Nottingham NG7 2RD, UK; seamus.garvey@nottingham.ac.uk
*   Correspondence: devans50.de@gmail.com

Featured Application: The work provides important data and information relating to future energy storage options and in particular the role CAES might play in load balancing and the integration of renewable energy technologies into electricity grids.

Citation: Evans, D.; Parkes, D.; Dooner, M.; Williamson, P.; Williams, J.; Busby, J.; He, W.; Wang, J.; Garvey, S. Salt Cavern Exergy Storage Capacity Potential of UK Massively Bedded Halites, Using Compressed Air Energy Storage (CAES). *Appl. Sci.* **2021**, *11*, 4728. https://doi.org/10.3390/app11114728

Academic Editors: Jorge Loredo and Javier Menéndez

Received: 31 March 2021
Accepted: 14 May 2021
Published: 21 May 2021

Publisher's Note: MDPI stays neutral with regard to jurisdictional claims in published maps and institutional affiliations.

Abstract: The increasing integration of large-scale electricity generation from renewable energy sources in the grid requires support through cheap, reliable, and accessible bulk energy storage technologies, delivering large amounts of electricity both quickly and over extended periods. Compressed air energy storage (CAES) represents such a storage option, with three commercial facilities using salt caverns for storage operational in Germany, the US, and Canada, with CAES now being actively considered in many countries. Massively bedded halite deposits exist in the UK and already host, or are considered for, solution-mined underground gas storage (UGS) caverns. We have assessed those with proven UGS potential for CAES purposes, using a tool developed during the EPSRC-funded IMAGES project, equations for which were validated using operational data from the Huntorf CAES plant. From a calculated total theoretical 'static' (one-fill) storage capacity exceeding that of UK electricity demand of ≈300 TWh in 2018, filtering of results suggests a minimum of several tens of TWh exergy storage in salt caverns, which when co-located with renewable energy sources, or connected to the grid for off-peak electricity, offers significant storage contributions to support the UK electricity grid and decarbonisation efforts.

Keywords: energy storage; exergy; CAES; salt caverns

1. Introduction

Current energy systems, relying primarily on fossil fuels (coal, oil, natural gas), produce carbon and greenhouse gases (C&GHG), contributing to the problem of global climate change. There is therefore, an increasing need to reduce C&GHG emissions. From initial targets of 80% reductions by 2050, in June 2019, the UK Government set a revised target of net zero emissions by 2050 [1], which was followed by the launch of the EU's 'European Green Deal' in December 2019 [2]. These aims will require significant effort across many industrial sectors that represent large emission sources, including electrical power generation.

Worldwide, transitioning from fossil fuel to cleaner, but intermittent, unpredictable, and inherently more variable mixed renewable energy sources (wind-power and solar photovoltaic [PV] plants) for electricity generation is enabling GHG emission reductions. However, if naturally variable renewable electricity sources comprise high percentages

(>80%) of the generated supply, the daily and seasonal variations in generation and capacity places greater challenges on power networks to meet transmission and distribution demands [3]. Alongside seasonal variation in electricity demand, issues then arise over security of supply, as power systems require balancing at various scales, ranging from second and minute reserves, to hourly, daily, weekly, and inter-seasonal (monthly) storage to meet and offset variability [3,4]. Therefore, patterns of demand not following such variations in electricity generation from renewable sources require fast-ramping, back-up generation, supported by reliable forecasting and, importantly, increased bulk, grid-scale storage capacity [3,4].

Electrical energy storage (EES) technologies are recognised as underpinning technologies to meeting these challenges, but they vary greatly in capacity, role, and costs. Some technologies provide short-term, small-scale energy storage options (e.g., batteries), whereas others represent load-levelling and longer-term utility scale and grid support through chemical and mechanical bulk energy storage technologies. The two largest and only current commercial, grid-scale, mechanical bulk energy storage technologies capable of providing fast ramp rates, good part load, and long duration are pumped hydroelectric storage (PHS) and compressed air energy storage (CAES) [5]. They are less economic or suitable as inter-seasonal storage options to balance longer term, seasonal fluctuations, or long-lasting wind shortages due to low volumetric energy storage densities ($\approx$0.7 and 2.40 kWh/m^3, respectively; see below) [6].

Despite extensive investigation and research into CAES technology from the 1960s [7,8], worldwide, commercially operational grid-scale CAES capacity is provided by just three salt cavern-hosted facilities: the conventional (diabatic) Huntorf, Germany (1978, 321-MW) [9], and McIntosh, USA (1991, 110-MW) CAES plants [7,8,10], and in November 2019, the small (1.75MW/7MWh+) plant at Goderich, Canada, which became the world's first commercial adiabatic CAES plant [11]. Sustained rapid growth in wind power and making it dispatchable has renewed interest in CAES [5,12]. Despite significant research and some extended tests [13,14], no porous rock CAES plants exist, which is due mainly to economic and geological factors that, prior to development as a realistic storage option at scale, must be overcome [3,12]. Nevertheless, offshore porous rock storage is advocated as having inter-seasonal potential for the UK [15].

Particularly pertinent, following the UK Government's October 2020 announced intention of becoming the world leader in green energy involving mainly increased offshore wind farm generation [16], we explore the prospects and possible capacity of salt caverns for UK CAES exergy storage in selected onshore and offshore massively bedded halite deposits (Figure 1). These offer large energy storage volumes to underpin affordable and energy-secured decarbonisation and the development of low-carbon energy system design, policy, and regulations. The method proposed here will also be applicable to other countries with storage potential identified in salt caverns, particularly in Eurasia, North and South America, and Sub-Saharan Africa [17].

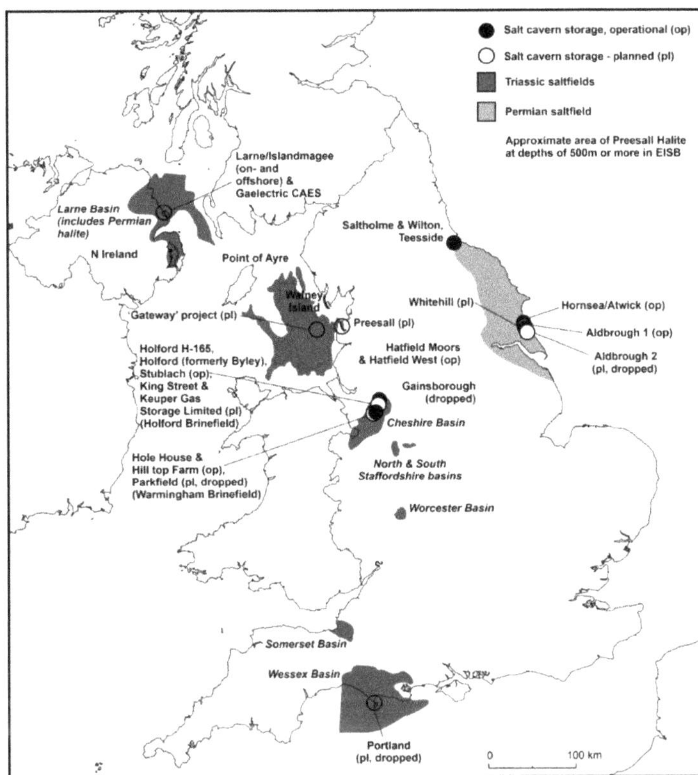

Figure 1. General outcrop map of the main halite basins studied onshore England and offshore East Irish Sea. Note area indicated in the East Irish Sea is that of the Triassic Preesall Halite at depths investigated (500–1500 m). Refer also S2, Table S1 for details on UGS facilities.

2. Mechanical, Bulk Electrical Energy Storage (EES), and the Potential of CAES

PHS is the most mature, proven bulk energy storage technology, whereby energy is stored in the form of the gravitational potential energy of water pumped from a lower to a higher elevation reservoir. Pumps are typically run by low-cost surplus, off-peak electrical power, and during periods of high/peak electrical demand, release of the stored water through turbines generates electric power. Used by electric power systems for load balancing, it reliably provides a large-scale and fast-responding storage option, with a current worldwide grid-connected capacity of ≈188 GW and representing ≈96% of the total global energy storage capability [18]. Significant potential for hydro-storage capacity may still exist in many other areas around the world [19]; however, ultimate development and capacity for PHS in most developed countries, including the UK, is considered limited and constrained by social, environmental, availability, and geographical considerations [5,20,21].

CAES, with a modest surface footprint and greater siting flexibility relative to PHS, represents a low-cost technology that is capable of a power output of over 100 MW. CAES is based on large quantities of electrical energy stored as high-pressure, compressed air in an underground storage 'reservoir' (currently salt caverns). During peak demand, air is withdrawn and used in the generation of electricity, and as with PHS, the release of power can be very quick. Worldwide, CAES capacity is currently around 431 MW [18], and CAES is viewed increasingly as offering bulk storage potential and a solution to levelling intermittent renewables generation (wind-power and solar photovoltaic [PV] plants), and capable of maintaining system balance (S1, Tables S1–S6) [3,5]. CAES technology has advantages over PHS, including a lower visible impact on the landscape and a greater scope for building CAES facilities nearer the centres of wind-power production, especially in parts of Europe and regions of the USA. CAES facilities in salt caverns already successfully provide minutes to hours reserve at Huntorf (Germany) and balancing out grid loads over a weekly cycle at McIntosh in the USA [4,9]. However, significant barriers to implementing large-scale CAES plants lie in identifying appropriate geological storage options and thus geographical locations, low round-trip efficiencies of CAES and the low volumetric energy density of compressed air (2.4 kWh/m^3) [6,22,23].

Energy in compressed air caverns is stored in the form of physical (mechanical) potential energy, whereas energy in compressed gases is chemical storage (chemical energy bonds). Consequently, the volumetric energy density of air is several orders of magnitude lower than that of gases such as hydrogen ($\approx$170 kWh/m^3) or natural gas ($\approx$1100 kWh/m^3) [4]. Accordingly, to make CAES economically viable requires very large volumes of air, which can only be achieved through high pressures and large volume storages. Geological storages at depth offer such storage conditions, with typical gas storage salt caverns, in particular, offering rapid cycling and high flow rates to provide ideal storage options. However, the lower volumetric energy density of air means that CAES plants are less suitable for long-term applications and storage because greater storage volume (increased cavern numbers) is required, increasing costs compared to gases with higher value.

Whilst geometrical volumes of compressed air caverns are comparable to those of conventional natural gas storage caverns, CAES operational pressure ranges (and thus storage volumes) will be considerably lower than for gas storage. This is because of the much higher cyclic pressure frequency rate together the with current technological development of compressors, heat storages, and turbines, meaning the operational pressures are also lower, being well below 100 bar [4]. Thus, commercial, central, grid-scale CAES plants will require deep underground (geological) storages such as those already used for natural gas, hydrogen, and the rare examples of already operational CAES plants.

Conventional (diabatic) CAES technology is based upon traditional gas-turbine plants requiring fossil fuel combustion and thus associated emissions during electricity generation, making it less attractive when compared with other EES technologies [24]. Nevertheless, the fitting of recuperators and advances in CAES technologies, particularly if advanced adiabatic or isothermal CAES technologies requiring no external source of energy to heat the withdrawn air eventually prove feasible, together with linking to renewables generation (including offshore wind), all offer the future prospect of improved cycle efficiencies, with the reduction and possibly elimination of emissions.

3. CAES-Geological Storage Options, Developments, and Restrictions

Bulk geological storage options and the technologies behind current and future electrical energy storages for compressed air are derived largely from tried and tested storage technologies developed for the underground storage of large volumes of high-pressure natural gas [4]. Most common geological options are porous rock formations (depleted gas fields and aquifers), or man-made (solution-mined) salt caverns. Where such options are not available conventionally mined, non-salt rock caverns and lined rock caverns represent alternatives, but they are significantly more expensive. These same options apply to potential CAES development (S1 and S1, Tables S1–S6).

As alluded to above, CAES has been considered for many decades [7,8] but to date, only three commercially operational CAES plants exist, at Huntorf [9], McIntosh [10], and most recently at Goderich [11]. Between 2012 and 2016, a small 2 MW isothermal CAES demonstration plant using a reconditioned former liquid hydrocarbon storage salt cavern and linked to wind generation, operated at Gaines, Texas, although it is not believed to be currently operating [25,26]. Salt caverns provide important high flexibility with respect to turnover frequency, as the open cavity enables very high flow rates permitting high injection/withdrawal rates required for rapid cycle storages. They also offer ideal conditions for compressed air storages because unlike porous reservoirs, the rock salt is inert to oxygen [4]. Thus solution-mined salt caverns are a likely first choice for CAES in the UK, and for CAES proposals linked with renewables, they are the overwhelming majority (S1, Table S1).

Many regions of the world lack suitable salt deposits, and so, the suitability of porous rock storage has long been and remains under investigation [12–14]. However, serious doubts exist over the likely development of porous rock storage (principally aquifers), with no CAES plants having operated commercially and only a few small test facilities having been constructed, with variable results (S1, Tables S2 and S3). The King Island project in California demonstrated the technical feasibility of using an abandoned natural gas reservoir for a 300 MW, 10 h CAES facility, with the reservoir capable of accommodating the flow rates and pressures necessary for the operation of the facility. Originally planned for opening around 2020, its progress appears stalled due to the high cost of a CAES facility relative to alternative energy storage technologies [27]. All test facilities encountered problems with one of more of the following: wells and economics, pressure anomalies, variations in reservoir quality and performance, formation of the 'air bubble' in the storage reservoir, and reaction between the oxygen of the injected air and minerals in the reservoir rock leading to oxygen depletion and/or potential for bacterial/micro-organism growth and porosity reduction. Proposed aquifer storage potential for the UK would be offshore [15], thereby increasing costs, which currently thus seems less likely than salt cavern storage.

Depleted field storages appear even more unlikely with a potential hazard posed by residual hydrocarbons in the depleted gas formation. Introducing compressed air presents the risk of ignition and explosion, both underground and during discharge [28].

Additionally, and although more expensive options, gas storages have and still operate in abandoned mines and unlined or lined, conventionally mined rock storages. Similar constructions could host CAES in regions lacking cheaper geological alternatives [7,8] and have been considered (S1, Tables S4–S6). Various CAES test facilities have operated briefly in Japan and Korea, and long-standing plans for CAES in a former limestone mine at Norton, Ohio were finally shelved in 2013 [29]. Small tests for adiabatic CAES are currently ongoing in an unlined Swiss tunnel [30] and a lined old mine working in Austria [31]. Whilst under consideration in, for example, USA, Mongolia, and Australia, such storages may be considered unlikely in the UK.

Non-geological CAES schemes offering storages of small volume. Though not considered here they include aboveground, or shallowly buried steel vessels or pipes [32,33], energy bags secured to the seabed [34], wind turbines linked with energy storage in supporting legs [35], or those in which power is converted directly from the rotor by means of gas/air compression within the rotor blades [36].

4. Materials, Exergy Storage Tool, and Methodology

This section outlines briefly the UK bedded halites, UGS sites together with the development of the model and the derivation of estimates for exergy storage (refer Figure 2), further details of which are provided in S1–S3.

4.1. Massively Bedded UK Halite Deposits Available

Important massively bedded halite deposits are developed in the UK and have been associated with, or identified as potential hosts for, large solution-mined natural gas storage caverns (Figure 1; S2, Table S1). The halite deposits considered extensive and thick enough for cavern construction occur in four main basins (with ages) [37]:

- The Northwich Halite Member of Cheshire Basin, onshore north-central England (Triassic)
- The Preesall Halite Member of the offshore East Irish Sea (EIS) (Triassic)
- The Dorset Halite Member of Wessex Basin, on- and extending offshore southern England (Triassic)

- The Fordon Evaporite Formation, on- and extending offshore Eastern England (Upper Permian, Zechstein [Z2]).

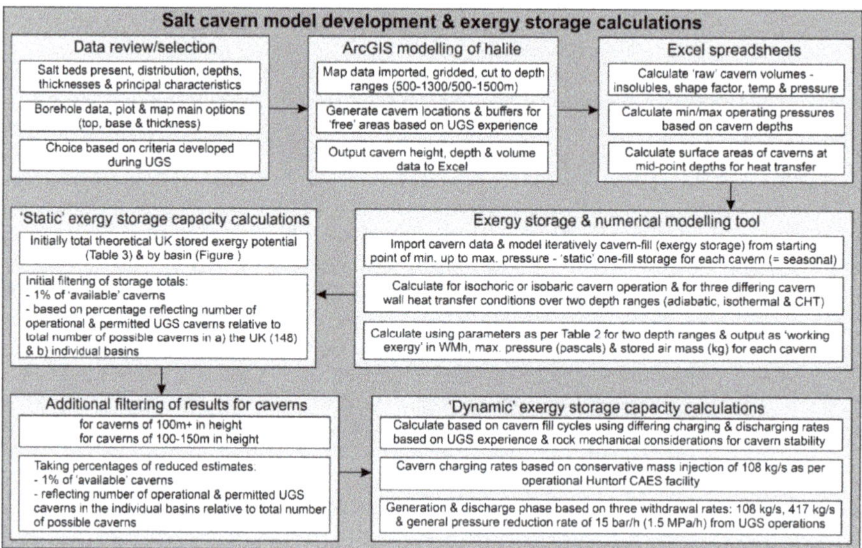

Figure 2. Workflow for the estimation of exergy storage provided by solution-mined salt caverns in the main halite-bearing basins of the UK.

These deposits offer important alternative energy storage capacity, and this study has assessed their potential for large-scale exergy storage through CAES. Differing from energy that is always conserved, exergy which takes its basis from the second law of thermodynamics, measures the loss of energy quality in every energy transformation process. Exergy tends to be destroyed during any conversion or storage processes, and therefore, exergy storage capacity quantifies the maximum useful work of the stored air that could be used in subsequent power generation. Exergy analysis is employed in applications with electricity output and power generation processes, and an exergy analysis tool was developed to estimate the exergy losses in energy conversions associated with a salt cavern-based CAES system. This permitted an estimate of the exergy storage capacity of the compressed air stored in a salt cavern for generating electricity during the discharging period [38]. Compared to conventional static thermodynamic exergy analysis, our developed tool also considers time-dependent factors that affect the overall electrical efficiency of a CAES system, such as dynamic internal air responses in the cavern and the coupled thermal effects of surrounding rocks [S3].

The Triassic and Permian bedded halite deposits in Northern Ireland have not been included here, as they are poorly defined and largely identified for UGS purposes [37]. Equally, the available Preesall Halite in NW England has also been identified for UGS and is not included here [37,39]. The Zechstein halite beds extend offshore from eastern England into the Southern North Sea, where due to halokinesis, they may attain great thicknesses. For various reasons, they have not been included in this study: they occur often far offshore and show significant changes in thickness over short distances, with some salt structures rising to shallow depths, even approaching close to sea bed, and are often in association with existing producing gasfields [40]. However, they should not be ruled out as CAES hosts, perhaps linked to the growing number of offshore windfarms. If existing hydrocarbon infrastructure (platforms, pipeline and cable routes, etc.) could be re-purposed, development costs, which are high for proposed gas storage caverns (A. Stacey, pers comm.), might be reduced significantly.

4.2. Exergy Storage Terminology—The Gas Storage Experience

The technology behind current and future storages for electrical energy based on compressed air, H_2, or SNG storages is derived largely from tried and tested storage technologies developed for the storage of natural gas [4]. A terminology has emerged to define operations and refer to the volumes of gas in an underground gas storage facility, which we adopt here when defining the exergy stored and explain below.

Underground gas storages generally operate by compressing the storage gas during injection and decompressing the gas again during withdrawal. The total gas storage capacity or volume is the maximum volume of natural gas that can be stored at the storage facility. This is governed by several physical factors such as the reservoir volume, engineering, and operational procedures including minimum and maximum pressure ranges and injection rates, which are determined from rock mechanical studies. The total storage volume comprises two elements:

- Working gas' volume, which represents the available gas that can be used between the maximum and the minimum operating storage pressures
- Cushion gas' volume, representing that below minimum operating pressure that is not available and which must remain permanently in the storage to provide the required minimum pressure to maintain the geomechanical stability of the storage. In the case of porous rock storage, it also provides some of the drive, but it is irretrievable, being effectively lost in the porosity.

The working gas volume represents the 'static', one-fill gas capacity and does not reflect multiple filling cycles. Thus, it is representative of a seasonal storage, similar to most traditional aquifer and depleted field storages. Of course, gas storages may be cycled many times during a year, which gives rise to what is described as a 'dynamic working gas volume' [39], which is greater than the static one-fill working gas volume.

Thus, exergy storage estimates are here referred to as the 'working exergy' (that available for work) and the 'cushion exergy' (that portion that must remain in the salt cavern/storage). The exergy tool was set up to calculate the static 'working exergy' (available) volume (see below). After introducing the static one fill 'working exergy' storage, we describe how, through a series of filters, attempts are made to derive realistic static 'working exergy' storage estimates from the total theoretical storage calculated (Figures 2 and 3a,b). These are based on cavern sizes and percentages of the total number of caverns, including that based upon the number of gas storage caverns in any particular basin (Figures 4–8).

However, as with gas storage caverns, the static 'working exergy' storage capacity is increased by multiple cavern-filling cycles. Therefore, also described and calculated are 'dynamic working exergy storage' capacity estimates, which are based upon multiple cavern cycles per year. The yearly cycle numbers are derived from different injection and withdrawal rates, which are informed by both CAES and UGS experience (S2, S3).

4.3. Exergy Storage Tool

The exergy storage system is represented by a thermal modelling tool developed during the EPSRC-funded IMAGES project [38] and augmented during this study (S3) to calculate stored exergy for individual caverns of known depths and size/volume, in two operational modes: constant volume, variable pressure (isochoric), and constant pressure, variable volume (isobaric) modes. The tool, equations for which were validated using operational data from the Huntorf CAES facility [38], considers three wall conditions to approximate and model the unsteady heat transfer (flux) between the injected air and cavern walls and models. Two cavern wall conditions represent idealistic and somewhat unrealistic, end-member models:

- Adiabatic boundary conditions in which heat flux into the surrounding rock mass is zero
- Isothermal boundary conditions in which heat flux is infinite with perfect conduction into and through the surrounding rock mass

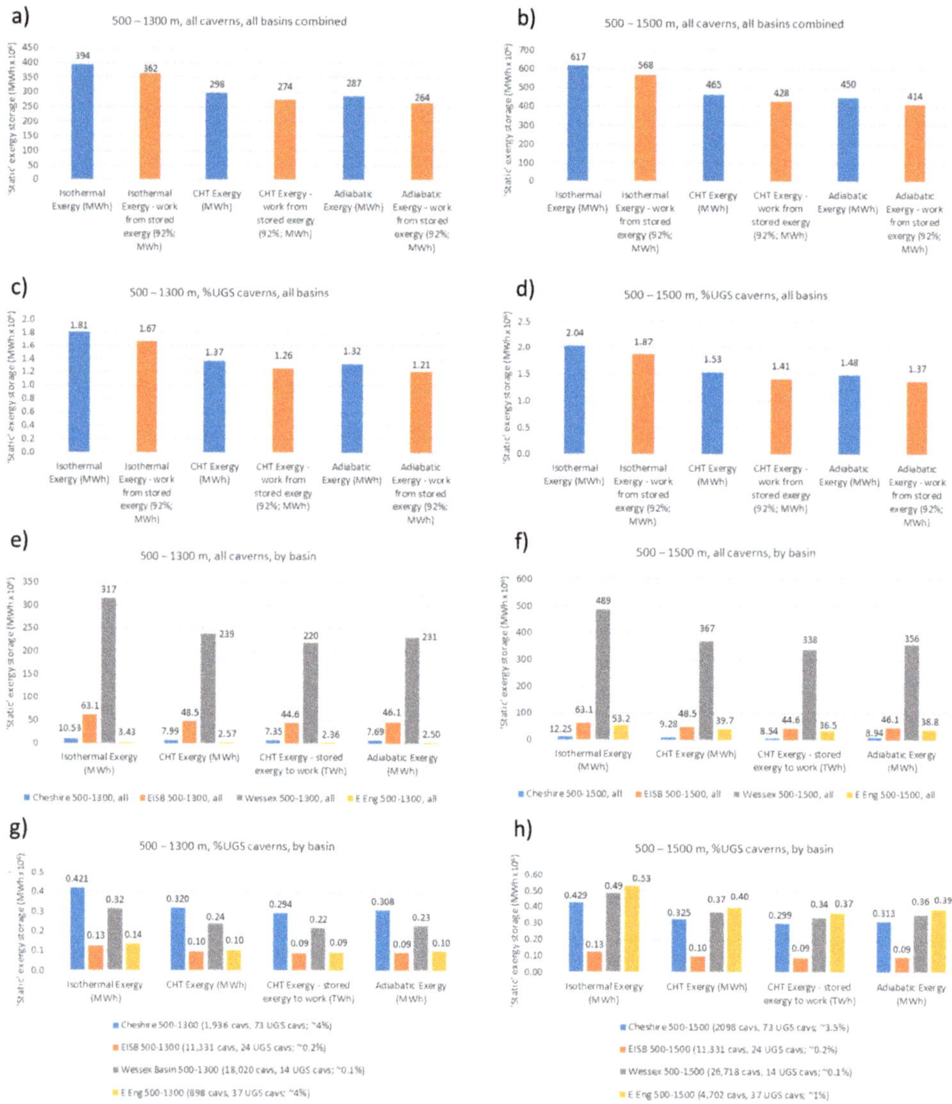

Figure 3. Plots of theoretical 'static' (one-fill) exergy storage estimates for the three thermal models for all potentially available caverns over the two depth ranges for all caverns with the basins studied. Parts (**a,b**) show graphs for combined totals from each basin for the two depth ranges, together with the estimated stored exergy to work for each thermal model. Parts (**c,d**) show graphs for the estimated stored exergy to work for each thermal model based upon percentages related to UGS numbers of the combined totals from each basin for the two depth ranges. Parts (**e,f**) show graphs breaking storage down by basin for the three thermal models, including stored exergy to work estimate for the CHT model also shown, with outlines data ranges being those pertinent to CHT model storage data presented in Figure 4. Parts (**g,h**) show graphs for estimates based upon a percentage related to the number of operation and/or planned UGS caverns in the basins.

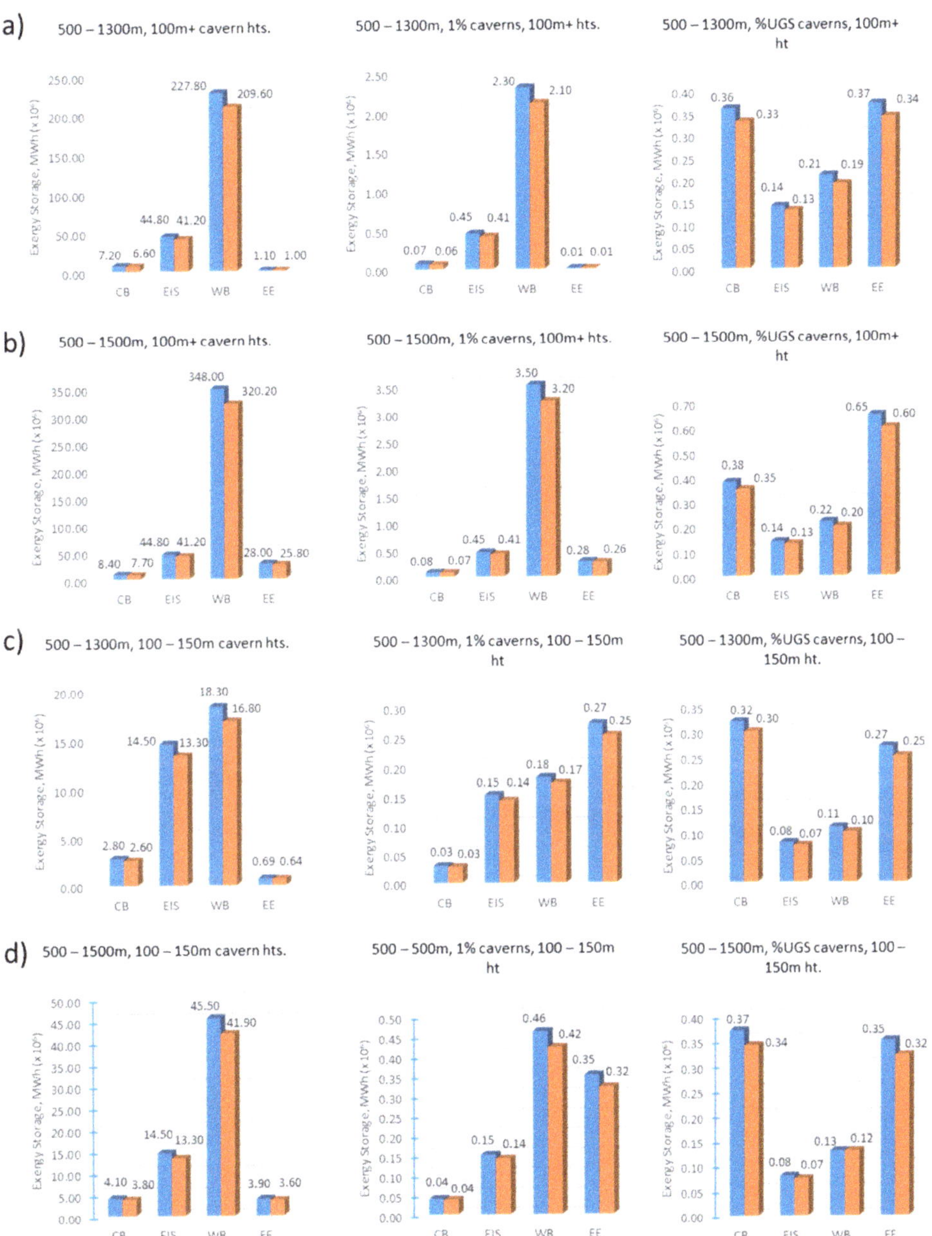

Figure 4. Plots of 'static' (one-fill) exergy storage estimates for the preferred CHT model, over the two depth ranges and cavern sizes (100 m+ and 100–150 m height) considered for CAES. Graphs for all potentially available caverns, 1% of available caverns and estimates based upon a percentage related to the number of UGS caverns in the basin. Parts (**a**,**c**) show data for the 500–1300 m depth range and parts (**b**,**d**) those data for the 500–1500 m depth range. Key common to all: blue = stored exergy, brown = stored exergy to work.

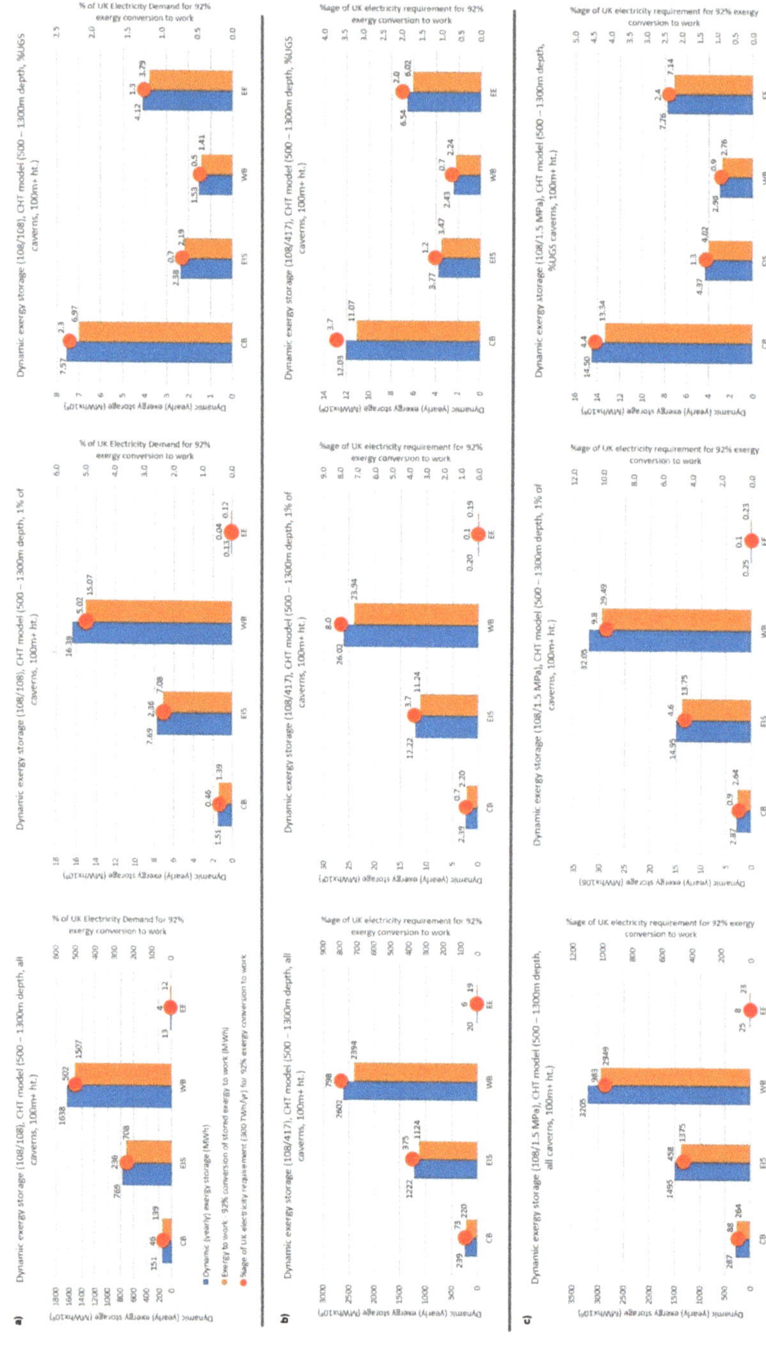

Figure 5. Plots of dynamic exergy storage and exergy to work estimates for the preferred CHT model, over the depth range 500–1300 m and cavern heights 100 m+ considered for CAES. Parts (a–c) show graphs for differing injection/withdrawal rates (108/108 kg/s and 108/417 kg/s) or fill and pressure reduction rates (108/108 kg/s/1.5 MPa/h) for all potentially available caverns, 1% of available caverns and estimates based upon the number of UGS caverns in the basins. Additionally shown, by basin, the percentage of UK electricity demand for 92% of stored exergy to work. Key common to all, see Figure 3.

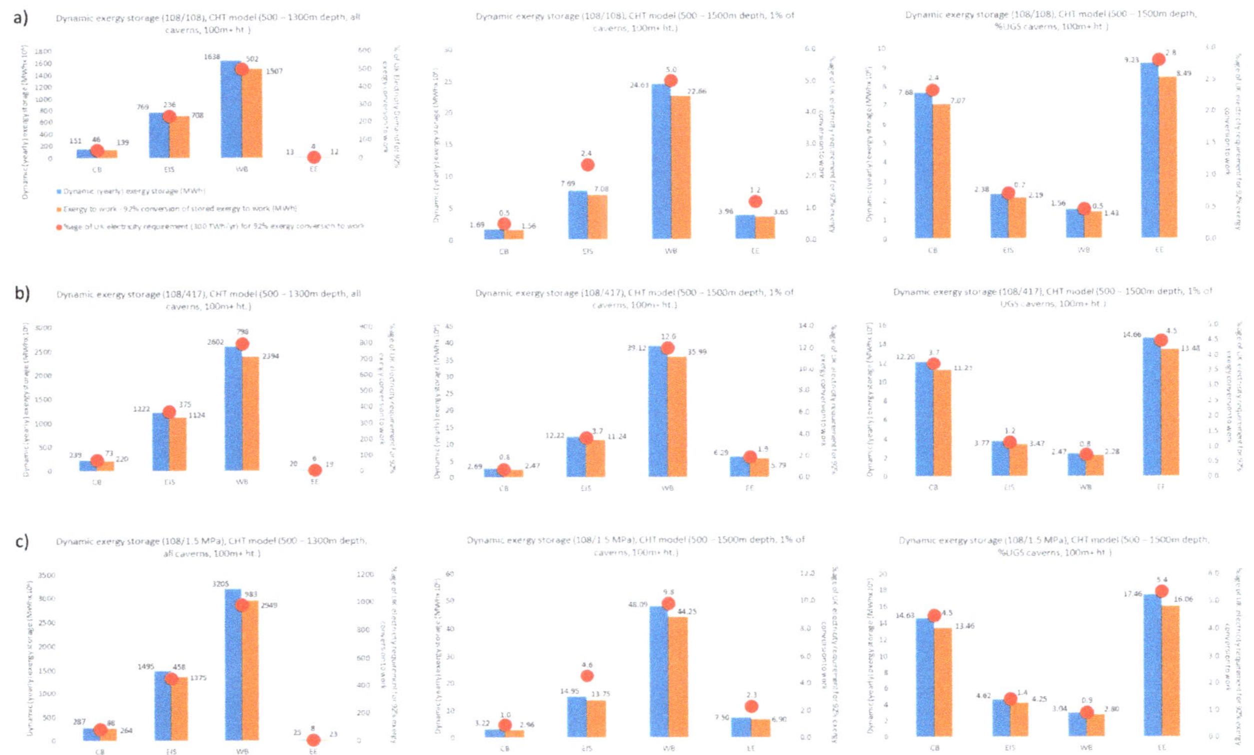

Figure 6. Plots of dynamic exergy storage and exergy to work estimates for the preferred CHT model over the depth range 500–1500 m and cavern heights 100 m+ considered for CAES. Parts (**a–c**) show graphs for differing injection/withdrawal rates (108/108 kg/s and 108/417 kg/s) or fill and pressure reduction rates (108 kg/s/1.5 MPa/h) for all potentially available caverns, 1% of available caverns, and estimates based upon the number of UGS caverns in the basins. Additionally shown, by basin, the percentage of UK electricity demand for 92% of stored exergy to work. Key common to all, see Figure 3.

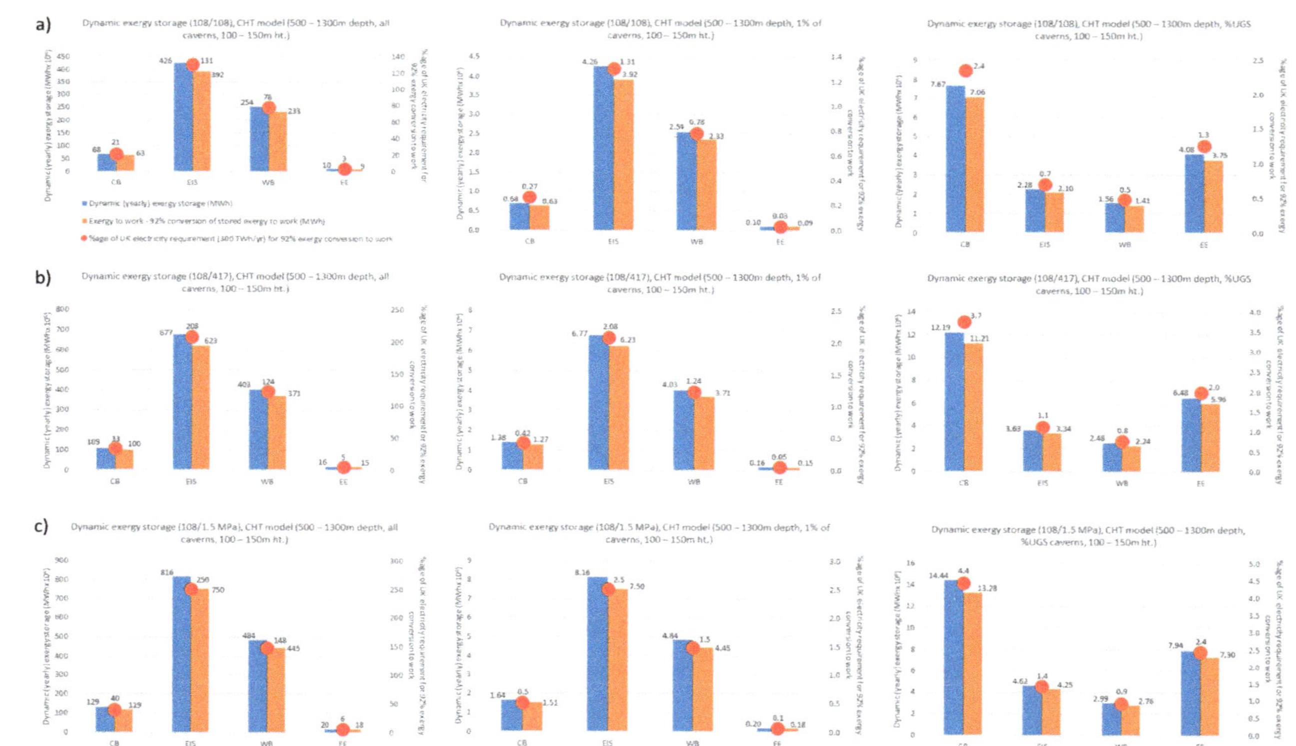

Figure 7. Plots of dynamic exergy storage and exergy to work estimates for the preferred CHT model, over the depth range 500–1300 m and cavern heights 100–150 m considered for CAES. Parts (**a**–**c**) show graphs for differing injection/withdrawal rates (108/108 kg/s and 108/417 kg/s) or fill and pressure reduction rates (108 kg/s/1.5 MPa/h) for all potentially available caverns, 1% of available caverns and estimates based upon the number of UGS caverns in the basins. Additionally shown, by basin, the percentage of UK electricity demand for 92% of stored exergy to work. Key common to all, see Figure 3.

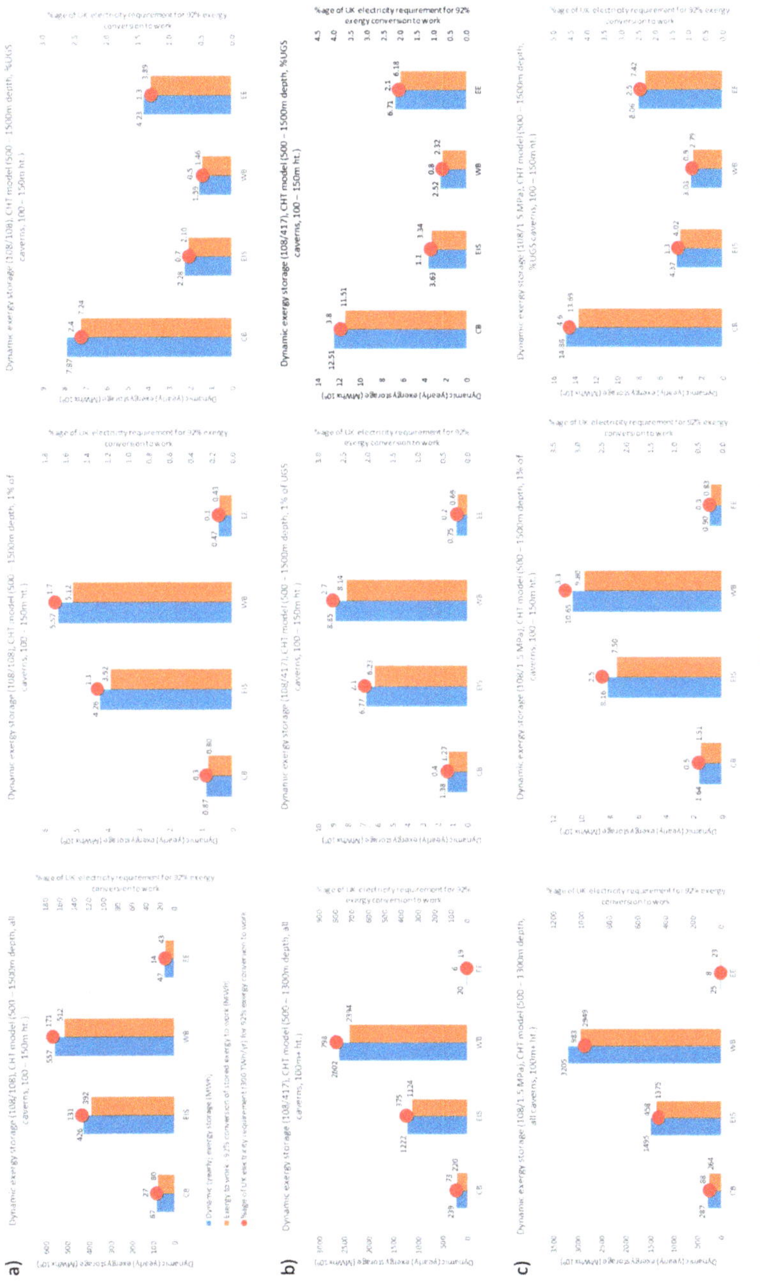

Figure 8. Plots of dynamic exergy storage and exergy to work estimates for the preferred CHT model, over the depth range 500–1500 m and cavern heights 100–150 m considered for CAES. Parts (**a**–**c**) show graphs for differing injection/withdrawal rates (108/108 kg/s and 108/417 kg/s) or fill and pressure reduction rates (108 kg/s/1.5 MPa/h) for all potentially available caverns, 1% of available caverns and estimates based upon the number of UGS caverns in the basins. Additionally shown, by basin, the percentage of UK electricity demand for 92% of stored exergy to work. Key common to all, see Figure 3.

In practice, realistic CAES cavern operation lies somewhere between the two end-member cases, and the convective heat transfer (CHT) wall condition for a practical (diabatic) cavern operational scenario was developed and is thought to more accurately represent actual storage conditions: during the cavern charging period, thermal energy of the air stored in the cavern is lost to the immediate surrounding rock mass, whilst the air temperature still increases due to the internal compression [38]. The two-end member scenarios produce slightly greater (isothermal) and smaller (adiabatic) exergy values, bracketing the CHT model (see Figure 3 and S2, Tables S3 and S4). Consequently, we have further refined the modelling tool for CHT conditions to implement their equations and predict the exergy stored when charging an uncompensated isochoric (constant volume, variable pressure) cavern or set of caverns. Results of this scenario are presented and discussed here.

Input parameters to the exergy modelling tool are summarised in S2, Table S2. Cavern surface areas and the calculation of heat transfer from the cavern void into the walls are necessary for CAES, estimates of which were derived relative to each cavern mid-point depth. They were calculated using the geothermal gradient for each specific basin, with an average annual surface temperature of 9.5 °C and pressure of 1 bar (14.5 psi). The tool imports the depths, volumes, temperatures, and min/max storage pressures calculated for each cavern and models iteratively, as well as the cavern-fill (exergy storage) from the starting point of the minimum to maximum permissible storage pressures. Results for the three differing cavern wall heat transfer models for each cavern over the two cavern depth ranges are output to a spreadsheet as the 'working exergy' storage in megawatt hours (MWh), together with the maximum pressure (pascals) and stored air mass (kg).

However, energy losses occur during generation, most notably through heat exchangers and in the turbines. From an energy and exergy analysis for 10 salt caverns of 100 m plus height in the Cheshire Basin, it was calculated that a full charge of all 10 caverns could store a net exergy of 25.32 GWh, of which ≈92% (23.19 GWh) could be converted to work via the turbines [41]. Therefore, alongside stored exergy estimates in Figures 3–8, we also present estimates of the stored exergy to work available, data behind which are provided in S2, Tables S3–S8.

4.4. Exergy Storage Assessment—Methodology

Figure 2 summarises the exergy storage assessment process. From borehole log data and map information in the public domain and held by the British Geological Survey, the tops, bases, and thicknesses of the halite deposits and major faults were mapped within each basin. These data were input to ArcGIS, which was used to obtain potential cavern locations, depths, and basic cavern parameters such as heights, diameters, available volumes, spacing and casing shoe depths based upon criteria applied to the design, development, and construction of gas storage caverns in the same strata [42]. The halite beds were evaluated over the depth ranges under consideration for CAES operations, with casing shoe depths (and thus pressures) in general between 500 and 1300 m [23] as well as up to 1500 m depth as at the proposed CAES plant at Larne, Northern Ireland [43]. Then, these basic cavern data were input to a modelling tool and used to estimate the exergy storage potential of prospective UK onshore and offshore East Irish Sea areas, using pressure and temperature ranges derived from gas storage investigations in these areas. Basic theoretical storage estimates are derived (Figure 3) that are only that and which, for various reasons, are clearly unrealistic totals. Most obviously, not all cavern locations will ultimately be available or suitable for cavern creation due to geological constraints, salt quality across the basin, together with economically and operationally viable cavern sizes. Therefore, a series of filters, based upon likely cavern height ranges, sizes, and differing storage operations, have been applied to derive more realistic CHT storage estimates for each basin. These can be compared to the annual UK electricity demand of 300 TWh [44].

In an attempt to obtain realistic assessments of the potential provided by the bedded halite resource, the total exergy storage estimates from each basin were filtered in a variety of ways:

- Taking 1% of the estimated exergy storage for the 'available' UK caverns
- Calculating the cavern storage estimates based on a percentage reflecting the number of operational or permitted UGS caverns in the UK (148) relative to the total number of possible caverns
- Filtering the caverns to include only those of greater than 100 m height
- Filtering the caverns to include only those of 100–150 m in height
- For the two filtered cavern height datasets, applying filters taking 1% of caverns and a percentage of the storage, using UGS cavern numbers relative to the total number of possible caverns in individual basins.

The figure of 1% is not based on industrial experience or previous studies. However, it provides a first pass understanding of the potential cavern numbers and storage capacities over the differing depth ranges, against which estimates set against the constructed or planned UGS cavern numbers in the two most developed basins can be evaluated (S2, Table S1): Cheshire Basin (73 caverns = 3.5–4%, for 500–1300 and 500–1500 m depths, respectively), Eastern England (37 caverns = 2–4%, respectively). Therefore, the figure of 1% is lower than these percentages and thus appears a reasonable gauge against which exergy storage estimates might be assessed initially (Figures 3 and 4). However, Figure 3 reveals very high storage estimates for the Wessex Basin, which is a potentially large region, but one in which the halite beds are less well characterised; halite beds were unknown in the area until oil and gas exploration began in the 1970s [37]. Consequently, further refinement of the estimates was attempted, reflecting more the degree of exploration and the proven potential and capabilities of the halite beds in each of the main halite basins. The greater numbers of storage caverns in the Cheshire Basin (73) and Eastern England (37) mean that these basins represent the most mature areas in terms of exploration and development of the halite beds. Thus, they potentially provide the more accurate and greater storage estimates when compared to using only the planned or permitted 24 and 14 cavern numbers for the lesser exploited EISB [45] and Wessex [46] basins, respectively. The latter two basins currently represent higher-risk target storage horizons, where in the case of the EIS, remoteness and its offshore location also increase CAPEX and OPEX costs of storage projects [45].

Additional efforts to derive realistic exergy storage estimates were also undertaken through filtering the storage outputs based upon cavern heights, with two sets of caverns assessed based on the experience of the maximum heights of gas storage caverns in the same salt beds, or those of proposed storage caverns in the Wessex Basin and the EIS. Firstly, caverns of 100 m and greater were selected, arising from the general sizes of UGS caverns developed or proposed in the same halite beds (refer S2, Table S1). Caverns smaller than $\approx$90 m in height are less economic to operate for gas storage purposes and are likely even more so for CAES due to the lower volumetric energy density ($\approx$2.4 kWh/m^3) of air in comparison to natural gas (1100 kWh/m^3) [23]. Importantly, those caverns in which diameters are much larger than cavern heights of a few tens of metres could be geomechanically less stable, with caverns thus requiring smaller diameters and thereby likely to also result in uneconomically small cavern volumes [47]. Secondly, very large (tall) caverns carry stability issues and operational limits for rapid cycle storage, and therefore, cavern heights were limited to 100–150 m. This is in part based upon the nature of halite beds in the Wessex Basin, where geophysical logs reveal that the insoluble content can comprise significant percentages of the Dorset Halite Member (DHM) [37] and are likely to significantly impact cavern volumes, stability, and location. This will likely limit areas of development to those with suitably clean halite for cavern construction. However, salt exploration boreholes for the construction of 14 gas storage caverns have proved a saliferous sequence 470 m thick in areas of the basin, with the main halite unit (referred to as 'S7') up to 140 m thick with a low insoluble content of $\approx$16.5% and in which it was assessed that caverns of 100 m in height could be constructed for the purposes of gas storage [46]. Elsewhere, the Winterborne Kingston Borehole in the NE of the basin proved halite beds to be 190 m thick [37,48]. Thus, constraining cavern heights to between 100 and 150 m was thought to be realistic for the UK in general and the Wessex Basin in particular. As previously, a percentage of potentially available caverns and volumes, based upon filtering for 1% and the percentage of UGS caverns relative to the UGS cavern numbers, was also extracted for the two cavern height ranges.

5. UK Salt Cavern Exergy Storage Capacity Estimates—Results and Commentary

We now summarise and present the exergy storage estimates and storage capacities (Figures 3–6) for salt caverns in four of the main UK halite-bearing sedimentary basins: onshore Cheshire, Wessex, eastern England, and the offshore East Irish Sea (Figure 1; S2, Tables S3–S8). Figure 3a,b illustrate the total theoretical UK 'static' (one-fill) exergy storage and work from stored exergy for the three models and two depth ranges considered. For the preferred CHT model conditions, the stored exergy to work available in caverns for the 500–1300 m depth range (274 MWh) would almost meet the annual UK electricity requirement of ≈300 TWh (Figure 3a), whilst for the depth range 500–1500 m, the stored exergy to work available from all three models would prove sufficient to meet UK electricity needs (Figure 3b). Taking just 1% of the potential caverns provides a 'static' exergy storage for the CHT scenario of between ≈3 and 4.7 TWh in the 500–1300 and 500–1500 m cavern depth ranges, respectively (Figure 3a,b, S2, Table S3). Cycled once a month, this could generate between 36 and 56.5 TWh of storage, or up to one-fifth of the UK electricity demand, illustrating the importance of this technology in providing a significant contribution to the UK's energy storage capacity and electricity supply. It should be noted that cavern numbers in eastern England are influenced strongly by depth, with much of the available halite and thus cavern volume being below 1300 m depth.

Taking a percentage of the 'static' exergy storage estimates derived from the numbers of operating or permitted UGS caverns (148) relative to the number of potential storage caverns (32,185 and 44,849), for the CHT scenario, exergy storage ranges from ≈1.37 TWh (500–1300 m depth range) to 1.53 TWh (500–1500 m) for the UK as a whole (Figure 3c,d, S2, Table S3). Figure 3e,f shows the influence of the less explored and characterised Wessex Basin halite beds on storage estimates, being significantly greater than other basins, suggesting perhaps 239–367 TWh of storage available and far greater than in the Cheshire Basin (8–9.3 TWh) or eastern England (2.6–39.7 TWh). Figure 3g,h show the effects on 'static' storage of applying a filter based upon the number of operational or planned UGS caverns in a basin, with the Wessex Basin storage reducing markedly to 0.24–0.37 TWh, similar to the Cheshire Basin (0.32 TWh) and eastern England (0.1–0.4 TWh).

To refine the estimates, the data were filtered to extract those caverns with heights of 100 m and greater and those caverns of 100–150 m in height, as described above. For caverns of 100 m and greater (Figure 4a,b, S2, Table S4), CHT 'static' exergy to work storage estimates for the basins range from between 6.6 (500–1300m) and 7.7 (500–1500 m) TWh in Cheshire to ≈210 to 348 TWh in the Wessex Basin, the latter skewing the estimates. Cycled once a month, this could generate between ≈79 and 95 TWh of storage in Cheshire and 12–300 TWh in eastern England. For caverns of 100–150 m in height, the results range from 2.6 and 3.8 TWh in the Cheshire Basin, to between 18.3 and 45.5 TWh in the Wessex Basin, the latter again being highest, although estimates appear more realistic than simply taking caverns of 100 m and greater, which takes much of the thick DHM interval as available. For reasons discussed above, cavern construction may not be feasible over much of the upper DHM across the basin. Cycled once a month, this could generate between ≈79 and 95 TWh of storage in Cheshire and 12–300 TWh in eastern England.

When the 'static' stored exergy to work estimates for each basin are assessed in relation to the numbers of operational or permitted gas storage cavern numbers in the basins over the two depth ranges (Figure 4c,d, S2, Table S4), then the potential exergy storage offered is highest in the Cheshire Basin (up to 0.30 TWh) and eastern England (0.25 to 0.32 TWh) areas, with much less estimated for the EIS (<0.1) and Wessex Basin (0.1 to 0.12 TWh). Cycled once a month this could generate between ≈3.6 TWh of storage in Cheshire and 3–3.84 TWh in eastern England.

Gas storage caverns are cycled more than once a year, and CAES caverns more than gas storage caverns, effectively increasing the 'static' working gas storage capacity and giving rise to a larger 'dynamic' working gas volume [39], or 'dynamic exergy storage', as considered here. At this stage, it is impossible to determine precise cavern depths, sizes, and temperatures and thus undertake detailed geomechanical and thermodynamic modelling for all potential cavern locations, volumes, storage pressures, operating scenarios, and cycle times. Therefore, we outline the processes behind an attempt to calculate the general 'dynamic exergy storage' potential for both the 100 m and greater, and the 100–150 m cavern sets. This was undertaken by estimating the number of cycles per year, which is based upon flow rates calculated from cavern fill and withdrawal rates in UGS and CAES operations and taking the average values for the outputs of maximum cavern pressure, exergy stored and air mass in caverns for CHT conditions from the exergy modelling tool (see S3 and S2, Table S5).

Thus, cavern emptying or withdrawal times were calculated for three scenarios based upon an injection phase (cavern charging), involving a conservative mass injection rate of 108 kg/s, as reported from the Huntorf CAES facility [9] and three differing withdrawal rates (generation or cavern discharge phase): 108 kg/s (equivalent to injection rate), 417 kg/s (from Huntorf [9]), and a general maximum pressure rate reduction of 15 bar/h (1.5 MPa/h) for gas storage operations [9,49]. For the latter, an approximate equivalent air mass withdrawal rate in kg/s was calculated to assess how realistic the rate might be for any particular scenario. For the higher flow rates, it may be that caverns would require more than one withdrawal well to achieve the air mass withdrawal rates. Then, the calculated injection and withdrawal rates were used to derive an estimate of the number of cycles per year and thus determine the 'dynamic' exergy capacity available for the different model categories (Figures 5–8, S2, Tables S6–S8).

Figures 5–8 illustrate the 'dynamic' exergy storage (and stored exergy to work) increases of the 100 m plus and 100–150 m cavern height subsets over the 'static' storage estimates presented in Figures 3 and 4. Dynamic exergy storage estimates based upon the Huntorf operational parameters (Figure 5a, Figure 6a, Figure 7a and Figure 8a, S2, Table S6) are lower than those using faster withdrawal rates, which increase the number of storage cycles possible (Figure 5b,c, Figure 6b,c, Figure 7b,c and Figure 8b,c, S2, Tables S7 and S8). The 'dynamic' exergy storage results illustrate more markedly the potentially significant contribution of exergy storage through CAES in salt caverns to the UK's energy storage capacity and electricity supply. This is highlighted by taking the Cheshire Basin as an example. Here, exergy storage estimates from the operational cycle based on 108 kg/s fill rates and 108 kg/s withdrawal rates suggest that caverns of 100 m and greater or 100–150 m in height over the two depth ranges have the potential to provide between ≈139–156 TWh (Figures 5a and 6a) and 63–80 TWh (Figures 7a and 8a) stored exergy to work respectively, meeting between ≈46–52% and 21–27% of the UK electricity demand respectively. For the operational cycle based on 108 kg/s fill rates and maximum withdrawal rates of 1.5 MPa/hr, stored exergy to work estimates range between ≈264–296 TWh (Figures 5c and 6c) and 119–151 TWh (Figures 7c and 8c) respectively, meeting between ≈88–98% and 40–50% of the UK electricity demand, respectively. Taking the estimates based on UGS cavern numbers for the two depth ranges and cavern sizes, stored exergy to work estimates could, using the most cycled operational mode (Figures 5c and 8c), provide between 13.3 and 13.7 TWh, respectively, meeting ≈4.5% of the UK electricity demand.

The other halite basins provide similarly important additional exergy storage and exergy to work support, with for example, just 1% of all caverns of 100 m and greater in the 500–1500 m depth range in the basins providing a further ≈34–65 TWh of work and meeting ≈12–22% of the UK electricity requirements, depending on the mode of operation and cycle numbers (Figure 6a–c, S2, Tables S6–S8). Whilst 1% of caverns 100–150 m height in the same depth range might provide a further ≈9.53–18.1 TWh of work, meeting ≈3.2–6.0% of the UK electricity requirements (Figure 8a–c, S2, Tables S6–S8). Relative to UGS cavern numbers in the basins, the figures for 100 m plus caverns in the depth range 500–1500 m might provide an additional ≈12.1–23.2 TWh of work, meeting ≈4–7.7% of the UK electricity requirements (Figure 6a–c, S2, Tables S6–S8), whilst 100–150 m height caverns might provide ≈7.5–14.2 TWh of work, meeting ≈1.5–4.7% of UK electricity requirements (Figure 8a–c, S2, Tables S6–S8). These data illustrate the potential importance of CAES and salt cavern storage to UK electricity demand and supply.

6. General Discussion

This study has attempted an estimate of the exergy storage (and stored exergy to work) potential of major bedded halite deposits of the UK onshore and offshore East Irish Sea areas. Storage would be using salt caverns constructed in the massively bedded halites and storage estimates are based on three thermodynamic models for the temperature and pressure variations within CAES caverns developed by ref [38]. Clearly, a number of significant assumptions and generalisations have been necessary when assessing entire sedimentary basins. However, current salt cavern hosted gas storage facilities prove that the UK halite beds studied are capable of hosting large, stable caverns for high-pressure gas storage. 'Static' theoretical storage volume is enough to meet the UK electricity demand of 300 TWh, although this is unrealistic. Various filters applied to the cavern storage data together with cycle numbers based upon gas storage operational parameters provide more realistic dynamic exergy storage and stored exergy to work estimates of at least 36 MWh, illustrating that salt caverns onshore and in the EIS could deliver significant EES and grid-scale support.

Estimates for future UK electrical energy storage capacity needs for a net-zero system in 2050 range from about 1 TWh in total [50] to the latest National Grid Future Energy Scenario (NGFES) estimates of about 200 GWh [51]. In both cases, the majority of capacity requirement will be for large-scale, long-duration energy storage, with CAES in all three NGFES net-zero scenarios contributing about 20–40 GWh. Currently, PHS accounts for the majority of the UK energy storage capacity, which has 2.8 GW power capacity and 27.6 GWh storage capacity. In 2019, the total energy discharged by PHS in the UK was 1.7 TWh, which is only about 1/70 of the total gas power generation. Due to the potential site-specific negative environmental and ecological impacts of PHS and the limited availability of favourable sites, further expansion of PHS capacity in the UK will be difficult. Lithium-ion battery storage and hydrogen are two promising technologies that may fulfil this required capacity. Lithium-ion batteries have attracted attention and undergone significant development in the last 5 years. However, the cost structure (high CAPEX of energy in $/kWh) renders it suitable only for mainly daily cycling applications, instead of energy storage operations at timescales greater than 10 h, even with a significantly reduced cost in future (e.g., $150/kWh) [52]. The design space for large-scale, long-duration electrical energy storage is plausibly set to be up to $20–40/kWh for balancing a grid with high-penetration (>90%) variable renewable energy generation [53,54]. Alternatively, hydrogen energy storage is at the other end of the storage spectrum, being particularly suitable for long-duration energy storage. Compared to technologies such as PHS, batteries, and CAES, hydrogen is still in the development phase of prototype or demonstration in order to validate its technical performance. Its cost reduction may require massive infrastructure construction (e.g., centralised electrolysis) that enables convenient transmission and distribution of hydrogen [54,55] and further research on currently less mature technologies such as high-temperature solid oxide or molten carbonate fuel cells that may enable low-cost scalable hydrogen production [56]. Either of these will add to the uncertainty in timescale and system-scale of the technology in decarbonising the power system. Although a diverse range of large-scale long-duration energy storage technologies are needed to deeply decarbonise electrical systems, technologies with relative high technology maturity and resource availability will help mitigate the risk and ensure an early and steady decarbonisation progress in the next decade, which may also help reduce the cost required for meeting the net-zero goal [57].

Amongst all the EES technologies, CAES is a relatively mature technology with large-scale conventional (diabatic) CAES having been commercially operational since 1978 at Huntorf, in systems of over 100 MW capacity and employing salt cavern storages. In this time, pilot ACAES plants have been considered, with a small (2 MW) pilot plant having operated between 2012 and 2016 in Texas [25,26], and the commissioning in 2019 of the world's first ACAES plant in Canada [11], also using salt caverns. Demonstration plants on the scale of 1–10 MW have been under appraisal (S1, Tables S1–S6) in Europe [30] and China, where there has also been a successful integration test of the world's first 100 MW CAES expander [58]. In comparison with other EES technologies, CAES has very low energy-storage costs ($3–6/kWh) [59], which makes it a cost-effective solution for long-duration grid-scale energy storage. The cost of CAES is described as low compared to all other energy storage technologies, which is evidenced by [59,60]. This includes the cost of hydrogen energy storage, amongst other energy storage technologies. The works are considered by the authors to be suitable resources for comparing the costs and other important performance methods of energy storage technologies as opposed to providing too much detail in this manuscript. Therefore, there exists the real possibility for the deployment of CAES to offer flexibility at a scale currently provided by fossil fuels in the system balance on various timescales from short duration (minute to hourly) to long duration (days/weeks). In contrast to the alternative large-scale storage technology, PHS, recent studies, and our analysis indicate that substantial exergy storage potential exists for CAES in the UK area. It is suggested that saline porous rocks (aquifers) in sedimentary basins of the UKCS area could provide inter-seasonal electricity storage amounting to approximately 160% of the UK's electricity consumption for January and February of 2017 [15]. However, this storage is offshore and distant to demand centres onshore. Additionally, whilst there has long been interest in the potential for CAES in porous rock formations [13,14], serious doubts exist over the likely development of porous rock CAES, with no plants having operated commercially and only a few, mostly small, test facilities having been constructed: a small 25 MW R&D CAES demonstration facility operated between 1987 and 1991 at Sesta, Italy [61], while an aquifer field test facility was built at Pittsfield, Illinois, USA and ran from 1982 through to mid-1984 [14,62]. Following eight years of investigations and research (2003–2011) funded by the US DOE at the Dallas Centre, Iowa USA, the Iowa Energy Storage Project, which aimed to develop a utility-scale, bulk energy storage facility linked to renewable wind energy, was shelved [12]. All of the above porous rock projects encountered problems with one of more of the following: pressure anomalies, variations in reservoir quality and performance, 'air bubble' formation in the reservoir, reaction between the oxygen of the injected air and minerals in the reservoir rock leading to oxygen depletion and/or potential for bacterial/micro-organism growth and porosity reduction. Aquifer storage for the UK, which would be remote offshore, thereby increasing costs, thus seems less likely than salt cavern storage, at least in the short term.

By contrast, our results illustrate the main halite-bearing strata of the UK onshore and East Irish Sea areas offer very significant CAES exergy storage possibilities and capacity, which being closer to demand could play a major role in grid support, load-levelling, and helping to meet the UK's annual electricity demand, which is currently at a level of $\approx$300 TWh [44]. Such resources in combination with renewable energy generation, particularly solar and wind, could replace the current flexible power generation at a national scale. Open-source data [63] illustrate that although the UK has achieved substantial carbon emission reductions in its power sector in the last decades by reducing the coal-based generation by almost 95%, from about 100 TWh in 2009 to 6 TWh in 2019, gas power is still an essential source in offering flexibility to maintain the second-by-second balance between the power supply (including intermittent renewable power) and varying demand. In 2019, gas power provided 114 TWh electricity that is 42% of all the electricity generated.

To decarbonise the gas power and provide the flexibility sacrificed, energy storage will play a significant role and the use of salt cavern-hosted CAES could underpin decarbonisation of the UK power system by offering large-scale flexibility over multiple timescales.

7. Conclusions

A study of the main halite-bearing strata of the UK onshore and East Irish Sea areas in which UGS caverns have been constructed or planned has been undertaken to assess their potential for the construction of salt caverns for CAES purposes and their exergy storage potential. Storage depths investigated are between 500 and 1500 m. Revisions to an earlier exergy modelling tool, equations for which were validated by operational data from the Huntorf CAES plant, have led to a series of exergy storage capacity estimates for three differing heat models. Both the 'static' one-fill exergy storage capacity and a series of 'dynamic' exergy storage capacities based on various fill and empty rates are derived. From a theoretical storage of over 300 TWh, more realistic storage estimates of many tens of TWh are achieved by way of filtering the estimates based on cavern dimensions and different storage cycles considering UGS projects and operational modes. Significant exergy storage capacity exists for CAES in salt caverns, which could provide important support to the UK electricity grid, requiring 300 TWh per year. As the contribution of intermittent renewables generation to the grid rises, it is suggested that salt cavern storages constructed onshore, rather than porous rock storages located offshore, are likely to be the main CAES storage technology available in the UK at least in the short term.

Supplementary Materials: The following are available online at https://www.mdpi.com/article/10.3390/app11114728/s1, S1, Tables S1–S6, S2, Tables S1–S8 and S3.

Author Contributions: Conceptualisation, D.E., J.B., J.W. (Jihong Wang) and S.G.; Formal analysis, D.E., D.P., M.D., P.W. and J.W. (John Williams); Funding acquisition, D.E., J.B., J.W. (Jihong Wang) and S.G.; Investigation, D.E., D.P., M.D., P.W. and J.W. (John Williams); Methodology, D.E., D.P., M.D., P.W., J.W. (John Williams) and W.H.; Project administration, D.E., J.W. (Jihong Wang) and S.G.; Software, D.P., M.D., P.W. and W.H.; Writing—original draft, D.E. All authors have read and agreed to the published version of the manuscript.

Funding: This research was funded from the UK Government's Engineering and Physical Science Research Council (EPSRC), UK grants EP/L014211/1 and EP/K002228/1.

Institutional Review Board Statement: Not applicable.

Informed Consent Statement: Not applicable.

Data Availability Statement: The data presented in this study are available on request from the corresponding author.

Acknowledgments: The authors thank three anonymous referees for comments on the manuscript all of which improved the final version. Financial support from the Engineering and Physical Science Research Council (EPSRC) funded Grand Challenge [Energy Storage] IMAGES project (EP/K002228) is gratefully acknowledged. This paper is published with permission of the Executive Director of the British Geological Survey (NERC).

Conflicts of Interest: The authors declare no conflict of interest. The funders had no role in the design of the study; in the collection, analyses, or interpretation of data; in the writing of the manuscript; or in the decision to publish the results.

References

1. BEIS. *UK Becomes First Major Economy to Pass NET Zero emissions Law New Target Will Require the UK to Bring all Greenhouse Gas Emissions to Net Zero by 2050*; Department for Business, Energy and Industrial Strategy (BEIS): London, UK, 2019.
2. EU. Climate Action and the European Green Deal. Available online: https://ec.europa.eu/clima/policies/eu-climate-action_en (accessed on 31 March 2021).
3. Elliott, D. A balancing act for renewables. *Nat. Energy* **2016**, *1*, 1–3. [CrossRef]
4. Crotogino, F.; Schneider, G.-S.; Evans, D.J. Renewable energy storage in geological formations. *Proc. IME Part A J. Power Energy* **2017**, *232*, 100–114. [CrossRef]

5. Succar, S.; Williams, R.H. Compressed Air Energy Storage: Theory, Resources, and Applications for Wind Power. Report Prepared by Princeton Environmental Institute, Princeton University, 8 April 2008. Available online: https://acee.princeton.edu/wp-content/uploads/2016/10/SuccarWilliams_PEI_CAES_2008April8.pdf (accessed on 17 May 2021).

6. Crotogino, F.; Donadei, S. Grid-scale energy storage in salt caverns. In Proceedings of the 8th International Workshop on Large-Scale Integration of Wind-Power into Power Systems as well as on Transmission Networks for Offshore Wind Farms, Bremen, Germany, 14–15 October 2009.

7. Budt, M.; Wolf, D.; Span, R.; Yan, J. A review on compressed air energy storage: Basic principles, past milestones and recent developments. *Appl. Energy* **2016**, *170*, 250–268. [CrossRef]

8. Evans, D.J.; Carpenter, G.; Farr, G. Mechanical Systems for Energy Storage—Scale and Environmental Issues. Pumped Hydroelectric and Compressed Air Energy Storage. In *Issues in Environmental Science and Technology No. 46 Energy Storage Options and Their Environmental Impact*; Hester, R.E., Harrison, R.M., Eds.; The Royal Society of Chemistry: London, UK, 2018; pp. 42–114.

9. Crotogino, F.; Mohmeyer, K.U.; Scharf, D.R. Huntorf CAES: More than 20 years of successful operation. In Proceedings of the Spring 2001 Meeting, Orlando, FL, USA, 15–18 April 2001.

10. PowerSouth. Compressed Air Energy Storage. *McIntosh Powerplant, McIntosh, Alabama. PowerSouth Energy Cooperative Brochure.* 2017. Available online: http://www.powersouth.com/wp-content/%20uploads/2017/07/CAES-Brochure-FINAL.pdf (accessed on 2 April 2018).

11. Peretick, K. Energy Storage in the Real World: Practical Applications and Considerations. Michigan Agency for Energy "Plugging Into Storage—Energy Storage Symposium", Held 25 March 2019. Available online: https://www.michigan.gov/documents/energy/Energy_Storage_Session_4-Energy_Storage_in_the_Real_World_Peretick_652341_7.pdf (accessed on 17 May 2021).

12. Schulte, R.H.; Critelli, N., Jr.; Holst, N.; Huff, G. *Lessons from Iowa: Development of a 270 Megawatt Compressed Air Energy Storage Project in Midwest Independent System Operator: A Study for the DOE Energy Storage Systems Program*; Sandia National Laboratories: Albuquerque, NM, USA, 2012. Available online: https://www.sandia.gov/ess-ssl/publications/120388.pdf (accessed on 17 May 2021).

13. General Electric. Economic & Technical Feasibility of CAES. General Electric Co. Final Report Prepared for the United States Energy Research and Development Administration Office of Conservation, Contract E (11-1)-2559. 1976. Available online: http://www.osti.gov/scitech/servlets/purl/7195898/ (accessed on 17 May 2021).

14. Allen, R.D.; Doherty, T.J.; Kannberg, L.D. *Summary of Selected Compressed Air Energy Storage Sites*; Report prepared for the U.S. Department of Energy; Pacific Northwest Laboratory for Battelle: Richland, WA, USA, 1985.

15. Mouli-Castillo, J.; Wilkinson, M.; Mignard, D.; McDermott, C.; Haszeldine, S.R.; Shipton, Z.K. Inter-seasonal compressed air energy storage using saline aquifers. *Nat. Energy* **2019**, *4*, 131–139. [CrossRef]

16. UK Gov. New Plans to Make UK World Leader in Green Energy—The Announcement Is Part of the Government's Commitment towards Net Zero Emissions by 2050 and Will Support 60,000 Jobs; Johnson, B.H., Sharma, H.A., Eds.; UK Government Press Release. 2020. Available online: https://www.gov.uk/government/news/new-plans-to-make-uk-world-leader-in-green-energy (accessed on 17 May 2021).

17. Aghahosseini, A.; Breyer, C. Assessment of geological resource potential for compressed air energy storage in global electricity supply. *Energy Convers. Manag.* **2018**, *169*, 161–173. [CrossRef]

18. United States Department of Energy Global Energy Storage Database. Available online: http://www.energystorageexchange.org/ (accessed on 5 December 2019).

19. Díaz-González, F.; Sumper, A.; Gomis-Bellmunt, O.; Villafáfila-Robles, R. A review of energy storage technologies for wind power applications. *Renew. Sustain. Energy Rev.* **2012**, *16*, 2154–2171. [CrossRef]

20. Rosenberg, D.M.; Bodaly, R.A.; Usher, P.J. Environmental and social impacts of large scale hydro-electric development: Who is listening? *Glob. Environ. Chang.* **1995**, *5*, 127–148. [CrossRef]

21. Renewables, S. *The Benefits of Pumped Storage Hydro to the UK, Scottish Renewables*; DNV GL: Glasgow, UK, 2016.

22. Crotogino, F.; Huebner, S. Energy storage in salt caverns-developments & concrete projects for adiabatic compressed air and for hydrogen storage. In Proceedings of the SMRI Spring Meeting, Porto, Portugal, 27–30 April 2008; p. 179.

23. Kepplinger, J.; Crotogino, F.; Donadei, S.; Wohlers, M. Present Trends in Compressed Air Energy and Hydrogen Storage in Germany. In Proceedings of the Solution Mining Research Institute (SMRI) Fall 2011 Technical Conference, York, UK, 3–4 October 2011; p. 13.

24. Denholm, P.; Kulcinski, G.L. Life cycle energy requirements and greenhouse gas emissions from large scale energy storage systems. *Energy Convers. Manag.* **2004**, *45*, 2153–2172. [CrossRef]

25. General Compression. Texas Dispatachable Wind 1, LLC. 2014. Available online: http://www.generalcompression.com/index.php/tdw1 (accessed on 8 November 2017).

26. Wilson, T.; Turaga, U. Oil Majors Pursue Energy Storage. 2016, Online Article by ADI Analytics, 29 July 2016. Available online: http://adi-analytics.com/2016/07/29/oil-majors-pursue-energy-storage-technology/ (accessed on 26 November 2016).

27. Medeiros, M.; Booth, R.; Fairchild, J.; Imperato, D.; Stinson, C.; Ausburn, M.; Tietze, M.; Irani, S.; Burzlaff, A.; Moore, H.; et al. *Technical Feasibility of Compressed Air Energy Storage (CAES) Utilizing a Porous Rock Reservoir*; Final Report; DOE-PGE-00198-1; Pacific Gas & Electric Company: San Francisco, CA, USA, 2018. Available online: https://www.osti.gov/servlets/purl/1434251 (accessed on 17 May 2021).

28. Grubelich, M.C.; Bauer, S.J.; Cooper, P.W. *Potential Hazards of Compressed Air Energy Storage in Depleted Natural Gas Reservoirs*; Sandia National Laboratories: Albuquerque, NM, USA, 2011.

29. Funk, J. First Energy Postpones Project to Generate Electricity with Compressed Air. Online Article, Posted 5 July 2013 at 1:44 PM. Available online: https://www.cleveland.com/business/index.ssf/2013/07/firstenergy_postpones_project.html (accessed on 17 May 2021).

30. Zanganeh, G.; Calisesi, Y.; Brüniger, R. *Demonstration of the Ability of Caverns for Compressed Air Storage with Thermal Energy Recuperation*; ALACAES SA Via Industria 10 CH-6710 Biasca; ALACAES: Lugano, Switzerland, 2016; 98p. Available online: https://www.aramis.admin.ch/Default?DocumentID=35281&Load=true (accessed on 17 May 2021).

31. Perillo, G. Report on Material Qualification Including Guidelines for Material Selection/Development. Design Study for European Underground Infrastructure Related to Advanced Adiabatic Compressed Air Energy Storage. RICAS2020 Deliverable D6.1 Report, 2016, 67p. Available online: http://www.ricas2020.eu/media/deliverable-d4-1-numerical-simulations/ (accessed on 17 May 2021).

32. EPRI. *Compressed Air Energy Storage Newsletter*; Electric Power Research Institute (EPRI): Palo Alto, CA, USA, 2010; p. 7.

33. EPRI. *Compressed Air Energy Storage Newsletter*; Electric Power Research Institute (EPRI): Palo Alto, CA, USA, 2012; p. 8.

34. Pimm, A.; Garvey, S.D.; de Jong, M. Design and testing of Energy Bags for underwater compressed air energy storage. *Energy* **2011**, *66*, 496–508. [CrossRef]

35. Li, P.Y.; Loth, E.; Simon, T.W.; Van de Ven, J.D.; Crane, S.E. Compressed Air Energy Storage for Offshore Wind Turbines. In Proceedings of the International Fluid Power Exhibition (IFPE), Las Vegas, NV, USA, 22–26 March 2011; p. 7.

36. Garvey, S.D. Integrating Energy Storage with Renewable Energy Generation. *Wind Eng.* **2015**, *39*, 129–140. [CrossRef]

37. Evans, D.J.; Holloway, S. A review of onshore UK salt deposits and their potential for underground gas storage. In *Underground Gas Storage: Worldwide Experiences and Future Development in the UK and Europe*; Evans, D.J., Chadwick, R.A., Eds.; The Geological Society: London, UK, 2009; pp. 39–80.

38. He, W.; Luo, X.; Evans, D.; Busby, J.; Garvey, S.; Parkes, D.; Wang, J. Exergy storage of compressed air in cavern and cavern volume estimation of the large-scale compressed air energy storage system. *Appl. Energy* **2017**, *208*, 745–757. [CrossRef]

39. Mott MacDonald. *Preesall Underground Gas Storage Facility: Geological Summary Report, 2014*; Mott MacDonald for Halite Energy Group: Altrincham, UK, 2014; p. 169.

40. Cameron, T.D.J.; Crosby, A.; Balson, P.S.; Jeffrey, D.H.; Lott, G.K.; Bulat, J.; Harrison, D.J. *United Kingdom Offshore Regional Report: The Geology of the Southern North Sea*; HMSO: London, UK, 1992.

41. Dooner, M.; Wang, J. Potential Exergy Storage Capacity of Salt Caverns in the Cheshire Basin Using Adiabatic Compressed Air Energy Storage. *Entropy* **2019**, *21*, 1065. [CrossRef]

42. Parkes, D.; Evans, D.J.; Williamson, P.; Williams, J.D.O. Estimating available salt volume for potential CAES development: A case study using the Northwich Halite of the Cheshire Basin. *J. Energy Storage* **2018**, *18*, 50–61. [CrossRef]

43. Haughey, C. Larne CAES: A Project Update In Gaelectric Energy Storage: The Missing Link. Gaelectric. 2015. Available online: http://www.gaelectric.ie/wpx/wp-content/uploads/2015/09/Gaelectric-Supplement-June-2015.pdf (accessed on 17 May 2021).

44. Dukes. Digest of United Kingdom Energy Statistics 2019: Chapter Five-Electricity. Digest of UK Energy Statistics (DUKES), Department for Business, Energy & Industrial Strategy (BEIS), Published 25 July 2019, p. 182. Available online: https://assets.publishing.service.gov.uk/government/uploads/system/uploads/attachment_data/file/820708/Chapter_5.pdf (accessed on 25 July 2019).

45. Gateway. *Gateway Gas Storage Project Offshore Environmental Statement: Non-Technical Summary*; Gateway Gas Storage: Edinburgh, UK, 2007; p. 282.

46. Wilke, F.H.; Wippich, M.G.E.; Zündel, F. *Outline Design of the Proposed Gas Storage Project at 'Upper Osprey' on the Isle of Portland, Dorset*; DEEP: Bad Zwischenahn, Germany, 2007; p. 67.

47. Charnavel, Y.; Durup, G. First Gaz de France Horizontal Salt Cavern Experiment. In Proceedings of the Spring 1998 Meeting, New Orleans, LA, USA, 19–22 April 1998; p. 14.

48. Rhys, G.H.; Lott, G.K.; Calver, M.A. *The Winterborne Kingston Borehole, Dorset, England*; Report of the Institute of Geological Sciences, Report Number CF/81-03; HMSO: London, UK, 1982.

49. Bérest, P.; Brouard, B.; Karimi-Jafari, M.; Pellizzaro, C. Thermomechanical aspects of high frequency cycling in salt storage caverns. In Proceedings of the International Gas Union Research Conference 2011, Seoul, Korea, 19–21 October 2011.

50. ETI. Catapult Energy Systems—Balancing Supply and Demand. Report funded by the Energy Technologies Institute (ETI), Published September 2019, p. 7. Available online: https://es.catapult.org.uk/reports/balancing-supply-and-demand/?download=true (accessed on 30 September 2019).

51. National Grid. Future Energy Scenarios. 2020, p. 124. Available online: https://www.nationalgrideso.com/document/173821/download (accessed on 17 May 2021).

52. Holladay, J.D.; Hu, H.; Kinf, D.L.; Wang, Y. An overview of hydrogen production technologies. *Catal. Today* **2009**, *139*, 244–260. [CrossRef]

53. Ziegler, M.S.; Mueller, J.M.; Pereira, G.D.; Song, J.; Ferrara, M.; Chiang, Y.M.; Trancik, J.E. Storage requirements and costs of shaping renewable energy toward grid decarbonization. *Joule* **2019**, *3*, 2134–2153. [CrossRef]

54. Albertus, P.; Manser, J.S.; Litzelman, S. Long-duration electricity storage applications, economics, and technologies. *Joule* **2020**, *4*, 21–32. [CrossRef]

55. Schmidt, O.; Gambhir, A.; Staffell, I.; Hawkes, A.; Nelson, J.; Few, S. Future cost and performance of water electrolysis: An expert elicitation study. *Int. J. Hydrog. Energy* **2017**, *42*, 30470–30492. [CrossRef]
56. Davis, S.J.; Lewis, N.S.; Shaner, M.; Aggarwal, S.; Arent, D.; Azevedo, I.L.; Benson, S.M.; Bradley, T.; Brouwer, J.; Chiang, Y.-T.; et al. Net-zero emissions energy systems. *Science* **2018**, *360*, 6396. [CrossRef]
57. Victoria, M.; Zhu, K.; Brown, T.; Andresen, G.B.; Greiner, M. Early decarbonisation of the European energy system pays off. *Nat. Commun.* **2020**, *11*, 1–9. [CrossRef]
58. Beijing Science and Technology Plan Project "Development and Demonstration of Large-Scale Advanced Compressed Air Energy Storage System" Passed the Acceptance Test. Available online: http://www.iet.cas.cn/xwdt/zhxw/201902/t20190211_5240114.html (accessed on 17 May 2021).
59. Luo, X.; Wang, J.; Dooner, M.; Clarke, J. Overview of current development in electrical energy storage technologies and the application potential in power system operation. *Appl. Energy* **2015**, *137*, 511–536. [CrossRef]
60. Mongird, K.; Viswanathan, V.; Balducci, P.; Alam, J.; Fotedar, V.; Koritarov, V.; Hadjerioua, B. An Evaluation of Energy Storage Cost and Performance Characteristics. *Energies* **2020**, *13*, 3307. [CrossRef]
61. Ter-Gazarian, A. Compressed air energy storage. In *Energy Storage for Power Systems*; Peter Peregrinus Ltd., on behalf of the Institution of Electrical Engineers: London, UK, 1994; p. 197.
62. EPRI. *Compressed Air Energy Storage Newsletter*; Electric Power Research Institute (EPRI): Palo Alto, CA, USA, 2011; p. 10.
63. Available online: http://www.gridwatch.templar.co.uk/download.php (accessed on 17 May 2021).

any isolating properties (like salt rock). Then, the shotcrete reinforcement, followed by the necessary installations, like drainage, are made. Final layer is the sealing lining. In order to ensure isolation and erosion, proven properties of sealing material are essential. A scheme of the LRC linings is shown in the Figure 1. In this technology, the most common lining material is steel [10,11]. It is justified to seek substitutional sealing materials beside stainless-steel, which will make the LRC storage more available and economical. One of the potential solutions is to use polymer-based concrete or epoxy resins, which are characterized by good sealing efficiencies. Research of gas permeability was done, including different methods, different polymer materials, and different gases. These studies include, inter alia, permeability of epoxy composites, using helium and CO_2 [12,13], permeability of N_2, O_2, CO_2, and H_2O through number of polymer materials (inter alia LDPE, HDPE, PVC, polypropylene) [14], and hydrogen permeability of high density polyethylene (HDPE) [15]. However, research related with hydrogen permeability through epoxy resins are very limited [16,17].

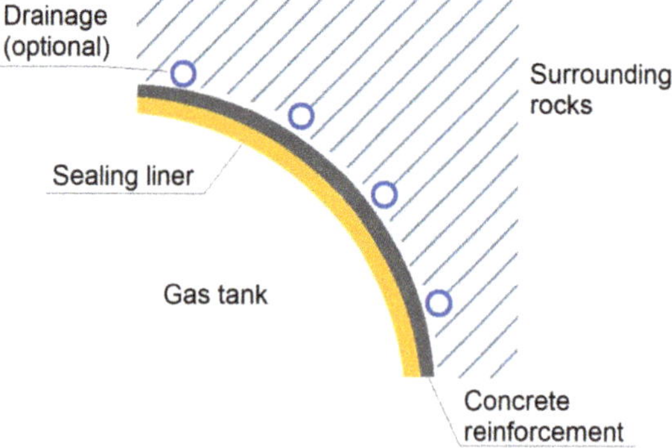

Figure 1. Scheme of the LRC linings.

The purpose of this work is to compare hydrogen permeability of different materials, which could be used as the sealing liner for hydrogen storage in LRC with focus on epoxy resins modified with various additives. Additionally, permeability of typical rocks that can be found around LRC, and rock salt, were also measured in order to compare results with available data. In this work, a setup combining Steady-State Flow Method and Carrier Gas Method was used for experiments. Samples were also examined under SEM to compare their structural properties.

2. Experimental Methodology and Materials

Setup used to investigate the hydrogen permeability was built in the Laboratory of Unconventional Gas and CO_2 Storage at the Silesian University of Technology. Setup is combining the Steady-State Flow Method and Carrier Gas Method. Setup can be switched between mentioned methods, depending on the permeability coefficient of investigated samples. A scheme of the setup is shown in the Figure 2. It consists of the gas cylinder, pressure regulation valves, sample holder, precise gas pressure transducers, and gas concentration detector.

The upstream side is a reference gas. A blend of 10% of hydrogen in methane was used as a reference gas on the upstream side. Gas is applied on the sample, held in the PVC sleeve with confining pressure of water. Tests were conducted at 1.0 MPa feed gas pressure. During the test, the downstream side was filled with carrier gas (helium) at the pressure of 100 kPa. A single gas detector for hydrogen, with sensitivity of 2–2000 ppm, was plugged on the end of the setup for gas concentration measurements made periodically. After the measurement is done, the same socket is used to plug the carrier gas (helium), the downstream side is filled with after vacuuming. There is also a back pressure valve, which can be used to adjust the pressure on the downstream side, when there is a flow of gas through the sample. Back pressure valve is giving the possibility to set a steady pressure gradient through the sample, by adjusting the pressure on the upstream and downstream side.

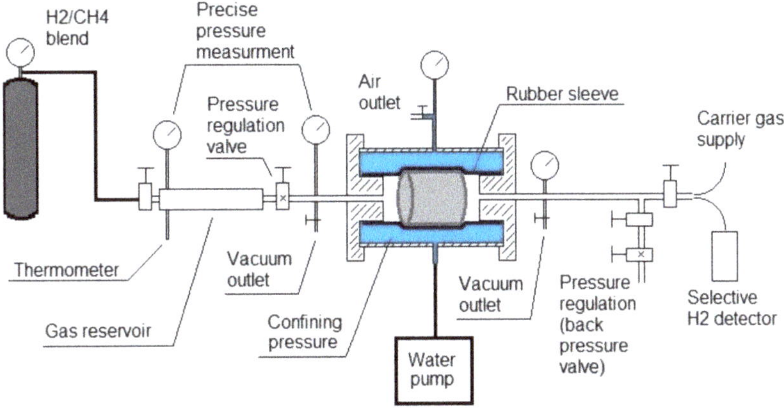

Figure 2. Steady-State Method/Carrier Gas Method setup.

Depending on the behavior of the sample, a proper method can be used. When a pressure increase on the downstream side is observed, there is a gas flow through the sample. It occurs, when the permeability of the sample is high enough to allow the gas to flow through the sample. Steady-State Flow Method is based on the pressure gradient in the sample, caused by the gas flow. Pressure gradient in the sample and gas pressure (together with temperature) in the reservoir are measured with precise pressure transducers (accuracy 0.5% FS). Filtration coefficient "k" can be calculated, using the Equations (1) and (2).

$$q = \left(\frac{\left(\frac{p_i \cdot V_{res}}{Z_i \cdot R \cdot T} \right) - \left(\frac{p_f \cdot V_{res}}{Z_f \cdot R \cdot T} \right)}{t} \right) \cdot 22.4 \cdot 10^{-3} \tag{1}$$

where:

q = gas flow, m^3/s
p_i, p_f = initial pressure (p_i) and final pressure (p_f) in reservoir, Pa
V_{res} = reservoir volume, m^3
Z_i, Z_f = gas compressibility factor at the initial (Z_i) and final (Z_f) pressure
R = gas constant (8.314463 m$^3 \cdot$Pa$\cdot$mol$^{-1} \cdot$K^{-1})
T = gas temperature in reservoir, K
t = time, s

$$k = \frac{2 \cdot q \cdot p_o \cdot L \cdot \mu}{A \cdot \left(p_i^2 - p_o^2 \right)} \tag{2}$$

where:

k = permeability coefficient, m^2
q = gas flow, m^3/s
p_i, p_o = average inlet (p_i) and outlet (p_o) gas pressure (pressure gradient), Pa
L = sample length, m
μ = gas viscosity, Pa$\cdot$s
A = sample cross area, m^2

When the gas permeability of the sample is low enough to prevent the gas flow through the sample, gas diffusion can be measured. In this case, Carrier Gas Method is used, which is based on the difference in gas concentration gradient through the sample. It is possible to measure it precisely with a single-gas hydrogen detector. Knowing the parameters of a reference gas on the upstream side, calculations of the permeability coefficient P are made, using a set of Equations (3)–(6). Using helium as a carrier gas, an ideal gas law can be assumed.

$$V = \frac{R \cdot T}{p} \tag{3}$$

where:

V = molar volume in test conditions, $m^3 \cdot mol^{-1}$
R = gas constant: 8.314463 $J \cdot mol^{-1} \cdot K^{-1}$
T = gas temperature, K
p = gas pressure, Pa

An amount of hydrogen diffused through the sample in a certain time can be calculated, using the Equation (4). In this case the volume of the downstream side of the setup needs to be determined. This was done by approximation of the inner volume of the pipes, based on the manufacturer's data and measurements of the setup. The volume of the downstream in the setup was 12.0 cm^3

$$N_{H_2} = \frac{c \cdot \left(\frac{N_A}{V}\right) \cdot v_{downstream}}{10^6} \tag{4}$$

where:

N_{H2} = amount of hydrogen, diffused through the sample
c = measured hydrogen concentration on downstream side, ppm
N_A = Avogadro's constant: $6.02214076 \cdot 10^{23}$, mol^{-1}
V = molar volume, $m^3 \cdot mol^{-1}$
$v_{downstream}$ = volume of downstream side of setup, m^3

By knowing the molar volume of one mole of gas in Standard Temperature and Pressure conditions (STP: 0 °C, 100 kPa), the volume of gas in STP can be calculated, using Equation (5).

$$V_{H_2} = \frac{N_{H_2} \cdot 22.414}{N_A} \tag{5}$$

where:

V_{H2} = volume of hydrogen diffused through the sample, cm^3STP
N_A = Avogadro's constant: $6.02214076 \cdot 10^{23}$, mol^{-1}
22.414 = mole volume of ideal gas in STP, cm^3

Using the calculation above, gas permeability coefficient can be calculated with Equation (6).

$$P_{H_2} = \frac{V_{H_2} \cdot l}{A \cdot t \cdot p} \tag{6}$$

where:

P_{H2} = hydrogen permeability coeficcient, barrer
V_{H2} = volume of hydrogen diffused through the sample, cm^3STP
l = sample length, cm
A = sample cross section area, cm^2
t = time, s
p = gas pressure (feed gas), cmHg (1 bar = 75 cmHg)

Permeability coefficient of hydrogen is given in Barrer unit [18], presented in Equation (7). Barrer unit is commonly used for presenting the permeability of membranes. However, it does not refer to the SI system, because of the pressure given in cmHg. Transforming the pressure into the SI unit (for example bar or Pa) is possible, however the results will not be comparable with the common literature data.

$$1 \, barrer = \frac{cm^3_{STP} \cdot cm}{cm^2 \cdot s \cdot cmHg} 10^{-10} \tag{7}$$

Hydrogen diffusion is calculated as a diffusion coefficient D, given in m^2/s, using the Fick's law [19] and following set of Equations (8)–(12), as well as values received from the Equations (3) and (4). For a gas diffusion through the polymer materials, the transport mechanism described by Fick's Law is accepted [20].

$$J = -D\frac{\partial \phi}{\partial x} \tag{8}$$

where:

J = flux, amount of hydrogen diffusing through the area in time, $mol \cdot m^{-2} \cdot s^{-1}$
D = diffusion coefficient, $m^2 \cdot s^{-1}$
ϕ = hydrogen concentration, $mol \cdot m^{-3}$
x = length of hydrogen path, m

$$D = -J\frac{x_2 - x_1}{\phi_{downstream} - \phi_{upstream}} \tag{9}$$

where:

D = diffusion coefficient, $m^2 \cdot s^{-1}$
J = flux, amount of hydrogen diffused through the area in time, $mol \cdot m^{-2} \cdot s^{-1}$
$\phi_{upstream}$ = concentration of hydrogen (upstream), $mol \cdot m^{-3}$
$\phi_{downstream}$ = concentration of hydrogen (downstream), $mol \cdot m^{-3}$
$x_2 - x_1$ = diffusion distance (sample length), m

For a non-porous polymer material, the tortuosity is equal to 1, thus the assumption can be made, that the hydrogen path is equal to sample length [21].

$$J = -\frac{\left(\frac{N_{H2}}{N_A}\right)}{A \cdot t} \tag{10}$$

where:

J = flux, amount of hydrogen diffused through the area in time, $mol \cdot m^{-2} \cdot s^{-1}$
N_{H2} = amount of hydrogen elements, diffused through the sample, from Equation (4)
N_A = Avogadro's constant: $6.02214076 \times 10^{23}$, mol^{-1}
A = sample cross area, m^2
t = time, s

$$\phi_{upstream} = \frac{c_{up}}{V \cdot 100} \tag{11}$$

where:

$\phi_{upstream}$ = concentration of hydrogen (upstream), $mol \cdot m^{-3}$
c_{up} = concentration of hydrogen in refenrence gas, %
V = molar volume of gas $mol \cdot m^{-3}$, from Equation (3)

$$\phi_{downstream} = \frac{\left(\frac{N_{H2}}{v_{downstream}}\right)}{N_A} \tag{12}$$

where:

$\phi_{downstream}$ = concentration of hydrogen (downstream), $mol \cdot m^{-3}$
N_{H2} = amount of hydrogen elements, diffused through the sample, from Equation (4)

$v_{downstream}$ = volume of downstream side of setup, m^3
N_A = Avogadro's constant: $6.02214076 \times 10^{23}$, mol^{-1}

Measurements of concentrations at the downstream side were done every 2–4 days and given in the diffusion ratio in time (ppm of H_2/24 h). Structure of the samples was investigated, using HRSEM SUPRA 35 (Carl Zeiss AG, Oberkochen, Germany) Scanning Electron Microscope, since the chemical composition of the samples was not the scope of this study; the Energy Dispersive Spectrometry was not used for this research.

For the purpose of the study, several materials that could be used as a liner in underground excavation were selected. Samples under investigation can be divided into three general groups: epoxy resins with different additives; concrete; polymer-concrete and rocks (mudstone and rock salt). Materials were selected based on their common availability, low cost, and ease of preparation. Resin samples were prepared by mixing resin and cured for at least 7 days. During the pot time, samples were held in the oven in a temperature of 30 °C to accelerate the venting of the samples. Relative low temperature was set to simulate the thermal conditions in the excavations. Higher temperatures would be more efficient in the samples venting but would not be related with the actual underground temperature conditions, where the resins might be applied. Concrete and polymer-concrete samples were prepared by companies, according to their recipes and general standards for curing the concretes. Concrete samples were cored using a diamond core drill of 2.54 mm (1 inch) diameter. In Table 1 detailed description of samples is given.

Table 1. Details of investigated samples.

Sample	Base	Physical Properties	Additives
Concrete "2"	CEM II *	Uniaxial strength: 16.0 MPa Water/cement ratio: n.a. Casted 3 years prior to experiment	Furnace slag (silica fumes) < 35%
Concrete "1"	CEM I *	Uniaxial strength: 17.0 MPa Water/cement ratio: 0.47 Casted 6 months prior to experiment	Limestone 7%
Commercial polymer-concrete "14-3"	Classified due to company's policy	Casted 3 month prior to experiment	Classified due to company's policy
Commercial polymer-concrete "G1"	Classified due to company's policy	Casted 3 month prior to experiment	Classified due to company's policy
Geopolymer	Classified due to company's policy	Uniaxial strength: 17.0–22.0 MPa Casted 3 month prior to experiment	Classified due to company's policy
Mudstone (Carbon)	Clay minerals	-	-
Salt rock (Permian) (before creep)	Sodium chloride	-	-
Salt rock (Permian) (after creep)	Sodium chloride	-	-
Epoxy resin	2.2-Bis (4-hydroxyphenyl) propane with epichlorohydrin resin-hardener ratio: 100:12	Viscosity: 15,000–30,000 mPa·s Epoxide number: 0.48–0.52 mol/100 g Chlorine content: <0.6% Pot time: 90 min.	Mechanical impurities: <0.03%
Epoxy resin +graphite (5% vol.)			Amorphous graphite < 50 μm
Epoxy resin +haloysite (5% vol.)			Grinded halloysite <125 μm
Epoxy resin +fly ash (5% vol.)			Sieved fly ash <125 μm
Epoxy resin +fly ash (30% vol.)			Sieved fly ash <125 μm

* CEM I, CEM II—cement types according to Polish Standard: PN-EN 197-1:2012, PN-B-19707:2013.

3. Results

Results of gas permeability of investigated samples are shown in Table 2. For the ease of comparison with other data, results are given in different units: permeability coefficient for hydrogen (P_{H2})($cm^3STP\cdot cm\cdot cm^{-2}\cdot s^{-1}\cdot cmHg^{-1}$ (Barrer)); diffusion coefficient ($m^2\cdot s^{-1}$); and filtration coefficient (m^2 and mD (milidarcy)). Blend of hydrogen (10%) in methane was used for the permeability tests. Results refer to this particular type of gas. For impermeable samples (10^{-11} $cm^3STP\cdot cm\cdot cm^{-2}\cdot s^{-1}\cdot cmHg^{-1}$ or lower), only hydrogen was permeating the sample, so these results refer to the pure hydrogen.

Table 2. Gas permeability of investigated samples.

Sample	Permeability Coefficient P_{H2}		Diffusion Coefficient D	Filtration Coefficient k
	($cm^3STP\cdot cm\cdot cm^{-2}\cdot s^{-1}\cdot cmHg^{-1}$)	Barrer	m^2/s	m^2 mD
Concrete "2"	4.170×10^{-4}	4.170×10^6	-	2.67×10^{-16} 0.2703
Concrete "1"	7.804×10^{-5}	7.804×10^5	-	4.99×10^{-17} 0.0505
Polymer-concrete "14-3"	3.414×10^{-5}	3.414×10^5	-	4.79×10^{-17} 0.0485
Polymer-concrete "G1"	6.214×10^{-5}	6.214×10^5	-	7.48×10^{-17} 0.0758
Geopolymer	9.897×10^{-5}	9.897×10^5	-	6.33×10^{-17} 0.0641
Mudstone (Carbon)	2.330×10^{-7}	2.330×10^3	-	2.13×10^{-19} 0.000216
Salt rock (Permian) (before creep)	4.815×10^{-7}	4.815×10^3	-	2.99×10^{-19} 0.000303
Salt rock (Permian) (after creep)	1.95×10^{-11}	0.195	1.586×10^{-12}	5.80×10^{-24} 1×10^{-8}
Epoxy resin	1.820×10^{-11}	0.182	1.479×10^{-12}	6.12×10^{-24} 1×10^{-8}
Epoxy resin +graphite (5% vol.)	2.350×10^{-11}	0.235	1.907×10^{-12}	7.13×10^{-24} 1×10^{-8}
Epoxy resin +haloysite (5% vol.)	3.220×10^{-11}	0.322	2.637×10^{-12}	9.78×10^{-24} 1×10^{-8}
Epoxy resin +fly ash (5% vol.)	1.770×10^{-11}	0.177	1.436×10^{-12}	4.61×10^{-24} 1×10^{-8}
Epoxy resin +fly ash (30% vol.)	1.774×10^{-11}	0.177	1.411×10^{-12}	5.24×10^{-24} 1×10^{-8}

As expected, tests showed a wide range of measured permeability values. The highest gas permeability coefficient has multi-grain materials, like concrete, polymer-concrete, and geopolymer. Filtration coefficient "k" of investigated concrete samples is 10^{-17} m^2 or higher. Lower gas permeability was observed for mudstone and salt rock before creep process. Both rocks have the filtration coefficient of 2.13×10^{-19} m^2 to 2.99×10^{-19} m^2, respectively. However, these rocks are still permeable for gases. In general, the lowest gas permeability has plain-structured materials, like epoxy resins and salt rock, after the creep process. Range of the filtration coefficient of these materials is of 10^{-24} m^2. Salt rock samples became impermeable after about 12 days of exposure to 2.0 MPa of confining pressure of water. Permeability dropped from 2.99×10^{-19} m^2 to 5.80×10^{-24} m^2. Hydrogen permeability of epoxy resin samples varied, depending on the additives in the resin. Admixture of halloysite or graphite powder caused the increase of hydrogen permeability. An exception is the addition of fly ash. Admixture of 30% of volume gave a similar, but slightly lower permeability, comparing to the pure epoxy resin sample. Significant share of fly ash in sample composition did not cause the deterioration of sealing properties.

Results were calculated for the steady-state diffusion. Usually, it took up to 3 weeks to achieve the steady-state diffusion. Increases in measured concentration of hydrogen per day of an example sample is presented in Figure 3. Because of the significant time required to complete the test for each sample, the hydrogen diffusion test was performed for a single gas pressure, which was 1.0 MPa. To verify the proper workings of the setup and sensitivity of the method, one test was extended and the feed gas pressure, after achieving steady-state diffusion, was increased. Pressure was increased after 24 days, from the value of 1.0 MPa to 1.7 MPa. After the pressure increase, several days were required to achieve the steady-state diffusion for a higher pressure. After that time, the measured concentration became stable again on a higher level. According to the Equations (3) and (4), increases of the pressure and permeability coefficient are linear. Pressure was raised by 70%, which gave the increase of measured concentrations by more than 60% as well (after achieving steady-state diffusion). This confirmed that the measuring methodology was correct. Slightly lower increase in concentration, in comparison with the pressure increase, may be explained by higher confining pressure (double the gas pressure, which was approx. 3.6 MPa). It can cause a compaction of the sample, which caused a slight decrease of the permeability. This phenomenon was described for concrete samples in [19], as well as for the polymer materials, where hydrostatic compression effect occurs [15]. However, the scale of permeability decreased, as well as mechanism responsible for it, is different.

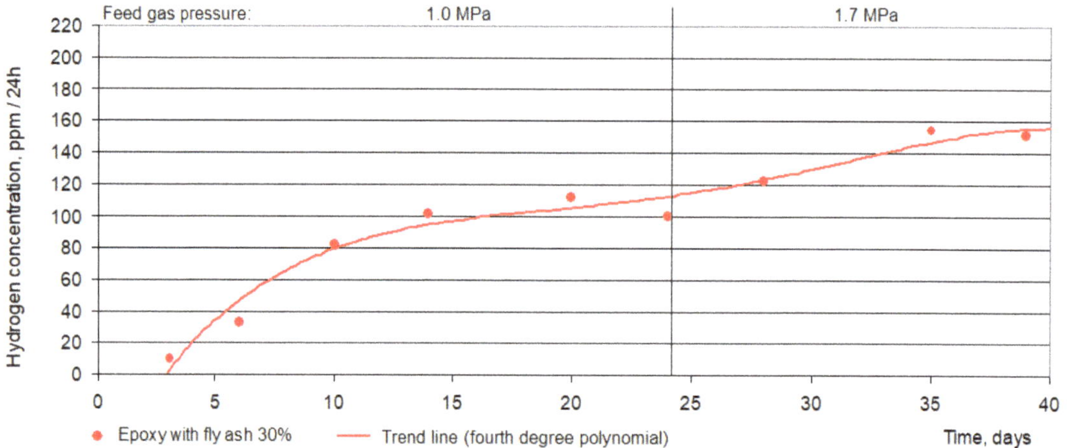

Figure 3. Increase of hydrogen concentration during the Carrier Gas Method test of epoxy resin with fly ash (30% vol.) with 1.0 MPa and 1.7 MPa feed gas pressure.

To compare the gas permeability results of different materials with the microstructure of the samples, SEM imaging of selected samples was done. Structures of the samples are shown in Figure 4. An evident difference is seen in microporous and plain materials. Dark areas representing pores are clearly visible in concrete and mudstone samples under similar magnification (Figure 4a,b). In the plain samples, voids appear mostly on the contact surface with grains of additives (Figure 4e).

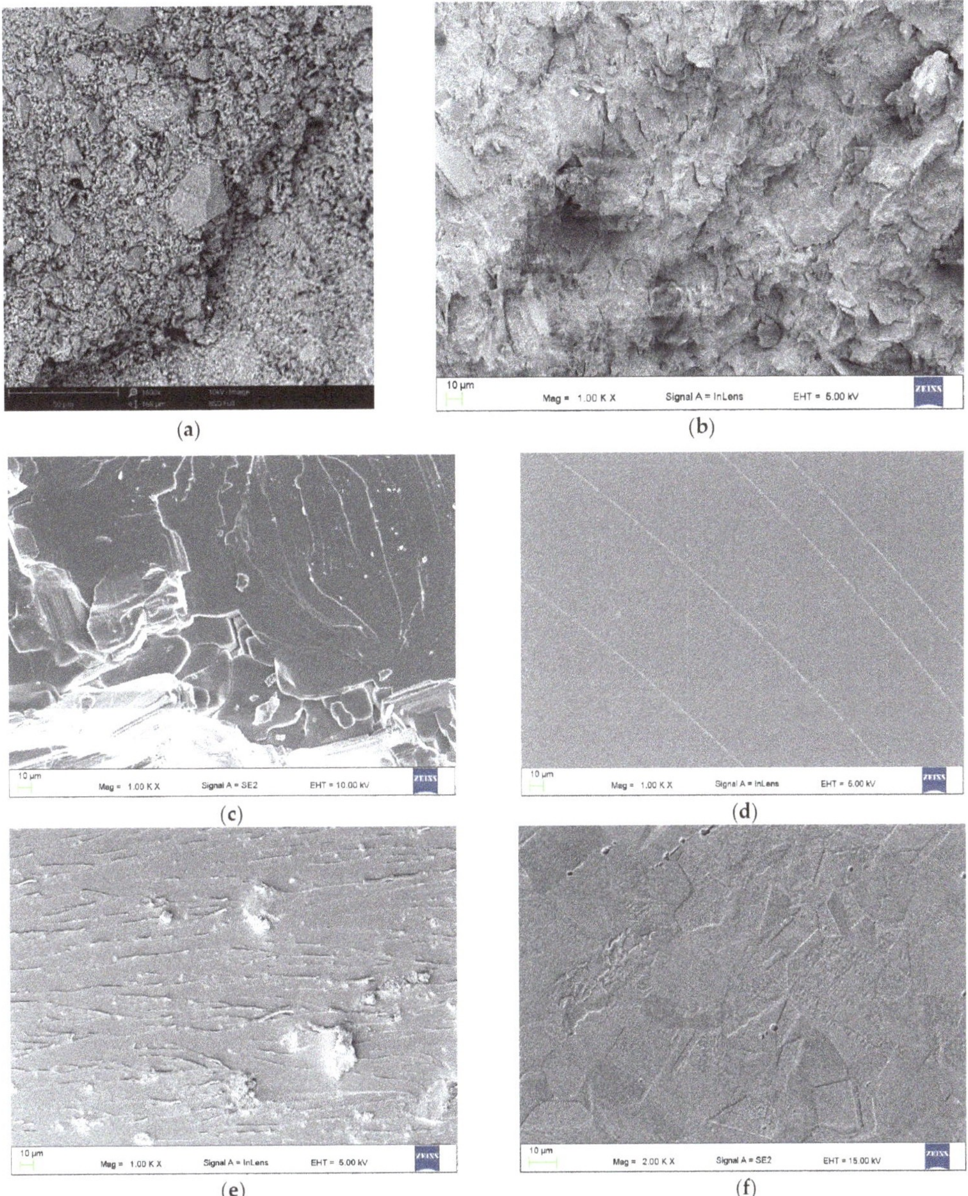

Figure 4. SEM imaging of the surface structure of investigated samples: (**a**) concrete "2" (year 2016); (**b**) mudstone (Carbon); (**c**) salt rock (Permian); (**d**) epoxy resin; (**e**) epoxy resin + graphite (5% vol.); (**f**) stainless-steel.

4. Discussion

Differences of the gas permeability of investigated materials is caused by the differences in structure of tested materials, which lead to the different mechanisms of gas migration. Multi-grain materials, such as concrete or mudstone, have many pores and voids within the structure. Fine grains can decrease the gas permeability because the intergranular porosity is lower while the compaction of the material is higher; this phenomenon was also observed in other research [22]. However, gas flow through the sample is still noticeable and is the main mechanism of gas migration in that kind of material. However, it is not essential data for the purpose of this paper. Because of the low gas sealing properties of the investigated concretes, this type of material is not suitable for sealing purposes of underground excavations. However, it can still be a reinforcement and base liner.

Plain materials, like salt rock, epoxy, or steel, do not have pores and voids in their structure. Gas is not flowing through the sample (there is no pressure gradient along the sample). Hydrogen particles can diffuse through the material by dislocations in crystal structure. It is a linear defect that induces the tensile stress. Gas elements can diffuse much easier through those kinds of zones [23]. Lattice diffusion is also possible, but in much higher temperatures, transcending Tamman temperature (in this temperature atoms in solid material acquire enough energy to make bulk reactivity and mobility significant, usually approximately half of the melting temperature) [24]. In this case, research temperature was much below the Tamman temperature.

Interesting results were obtained with the fly ash epoxy sample. Graphite and halloysite are slightly increasing hydrogen permeability of the samples, while fly ash do not affect the sealing properties. Fly ash used for this research was not grinded but only sieved. Since fly ash from power plants is easily available, it could be a cost-effective filler by reducing the amount of epoxy resin in the sealing liner.

Obtained results of epoxy resin hydrogen permeability is slightly lower than presented in literature. Other research, presented collectively in [16], showed the hydrogen permeability of epoxy resin of approximately 1.0 Barrer, while performed research of investigated samples gave the permeability coefficient of approximately 0.2 Barrer. Results of hydrogen diffusion coefficient presented in [17], which was 6.9×10^{-11} m^2/s, was also higher than the obtained 1.5×10^{-12} from this research. Lower results may be caused by numerous factors, influencing on the gas permeability in polymeric materials like polymer chemistry (epoxide number), free volumes (crystallinity, orientation of molecules), porosity or voids (air inclusions), and fillers presence [16].

Permeability coefficient of 316SS steel was calculated using the Equations (3)–(6), based on the literature data [25]. Permeability of 316 L steel were taken from [26,27]. Because of the sensitivity and construction of the setup, investigation of the steel samples was not possible. Permeability of steel is orders of magnitude lower than materials the setup is meant for. To investigate the steel or alloys, much thinner samples need to be used. The Setup is not designed for that kind of samples, but only for a core-shape samples. Calculated hydrogen permeability P_{H2} of 316SS steel was 4.6×10^{-17} cm^{3}STP·cm·cm^{-2}·s^{-1}·cmHg^{-1} (4.6×10^{-7} Barrer), and given [26,27], hydrogen diffusion coefficient D of 316L steel was in range from 10^{-13} to 10^{-15} m^2/s.

5. Conclusions

Hydrogen permeability of investigated epoxy resins are similar to the permeability of salt rock in certain pressure conditions (after the salt creep). However, storage of pressurized gas in salt caverns is causing that kind of constant pressure anyway, which will lead to drop of salt permeability in time. Obtained results of epoxies are very promising for the purpose of use as the sealing liners in Lined Rock Caverns (LRC) underground gas storage, particularly for the storage of hydrogen. Epoxy resin can be used as a substitution of stainless-steel, because of the satisfying mechanical properties and extraordinary adhesive properties, which will be reinforcing the general construction, and thanks to the low density, the construction will not be overloaded at the same time. Despite the orders of magnitude higher of hydrogen permeability of epoxies, the hydrogen loss through this material is still acceptable, compared to the steel. For example, using a 3 cm thick epoxy liner in storing the 20/80 hydrogen/methane blend at 1.0 MPa, the loss of hydrogen in the period of 30 days is only approximately 0.003–0.005% [28]. It is fully acceptable, taking the other properties and economy of epoxies. Only applying the liquid epoxy on the surface of cavern may cause some technical problems. Despite the better sealing properties of stainless-steel, the hydrogen permeability coefficient of investigated epoxy resin is still satisfying. Epoxies are also less susceptible to hydrogen erosion, which is making them a suitable material for hydrogen storage and cost-effective substitution for steel.

Author Contributions: Conceptualization D.G. and M.L., methodology D.G., investigation D.G., data curation D.G., formal analysis D.G. and M.L., writing—original draft preparation D.G., writing—review and editing M.L., supervision M.L. All authors have read and agreed to the published version of the manuscript.

Funding: This research was funded by Own Scholarship Fund of the Silesian University of Technology in year 2019/2020, grant number: 19/FSW18/0003-03/2019.

Institutional Review Board Statement: Not applicable.

Informed Consent Statement: Not applicable.

Data Availability Statement: All analyzed data in this study has been included in the manuscript.

Conflicts of Interest: The authors declare no conflict of interest.

References

1. Paul, D.; Ela, E.; Kirby, B.; Milligan, M. *The Role of Energy Storage with Renewable Electricity Generation*; National Renewable Energy Laboratory: Golden, CO, USA, 2010.
2. *Renewable Energy Progress Report*; European Comission: Brussels, Belgium, 2019.
3. Matos, C.R.; Carneiro, J.F.; Silva, P.P. Overview of Large-Scale Underground Energy Storage Technologies for Integration of Renewable Energies and Criteria for Reservoir Identification. *J. Energy Storage* **2019**, *21*, 241–258. [CrossRef]
4. Apostolou, D.; Enevoldsen, P. The past, present and potential of hydrogen as a multifunctional storage application for wind power. *Renew. Sustain. Energy Rev.* **2019**, *112*, 917–929. [CrossRef]
5. Bailera, M.; Lisbona, P.; Romeo, L.M.; Espatolero, S. Power to Gas projects review: Lab, pilot and demo plants for storing renewable energy and CO_2. *Renew. Sustain. Energy Rev.* **2017**, *69*, 292–312. [CrossRef]
6. Altfeld, K.; Pinchbeck, D. Admissible Hydrogen Concentrations in Natural Gas Systems. *Gas Energy* **2013**, *3*, 36–47.
7. Kuczyński, S.; Łaciak, M.; Olijnyk, A.; Szurlej, A.; Włodek, T. Thermodynamic and Technical Issues of Hydrogen and Methane-Hydrogen Mixtures Pipeline Transmission. *Energies* **2019**, *12*, 569. [CrossRef]
8. Panfilov, M. Underground and pipeline hydrogen storage. In *Compendium of Hydrogen Energy*; Elsevier BV: Amsterdam, The Netherlands, 2016; pp. 91–115.
9. Gabrielli, P.; Poluzzi, A.; Kramer, G.J.; Spiers, C.; Mazzotti, M.; Gazzani, M. Seasonal energy storage for zero-emissions multi-energy systems via underground hydrogen storage. *Renew. Sustain. Energy Rev.* **2020**, *121*, 109629. [CrossRef]
10. Tengborg, P.; Johansson, J.; Durup, J.G. Storage of highly compressed gases in underground Lined Rock Caverns—More than 10 years of experience. In Proceedings of the World Tunnel Congress 2014—Tunnels for a Better Life, Foz do Iguacu, Brazil, 9–15 May 2014.
11. Natural Gas-Extraction to End Use. *Natural Gas-Extraction to End Use*; IntechOpen: London, UK, 2012; pp. 159–180.
12. Van Rooyen, L.J.; Karger-Kocsis, J.; Kock, L.D. Improving the helium gas barrier properties of epoxy coatings through the incorporation of graphene nanoplatelets and the influence of preparation techniques. *J. Appl. Polym. Sci.* **2015**, *132*. [CrossRef]
13. Zhang, Q.; Wang, Y.C.; Bailey, C.G.; Istrate, O.M.; Li, Z.; Kinloch, I.A.; Budd, P.M. Quantification of gas permeability of epoxy resin composites with graphene nanoplatelets. *Compos. Sci. Technol.* **2019**, *184*, 107875. [CrossRef]
14. Zeman, S.; Kubík, L. Permeability of Polymeric Packaging Materials. *Tech. Sci.* **2007**, *10*, 33–34. [CrossRef]
15. Fujiwara, H.; Ono, H.; Onoue, K.; Nishimura, S. High-pressure gaseous hydrogen permeation test method -property of polymeric materials for high-pressure hydrogen devices (1)-. *Int. J. Hydrogen Energy* **2020**, *45*, 29082–29094. [CrossRef]
16. Maxwell, A.S.; Roberts, S.J. *Review of Data on Gas Migration through Polymer Encapsulants: Report to NDA–Radioactive Waste Management Directorate*; Serco Ltd.: Osfordshire, UK, 2008.
17. Prewitz, M.; Gaber, M.; Müller, R.; Marotztke, C.; Holtappels, K. Polymer coated glass capillaries and structures for high-pressure hydrogen storage: Permeability and hydrogen tightness. *Int. J. Hydrogen Energy* **2018**, *43*, 5637–5644. [CrossRef]
18. Hagg, M.B. Gas Permeation Unit (GPU). In *Encyclopedia of Membranes*; Drioli, E., Giomo, L., Eds.; Springer: Berlin/Heidelberg, Germany, 2015.
19. Fick, A. On liquid diffusion. *J. Membr. Sci.* **1995**, *100*, 33–38. [CrossRef]
20. Choudalakis, G.; Gotsis, A. Permeability of polymer/clay nanocomposites: A review. *Eur. Polym. J.* **2009**, *45*, 967–984. [CrossRef]
21. Webb, S.W. 2–Gas Transport Mechanisms. In *Gas Transport in Porous Media*; Ho Clifford, K., Webb Stephen, W., Eds.; Springer: Berlin/Heidelberg, Germany, 2006; ISBN 978-1-4020-3962-1.
22. Gajda, D.; Liu, S.; Lutyński, M. The concept of hydrogen-methane blends storage in underground mine excavations–gas permeability of concrete. In Proceedings of the XVI International Forum-Contest of Students and Young Researchers "Topical Issues of Rational Use of Natural Resources", St. Petersburg, Russia, 17–19 June 2020. paper accepted.

23. Zajęcia Dydaktyczne (Didactics Materials). Available online: http://home.agh.edu.pl/~{}grzesik/ (accessed on 10 March 2021).
24. Anandprakash, K.P. Effect of Tammann Temperature and Relative Humidity on Lead Chromate and Magnesium-Based Compositions. *Def. Sci. J.* **1998**, *48*, 303–308. [CrossRef]
25. Henager, C.H. Hydrogen Permeation Barrier Coatings. In *Materials for the Hydrogen Economy*; Jones, R.H., Thomas, G.J., Eds.; CRC Press: Boca Raton, FL, USA, 2007; pp. 181–190.
26. Duportal, M.; Oudriss, A.; Feaugas, X.; Savall, C. On the Estimation of the Diffusion Coefficient and Distribution of Hydrogen in Stainless Steel. *SSRN Electron. J.* **2020**. [CrossRef]
27. Kim, Y.; Kim, Y.; Kim, D.; Kim, S.; Nam, W.; Choe, H. Effects of Hydrogen Diffusion on the Mechanical Properties of Austenite 316L Steel at Ambient Temperature. *Mater. Trans.* **2011**, *52*, 507–513. [CrossRef]
28. Gajda, D. Epoxy resin for sealing the underground hydrogen storage reservoirs. Presented at the 5th International Conference on Energy Harvesting, Storage and Transfer (EHST'21), Niagara Falls, ON, Canada, 21–23 May2021. paper accepted.

applied sciences

Article

Hydrodynamical and Hydrochemical Assessment of Pumped-Storage Hydropower (PSH) Using an Open Pit: The Case of Obourg Chalk Quarry in Belgium

Angélique Poulain [1,*], Estanislao Pujades [2,*] and Pascal Goderniaux [3]

[1] UMR EMMAH Environnement Méditerranéen et Modélisation des Agro-Hydrosystèmes, University of Avignon, 84000 Avignon, France

[2] Institute of Environmental Assessment and Water Research (IDAEA), Severo Ochoa Excellence Center of the Spanish Council for Scientific Research (CSIC), Jordi Girona 18–26, 08034 Barcelona, Spain

[3] Geology and Applied Geology, Polytech Mons, University of Mons, 7000 Mons, Belgium; Pascal.Goderniaux@umons.ac.be

* Correspondence: poulain.angelique@outlook.fr (A.P.); estanislao.pujades@idaea.csic.es (E.P.)

Abstract: Pumped storage hydropower (PSH) enables the temporary storage of energy, including from intermittent renewable sources, and provides answers to the difficulties related to the mismatch between supply and demand of electrical energy over time. Implementing a PSH station requires two reservoirs at different elevations and with large volumes of water. The idea of using old, flooded open-pit quarries as a lower reservoir has recently emerged. However, quarries cannot be considered as impervious reservoirs, and they are connected to the surrounding aquifers. As a result, PSH activities may entail environmental impacts. The alternation of the pumping–discharge cycles generates rapid and periodic hydraulic head fluctuations in the quarry, which propagate into the surrounding rock media forcing the exchange of water and inducing the aeration of groundwater. This aeration can destabilize the chemical balances between groundwater and minerals in the underground rock media. In this study, two numerical groundwater models based on the chalk quarry of Obourg (Belgium) were developed considering realistic pumping–discharge scenarios. The aim of these models was to investigate the hydrodynamic and hydrochemical impact of PSH activities on water inside the quarry and in the surrounding rock media. Results showed that (1) water exchanges between the quarry and the adjacent rock media have a significant influence on the hydraulic head, (2) the frequency of the pump–discharge scenarios influence the potential environmental impacts, and (3), in the case of chalk formations, the expected impact of PSH on the water chemical composition is relatively limited around the quarry. Results highlight that those hydrogeological and hydrochemical concerns should be assessed when developing a project of a PSH installation using a quarry as a lower reservoir, considering all particularities of the proposed sites.

Keywords: pumped storage hydropower; energy storage system; hydropower; quarry; open pit; numerical modelling; environmental impacts; hydrochemistry

Citation: Poulain, A.; Pujades, E.; Goderniaux, P. Hydrodynamical and Hydrochemical Assessment of Pumped-Storage Hydropower (PSH) Using an Open Pit: The Case of Obourg Chalk Quarry in Belgium. *Appl. Sci.* **2021**, *11*, 4913. https://doi.org/10.3390/app11114913

Academic Editors: Jorge Loredo and Javier Menéndez

Received: 9 April 2021
Accepted: 21 May 2021
Published: 27 May 2021

Publisher's Note: MDPI stays neutral with regard to jurisdictional claims in published maps and institutional affiliations.

check for **updates**

1. Introduction

The development and use of renewable energies involve the necessity to temporarily store the energy because of their intermittence [1]. In this context, pumped storage hydropower (PSH) appears an efficient way to store and produce large amounts of electricity that can be used in combination with intermittent renewable energies [1,2]. PSH consists of two large water reservoirs, with a difference in elevation between them. The general principle is to pump water from the lower reservoir to the upper one when the demand of electricity is low. Later, during peak demand, water stored in the upper reservoir is discharged into the lower reservoir through turbines to produce electricity. The flexibility of this system allows the regulation of the supply and demand of electricity at the daily scale. Usually, the reservoirs are artificial, which has the advantage of offering a certain freedom in terms of plant sizing, volumes, and/or location. However, these constructions are also extremely expensive. In addition, in countries with a gentle relief, such as Belgium, potential locations allowing a significant difference in level between the two reservoirs are sometimes difficult to find. Therefore, the possibility of using old flooded mines and quarries for constructing PSH plants [3] and regulating local power grids [4–6] has been considered. The implementation of these systems would allow a better management of intermittent local renewable energy production by coupling them with wind and/or photo-voltaic systems [2]. The flooded mines and quarries are in continuous interaction with the water contained in the surrounding aquifer systems, both quantitatively and qualitatively. Therefore, pumping and discharging large volumes of water into the lower reservoir can (1) modify the hydraulic head in the quarry and the adjacent aquifer systems impacting on the environment [7] and on the system efficiency [8] and (2) alter the hydrogeochemical balances, and thus, the water quality [9]. The induced hydrodynamic and hydrochemical changes must be compatible with the natural functions observed and established on and around the PSH site. Similarly, these changes should not negatively affect other activities undertaken near to the quarry, such as drinking water pumping stations. Thus, preliminary hydrodynamic and hydrochemical studies must be developed to avoid unexpected and undesired impacts.

To date, few studies have been carried out about water hydrochemistry issues in former quarries that are now flooded. Most of them are mainly focused on the presence of heavy metals in the extraction zone [10]. The issue of groundwater pollution, in the context of rock or mineral extraction sites, is better studied in the context of underground mines [11] since many studies have been focused on problems related with mine water acidification [12] as well as other more specific problems. However, investigations focused on hydrogeochemical issues in the specific context of PSH using quarries have not been conducted. Only [8] and [9] developed a numerical study focused on hydrochemical issues related with PSH. However, these previous investigations are limited since they were based on deep, underground coal mines. The results show that the presence of pyrite and calcite significantly influences the evolution of the hydrochemical properties of the water contained in the mine during the pumping–discharge operations. These previous works are exclusively numerical and based on hypothetical cases. To date, very few case studies of PSH using underground works have been investigated. One of them is [13], which investigated the possibility of recycling some USA underground iron mines for PSH.

In this context, the main objective of this study was to quantify and assess the hydro-dynamic and hydrochemical impacts induced by PSH using a quarry as the lower reservoir. This objective was reached by developing two numerical models. One of the numerical models aimed at investigating the hydrodynamic behavior induced by PSH, whereas the other one aimed at studying hydrogeochemical issues produced by PSH. Both models were based on the former quarry of Obourg, located in the chalk aquifer of the Haine Valley (Mons city area, Belgium).

2. Materials and Methods

2.1. Study Site

Several chalk quarries are present in the Walloon region (Belgium), including active and abandoned ones. The chalk open pit, which was studied in this article, is representative of those quarries. It is located close to the Obourg locality, about 4.5 km north-east of Mons city (South-West Belgium). The study site is composed of five quarries located in chalk geological formations. The two most easterly quarries are still in operation, whereas the three located to the west are no longer used. One of those abandoned quarries is studied to be used for PSH (Figure 1). The considered quarry has a surface area of 0.34 km^2. The upper reservoir, with a volume of 1 million m^3 ($100 \times 1000 \times 10$ m), would be built north of the quarry, close to the E19 motorway. The difference in altitude between the upper reservoir and the quarry would be 40 m as shown in Figure 1.

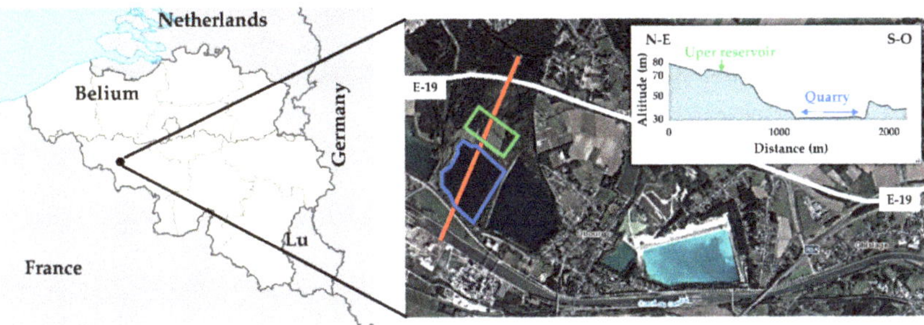

Figure 1. View of the Obourg chalk quarries. The red line, intersecting the studied open pit, gives the location of the altimetric profile.

2.2. Geological and Hydrogeological Context

The Mons sedimentary basin has a syncline-shaped (Figure 2) structure and is composed of Mesozoic and Cenozoic deposits. This basin has been affected by subsidence events since the end of the Paleozoic. The Obourg chalk quarries are in the northern part of the Mons Basin where the Cretaceous chalk formations are exploited. These geological formations are included in the "Chalk Group" and represent a major aquifer system called the "Haine Valley Chalk Aquifer" or the "Mons Basin Chalk Aquifer." The thickness of this aquifer is variable and can reach up to 250 to 300 m. The chalk aquifer is bounded at the north, east, and south-west by schisto-sandstone formations from the Upper Carboniferous [14]. In the South, the aquifer is limited by the Lower Devonian deposits, which include several aquifer systems. The chalk aquifer overlies low-permeability marly geological formations [14]. The chalk aquifer is characterized by a relatively high hydraulic conductivity on a macroscopic scale. This high hydraulic conductivity is the result of the fracture network, consisting of diaclases, stratification joints, and faults. The chalk has a high value of total porosity that can be divided in a matrix porosity, which allows the storage of large quantities of water and a fracture porosity, which allows preferential flows [15–20]. The Obourg quarries are located in the vicinity of drinking water abstraction stations, pumping in the same aquifer, and located about 1.5 km in the south-west direction. This groundwater abstraction complex, using pumping wells, is one of the most important in the Walloon region.

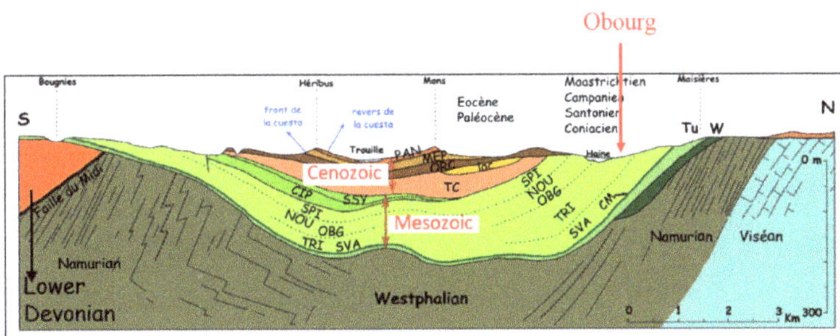

Figure 2. Modified south–north geological section of the Mons Basin and projection of the Obourg quarry location [21].

2.3. Hydrochemical Context

Chalk is generally composed of high percentage values (60–95%) of calcite ($CaCO_3$) [22]. Specifically, the chalk of the Trivières formation is composed of more than 92% $CaCO_3$. These chalk geological formations also contain some slightly ferruginous beds, as well as some phosphate nodules [14]. The hydrochemical composition of the groundwater is mainly explained by the water–rock interactions and, in particular, by the different alteration processes inducing dissolution/precipitation reactions. In the chalk aquifer, the chemistry of the groundwater is related to the dissolution of $CaCO_3$ in the presence of dissolved CO_2. The dissolution of $CaCO_3$ is governed by a series of acid–base equilibria as follows:

$$CO_2 + H_2O \leftrightarrows H_2CO_3 \tag{1}$$

$$H_2CO_3 \leftrightarrows H^+ + HCO_3^- \tag{2}$$

$$HCO_3^- \leftrightarrows H^+ + CO_3^{2-} \tag{3}$$

$$H_2O \leftrightarrows H^+ + OH^- \tag{4}$$

$$CaCO_3 + 2H^+ \leftrightarrows Ca^{2+} + H_2O + CO_2 \tag{5}$$

In accordance with these equilibrium reactions (1–5), most of the dissolved elements in the chalk aquifers' groundwater are Ca^{2+} and HCO_3^-. The HCO_3^- ion is the predominant form. Other major ions in chalk aquifers' groundwater are Mg^{2+}, Na^+, Cl^-, SO_4^{2-}, and Fe^{2+} [23–25].

Table 1 summarizes the average concentration of ions in the groundwater of the chalk aquifer of the Mons Basin. Note that a great disparity in concentrations can be observed depending on the location. Groundwater in the study site has high concentrations of iron and manganese, which under reducing conditions are present in the forms Mn^{2+} and Fe^{2+}.

Table 1. Average concentration of ions in the chalk aquifer of the Mons Basin [26].

Ion	Concentration (mg/L)
Ca^{2+}	142.2
HCO_3^-	350.1
Mg^{2+}	11.3
Na^+	26.2
Cl^-	41.4
SO_4^2	111.5
Mn^{2+}	0.82
Fe^{2+}	2.62

These two ions (Mn^{2+} and Fe^{2+}) are involved in redox reactions, shown below (Equations (6) and (7)), involving the O_2 dissolved in the water. The aeration of the water during the pumping-turbine cycles in the upper reservoir induces the increase in the equivalent partial pressure of O_2. The presence of dissolved O_2 leads to oxidation of Fe^{2+} and Mn^{2+} to FeOOH (goethite, iron hydroxide) and MnO_2 (pyrolusite, manganese oxide), respectively.

$$4Fe^{2+} + O_2 + 8OH^- + 2H_2O \leftrightarrows 4Fe(OH)_{3(s)} \tag{6}$$

$$Mn^{2+} + \frac{1}{2}O_2 + H_2O \leftrightarrows MnO_{2(s)} + 2H^+ \tag{7}$$

Concerning the major ions observed in the study site, the presence of magnesium may indicate the existence of dolomite ($CaMg(CO_3)_2$) [22]. The presence of sulphates and iron may be explained by the result of pyrite and marcasite oxidation, although these minerals were not observed extensively in the aquifer. Iron oxidation benches are however regularly visible within the chalk layers. Groundwater near the Obourg quarry was characterized by an alkalinity of 26.5 meq/L, which means that it is relatively difficult to change the pH of the solution. The pH value was equal to 7.24. The presence of dissolved O_2 was considered to be zero, and the partial pressure corresponding to the concentration of dissolved CO_2 was 10–2 bar.

3. Model Development

3.1. Groundwater Flow Model

3.1.1. Model and Spatial Discretization

The numerical models were developed using the finite difference code Modflow [27]. The modelled area was equal to 32.8 km². Two hydrogeological units were represented in the numerical flow model (Figure 3): (1) the chalk aquifer that covered the whole modelled area, reaching a thickness of 300 m in the southern part of the model, and (2) a shallower aquifer located in the southern area and overlying the chalk aquifer. This shallower aquifer is made up of marine and fluvial–alluvial sediments from the Cenozoic age. They consist of sand, silt, and clay, structured in alternating permeable and low-permeable formations.

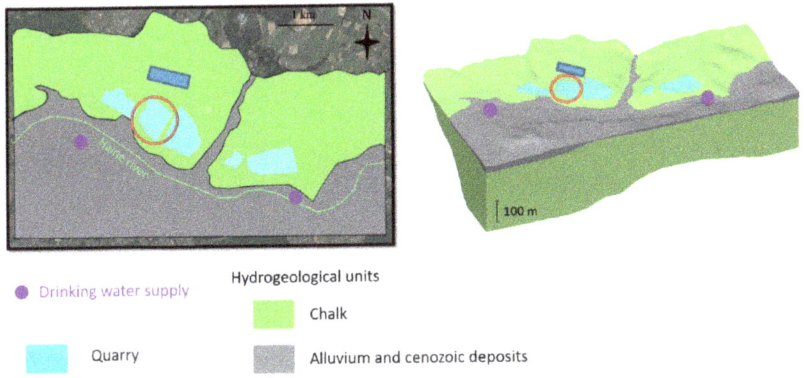

Figure 3. Plan view and 3D view of the hydrogeological model. The red circle highlights the open pit chalk quarry where PSH operations were studied.

The modelled area was discretized using irregular rectangular cells. The dimensions of the cells were 10 × 10 m in the quarry and increased by a factor of 1.05 in the direction of the boundaries. The mesh was divided vertically in three layers of cells representing the hydrogeological units. The first layer represented the shallower aquifer, made of marine and fluvial–alluvial sediments, whilst the two lower layers represented the chalk aquifer. The central layer was characterized by a constant thickness of 40 m, corresponding to the thickness of the quarry. The Obourg quarry, which was used for PSH, as well as the other quarries, were therefore contained in this layer. This configuration allowed the expected variations of the water table in the quarry and in the aquifer to be contained in the central layer. This enabled us to avoid numerical problems related with the saturation and desaturation of the cells. The lower layer was characterized by a variable thickness, ranging from 10 m in the north to 300 m in the south, forming a beveled layer.

3.1.2. Boundary Conditions

The northern limit of the model corresponds to the interface between the chalk aquifer and low-permeability deposits in contact with the base of the chalk aquifer (Figure 3). This limit is represented by a zero flow Neumann boundary condition (BC). The southern, western, and eastern boundaries were implemented in the model using Dirichlet BCs whose values correspond to the mean values of the piezometric head of the chalk aquifer at these locations [14] (Figure 3). Those boundaries were chosen sufficiently far around the quarry not to interfere with the influence of the PSH operations. The distance and location of these boundaries were determined from the results of a generic study about the impact of PSH using quarries as lower reservoirs [7]. The base of the model corresponds to the lower limit of the chalk aquifer. The "Haine River" is a draining stream along the considered section. It is represented by a potential-dependent flow BC (i.e., Fourrier BC), which only allows groundwater flowing out the model. The value of the conductance factor, which governs the Fourrier BC, was high enough to keep the piezometric head in equilibrium with the river. Two major groundwater pumping stations in the chalk aquifer were included in the modelled area. They were implemented in the chalk layer of the model using Neuman BCs, with prescribed flow rates equal to 5,234,976 and 883,008 m^3/year. A recharge of 300 mm/year was applied to the top of the model.

3.1.3. Parametrization

The hydraulic properties were distributed according to the hydrogeological units. They were determined by calibrating the model by fitting the measured piezometric head in different piezometers in the modelled area. In some locations however, few measured piezometric values were available. Calibrated parameter values were in accordance with the commonly accepted intervals for the different hydrogeological units considered in the model [14]. The quarry was represented by adopting a high hydraulic conductivity and porosity of 1. This hydraulic conductivity was high enough to simulate a homogeneous hydraulic head inside the quarry. The hydraulic properties finally used in the model simulations are summarized in Table 2.

Table 2. Calibrated hydraulic parameters values used in the groundwater flow model of the Obourg quarries.

	Alluvium	**Chalk**	**Quarry**
Hydraulic conductivity [m/s]	10^{-4}	10^{-5}	10^{-1}
Specific yield [%]	30	16	100

3.1.4. Simulated Pump–Storage Operations Scenarios

Four pumping–discharge scenarios (Figure 4) have been considered (Smartwater project, 2018; [4,7]). The pumping–discharge scenarios, which were implemented as boundary conditions in the models, are based on real electricity data. These scenarios describe the emptying and filling of the upper reservoir and the alternating phases of pumping, discharge, and no-activity over a period of fourteen days. In Figure 3, negative flow rate values correspond to pumping operations in the quarry and filling of the upper reservoir. Positive flow rate values correspond to discharge steps of the upper reservoir through the turbines and injection in the quarry. Three of the pumping–discharge scenarios are based on the electricity market daily data at different seasons of the year (spring, summer, and winter). The pumping and discharge periods take approximately five hours. The succession of the different phases (pumping/discharge/no-activity) is faster in winter, intermediate in spring, and slower in summer. The last scenario was generated randomly, with a change of phase (i.e., pumping/discharge/no-activity) every 15 min, following a uniform law. This scenario has been developed in line with PSH stations potentially connected to renewable and intermittent energy sources, whose production management would probably require higher frequencies, with periods shorter than an hour. The pumping–discharge scenarios were used as input data for the numerical model and were implemented by prescribing the flow rates (i.e., Neuman boundary conditions) in the cells corresponding to the quarry. The pumping and discharge flow rates were chosen according to the volume of the upper reservoir ($\approx$1 million m^3) to satisfy that it can be filled or emptied during one pumping/discharge cycle. Thus, for an example cycle of 4.8 h, the pumping/discharge rate was 55.56 m^3/s.

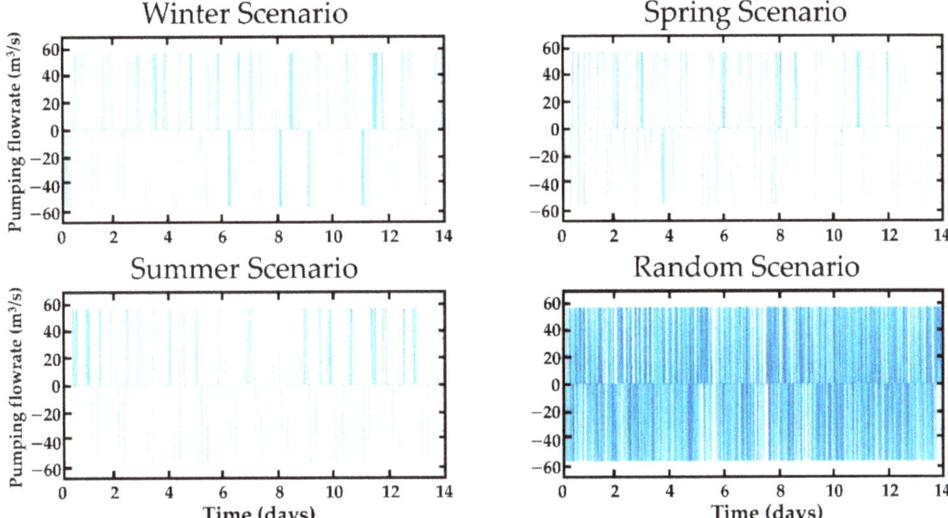

Figure 4. Pumping-discharge scenarios developed and used as inputs to the numerical flow model. Negative flow rate values correspond to pumping operations in the quarry and filling of the upper reservoir. Positive flow rate values correspond to discharge steps of the upper reservoir through the turbines and injection in the quarry.

3.2. Groundwater Hydrochemical Model

3.2.1. Conceptual Model

To simulate the hydrochemical evolution of the water contained in the upper reservoir, in the quarry, and in the chalk aquifer (Figure 5), several hypotheses were established concerning the aquifer and the hydrochemical characteristics of the groundwater and the pump–storage operations. The porous medium was considered as homogeneous. The hydraulic conductivity and the drainage porosity were identical at each point of the rock media. Values were identical to those used for the flow groundwater model described in Section 3.1. The rock was considered to be composed of 92% calcite, reflecting the composition of the chalk of the Trivières geological formation. The pumping and discharge flowrates were derived from the pumping–storage scenarios described in the groundwater flow model and consisting of regular successive pumping and discharge slots, the period of which was 4.8 h for 14.6 days. The volume of pumped and then discharged water during each cycle was 125,000 m^3. It is considered that chemical equilibrium with the atmosphere was reached in the upper reservoir before each discharge phase assuming partial pressures for O_2 and CO_2 of $10^{-0.7}$ bar and $10^{-3.5}$ bar, respectively. In the initial state, the water present in the quarry was considered to be in equilibrium with the groundwater in the aquifer. In the quarry, the equilibrium of the first few meters of water with the atmosphere was neglected. The temperature and density of the water were considered to be constant and equal to 12 °C and 1000 Kg/m^3, respectively. Despite variations in water temperature and density, which were expected to be negligible within the quarry, further investigations by using numerical codes with more capabilities should be developed to establish their role in the system's behavior. The used numerical code does not allow temperature variation within the same simulation.

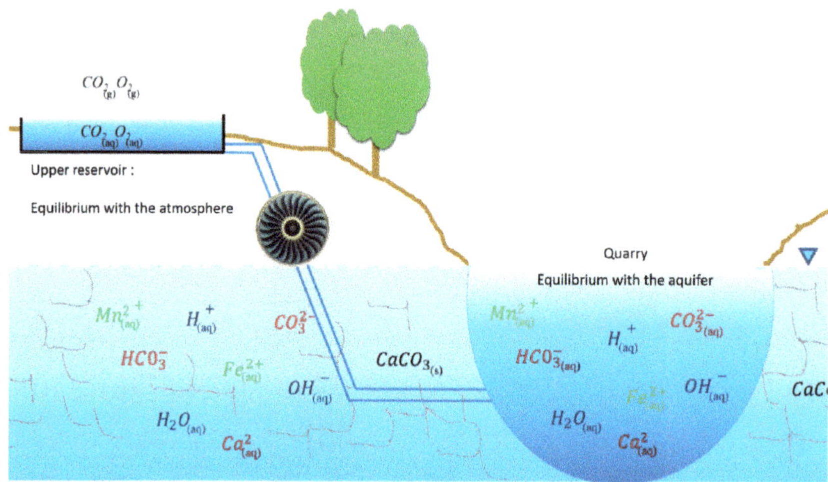

Figure 5. Schematization of the hydrogeochemical model.

3.2.2. Numerical Model and Boundary Conditions

The model was three-dimensional and was developed using the finite difference code PHAST [28]. PHAST aims to simulate groundwater flow, solute transport, and the hydrogeochemical reactions. PHAST couples HST3D [29], which computes flow and solute transport processes, with PHREEQC [30], which simulates the hydrogeochemical reactions.

Taking advantage of the symmetry of the problem, only one quarter of the system was modeled, allowing a significant reduction in computation time. As for the groundwater flow model, the 2-km long domain was discretized with irregular rectangular cells. The size of the cells was 10 m × 10 m in the quarry and increased progressively by a factor of 1.05 towards the boundaries. The piezometric head was prescribed (Dirichlet BCs) at the outer boundaries of the model. The size of the modelled zone was chosen so that these limits were sufficiently far from the quarry to not influence the effects of pumping–discharge operations. This choice was based on the results of the groundwater flow model developed (Section 3.1). Note that the prescribed head at the outer boundaries was uniform, and thus, the regional piezometric gradient was not represented. The pumping and discharge flows were simulated by implementing a Neumann BC in the center of the quarry. The hydrochemical characteristics of the water released into the quarry during the discharge phases depended on the results of the routine relating to the upper reservoir and were readapted to each new pump–discharge cycle.

3.2.3. Parameterization

The quarry was assimilated as a linear reservoir with a high hydraulic conductivity and a porosity of 100%. Hydraulic conductivity and drainage porosity values assigned to the elements representing the chalk aquifer were the same as those implemented in the groundwater flow model (Section 3.1) and were equal to 10^{-5} m·s^{-1} and 0.16, respectively. In this case, we considered the total porosity because the groundwater exchanges and the potential environmental impacts increased with high porosities [7,31]. Concerning the solute transport parameters, the longitudinal and transverse dispersivities were 10 m and 1 m, respectively, and the tortuosity was equal to 1. The longitudinal and transverse dispersivities adopted to simulate the water behavior in the quarry were equal to 100 m in order to assimilate it to an environment with high degrees of mixing.

4. Results

4.1. Hydraulic Head Fluctuations in the Quarry and Water Exchanges with the Aquifer

Figure 6 shows the simulated hydraulic head variations in the quarry and water exchanges between the quarry and the aquifer as a function of time and pumping–discharge operations. Positive and negative values for the exchange flowrate relate to groundwater that flowed towards the quarry and water that flowed from the quarry to the chalk aquifer, respectively. These exchange flow rates were globally inversely correlated to the hydraulic head in the quarry. When the hydraulic head in the quarry increased as a result of a water discharge from the upper reservoir, the exchange rates decreased to negative values, reflecting that water flows from the quarry to the aquifer. The opposite behavior was observed during the pumping phases when the hydraulic head decreased in the quarry.

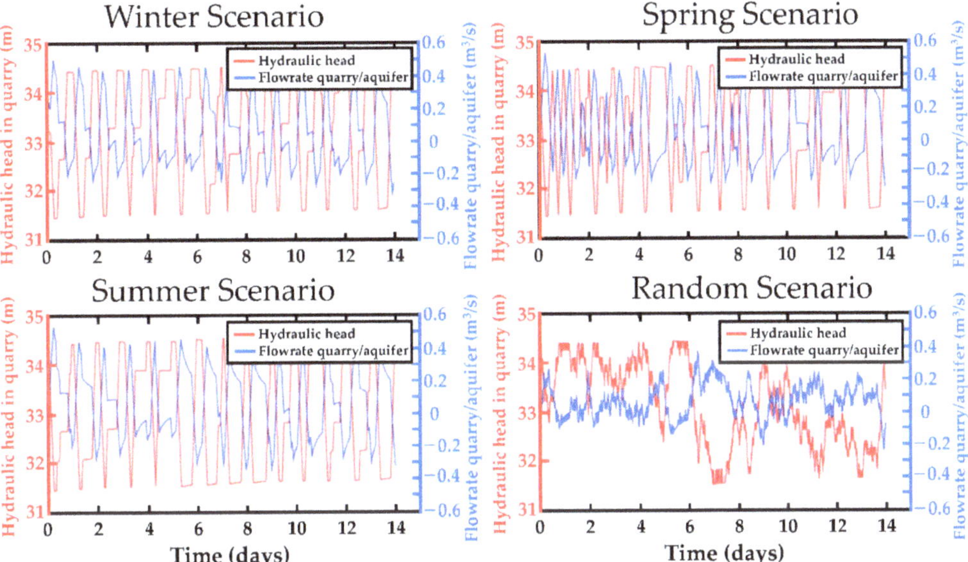

Figure 6. Hydraulic head variations in the quarry (red line) and flowrate exchange between the quarry and the chalk aquifer (blue line). Positive and negative values for the exchange flowrate relate to groundwater that flowed towards the quarry and water that flowed from the quarry to the chalk aquifer, respectively.

During no-activity phases, exchange rates were positive when the hydraulic head in the quarry was located at a lower elevation than the surrounding piezometric level in the aquifer, and vice-versa. During no-activity phases, the hydraulic head in the quarry continued to vary slightly since water continued to be exchanged between the quarry and the chalk aquifer until an equilibrium was reached. For example, at the end of a pumping period, the hydraulic head in the quarry was lower than the surrounding piezometric level in the aquifer. Consequently, the chalk aquifer was drained by the quarry. The exchange rates were positive but decreased slowly, and the hydraulic head in the quarry increased slowly.

In the long term, the average elevation around which the hydraulic head fluctuates tended to be equal to the elevation of the hydraulic head in natural conditions. This fact was better observed in the regular scenarios than in the random one because of its more chaotic behavior (Figure 3). This phenomenon of gradual increase in the hydraulic head occurred because the simulated scenarios started with a pumping phase. During this initial phase, the hydraulic head in the quarry fell below the elevation of the hydraulic head in natural conditions. As a result, groundwater inflows from the chalk aquifer to the quarry contributed to increase the hydraulic head. This behavior differed to that expected in the case of an isolated lower reservoir where the hydraulic head remained constant and was not influenced by the groundwater exchanges. During discharge phases, when the volume of water previously pumped is discharged into the quarry, the hydraulic head in the quarry increased above its natural elevation, and thus, water was forced to flow from the quarry to the chalk aquifer. However, total water inflows from the aquifer to the quarry were generally greater than losses, as long as the average hydraulic head was below the natural level. Water inflows and outflows were not equilibrated until the hydraulic head oscillated around its natural elevation. This behavior must be considered, especially if there are constraints concerning the maximum hydraulic head to be reached in the quarry such as stability problems related with the elevation of the hydraulic head. Moreover, the gradual increase in the average hydraulic head was also critical because it may affect the balance between energy stored and produced by the system since the efficiency of the pumps and turbines depends on the difference in the hydraulic head between the quarry and the upper reservoir [8].

Table 3 summarizes the results from the simulations regarding the quarry–aquifer exchanges and the hydraulic head fluctuations in the quarry. The amplitude of the hydraulic head fluctuations was about 3 m, despite the large upper reservoir considered. Natural piezometric head fluctuations in the aquifer were 1 to 4 m depending on where you were in the aquifer. Concerning the maximum water exchange rates between the quarry and the chalk aquifer, they reached up to 0.53 m^3/s. Slight differences in results were observed for the different pumping–discharge scenarios considered. The maximum variations in the hydraulic head in the requested quarry were less important for the random scenario than for the seasonal scenarios. The difference between them was about 20 cm. This was related with the higher frequency of pumping–discharge cycles of the random scenario since pumped and discharged volumes of water per phase are less important. As a result, fluctuations of the hydraulic head in the quarry as well as the exchange rates with the aquifer were smaller than in the seasonal scenarios. The maximum water exchange rates, which were similar for the three seasonal scenarios, were about 0.5 m^3/s, whilst they were about 27% lower for the random scenario.

Table 3. Summary of the results concerning the quarry–groundwater interactions and hydraulic head fluctuations in the quarry during the simulations. These data concern the minimum and maximum hydraulic head in the quarry, the maximum variation in the hydraulic head in the quarry, and the maximum exchange flowrate between the quarry and the aquifer for each scenario.

	Min. Hydraulic Head [m]	Max. Hydraulic Head [m]	Max. Hydraulic Head Variations [m]	Max. Exchange Flowrate [$m^3 \cdot s^{-1}$]
Winter	31.44	34.59	3.14	0.53
Spring	31.43	34.55	3.12	0.50
Summer	31.43	34.61	3.18	0.53
Random	31.50	34.44	2.93	0.37

4.2. Piezometric Head Fluctuations in the Aquifer

Figure 7 shows the zone of influence of pumping–discharge operations at the end of the simulations (after 14 days) for all simulated scenarios.

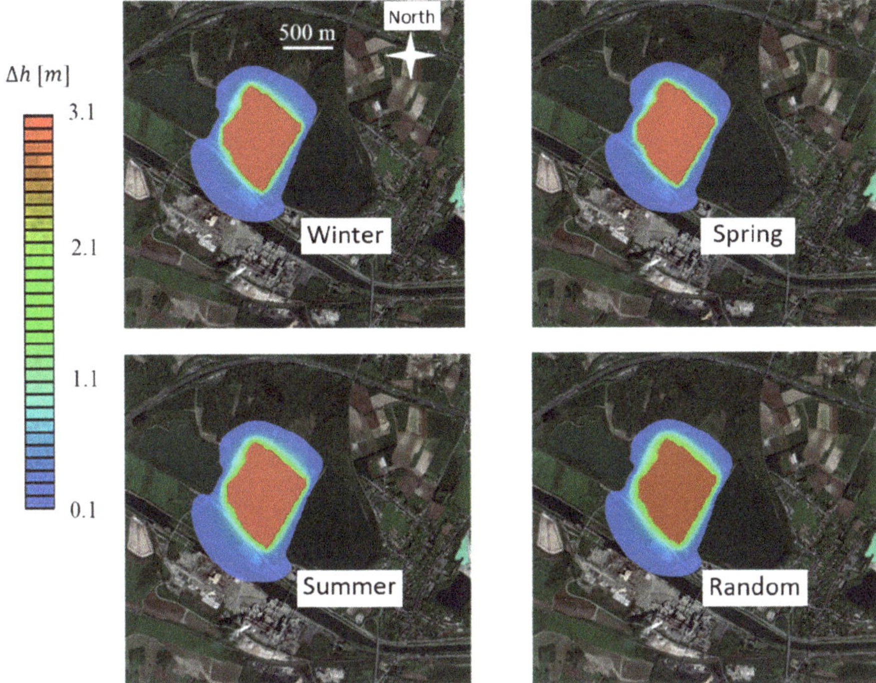

Figure 7. Zone of influence of pumping–discharge operations around the Obourg quarry at the end of the simulation time (14 days). A threshold of 0.1 m for piezometric head variations was assumed to delimitate the area of influence. If piezometric head fluctuations were lower than this threshold, it was considered as outside the zone of influence of PSH operations in the aquifer.

This zone of influence corresponds to an area in which the fluctuations of the piezometric head were larger than 0.1 m. The typical summer, winter, and spring scenarios were a priori considered as low frequency scenarios in comparison with the random scenario, and thus, greater distances of influence were expected. However, the maximum distance of influence was similar in all scenarios. For all of them, the induced variations of the piezometric head (Δh) were less than 0.1 m within a maximum distance of 380 m around the quarry walls. This similarity between scenarios is probably explained by the fact that the pumping–discharge scenarios were not sinusoidal and mono-frequent but contained several frequencies. This effect is particularly visible on the hydraulic head fluctuations in the quarry simulated for the random scenario (Figure 4). The lower frequencies control the distance of influence [7]. In addition, the presence of large water bodies around the quarry limited the extension of the zone of influence. In other words, the quarries located at both sides of the central Obourg quarry act as a kind of buffer.

The results reflect the importance of considering these interactions to maximize the efficiency and to minimize the potential environmental impacts. Results obtained with the groundwater flow model show the preponderant influence of the lower frequency of the pumping–discharge cycles. The lower the frequency, and thus the longer the pumping–discharge cycles, the greater the quarry–aquifer interactions. Realistic pumping–discharge scenarios (winter, spring, and summer) used as input to the models had a minimum period of 5 h. If PSH stations are connected to renewable and intermittent energy sources, the management of this production would probably require higher frequencies, with periods shorter than an hour. The random scenario was developed accordingly. Concerning the distances of influence, however, studying low frequencies allows consideration of the safety side and the potentially largest zones of influence [7].

4.3. Evolution of Water Chemistry in the Upper Reservoir

The following results are related to the hydrochemical context described at 2.3. Figure 8 shows the main results simulated with the hydrochemical groundwater model concerning the hydrochemical evolution of the water in the upper reservoir. Figure 7 displays the evolution of pH and the precipitated quantities of calcite, pyrolusite, and goethite.

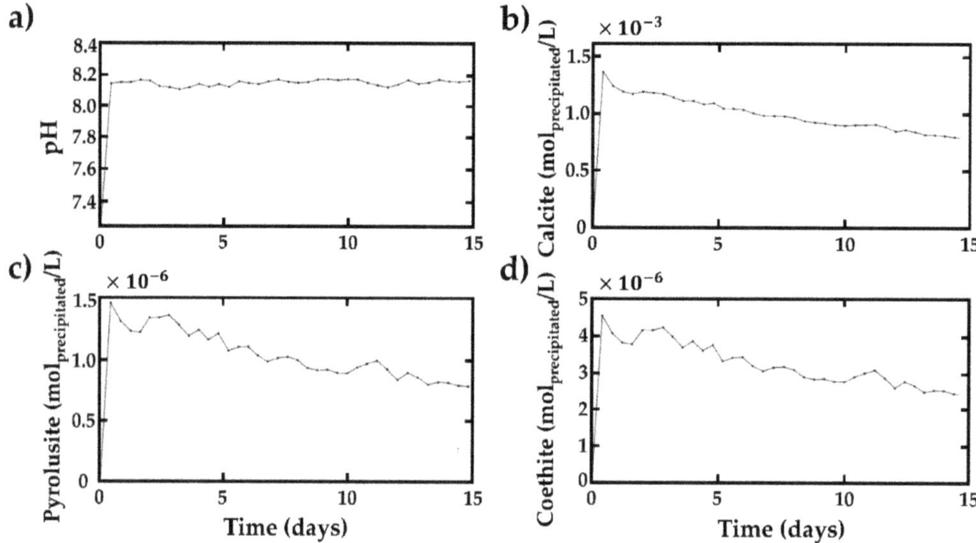

Figure 8. Evolution of hydrogeochemical variables during pumping–injection cycles in the quarry. (**a**) Evolution of pH. (**b**) Evolution of the quantity of Ca^{2+} in moles per liter. (**c**) Evolution of the quantity of Mn^{2+} in moles per liter. (**d**) Evolution of the quantity of Fe^{2+} in moles per liter.

In the upper reservoir, the chemical equilibrium of the water with the atmosphere induced an increase in the concentration of dissolved O_2 and a decrease in the concentration of dissolved CO_2. In general, the increase in the dissolved O_2 caused the precipitation of pyrolusite and goethite (Equations (6) and (7)), whereas the decrease in dissolved CO_2 caused the precipitation of calcite (Equations (1) to (5)). The precipitation of these elements occurred from the first pumping–discharge cycle, during which the amount of precipitated minerals reached a maximum (1.46×10^{-6}, 4.55×10^{-6}, and 1.4×10^{-3} moles of precipitate/L of pyrolusite, goethite, and calcite, respectively) as shown in [9] in the scenario with only calcite. Precipitation continued during the following cycles, but the amount of precipitated minerals decreased gradually as the water pumped into the upper reservoir became depleted in calcium, manganese, and iron. The precipitated quantities were significant, especially at the beginning of the PSH activities. The quantity of precipitated calcite was three orders of magnitude higher than those of goethite and pyrolusite. Precipitated amounts of goethite and pyrolusite were similar, although slightly higher for goethite. Indeed, the iron concentration (4.7×10^{-6} mol/L) in groundwater was initially higher than that of manganese (1.5×10^{-6} mol/L). In addition, iron oxidation is preferential and has faster kinetics. Theoretically, calcite, pyrolusite, and goethite should form deposits in the upper reservoir, which would require periodic cleaning tasks. In practice, however, given the flows involved, turbulence in the upper reservoir could reduce these deposits. Overall, the volume of precipitated minerals can be relatively important and periodical cleaning tasks could be needed, which would affect the global efficiency of the PSH plant [9,32]

pH values in the upper reservoir increased drastically during the first pumping–discharge cycle as a result of CO_2 degassing and calcite precipitation. After the first cycle, pH remained relatively constant throughout the following cycles, oscillating between 8.16 and 8.18. Note that incrustations were promoted under these values of pH, in accordance with calcite precipitation, and potentially needed cleaning operations.

4.4. Hydrochemical Evolution of the Water in the Quarry

Concentrations of Ca^{2+}, Fe^{2+}, and Mg^{2+} tend to decrease during the pumping–discharge cycles, whereas the average pH increases (Figure 9). The evolution of pH in the quarry was related to that in the upper reservoir where the pH increased abruptly during the first cycle. During discharge phases, the pH of the released water was higher than that of the water in the quarry, and thus, the pH in the quarry tended to increase during each discharge phase. Conversely, during the pumping periods, water from the chalk aquifer, which had lower values of pH than that in the quarry and upper reservoir, inflowed to the quarry causing its pH to decrease. Overall, the pH of the water in the quarry gradually increased towards an average value between the pH of the aquifer and the pH of the upper reservoir through cumulative effects [33]. The volumes of groundwater and chalk rock present in the surrounding aquifer stabilized the chemical equilibrium of the groundwater. This also explains the pH evolution in the quarry towards intermediate values compared to the upper reservoir.

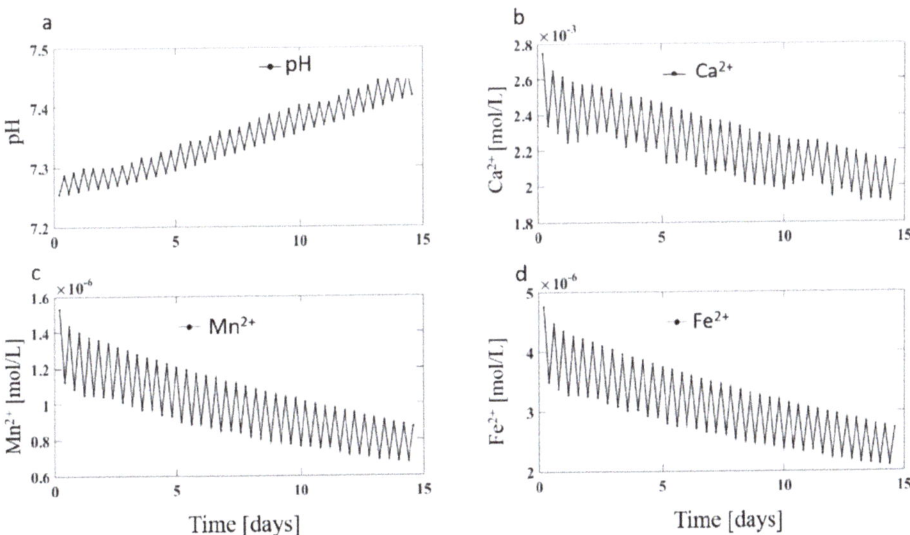

Figure 9. Evolution of hydrogeochemical variables during pumping–injection cycles in the quarry. (**a**) Evolution of pH. (**b**) Evolution of the quantity of Ca^{2+} in moles per liter. (**c**) Evolution of the quantity of Mn^{2+} in moles per liter. (**d**) Evolution of the quantity of Fe^{2+} in moles per liter.

The opposite behavior was observed for Ca^{2+}, Fe^{2+}, and Mg^{2+}. In this case, as a result of calcite, pyrolusite, and goethite precipitation, their concentrations were lower in the upper reservoir than in the chalk aquifer. Thus, concentrations of Ca^{2+}, Fe^{2+}, and Mg^{2+} increase during pumping phases (water flows from the aquifer towards the lower reservoir) and decrease during discharge phases (the quarry is filled with water from the upper reservoir).

4.5. Hydrochemical Evolution of the Groundwater in the Chalk Aquifer

Regarding the hydrochemical evolution in the aquifer, pH values tended to increase, as in the quarry, but only within a zone limited to the first 20 meters around the quarry (Figure 10).

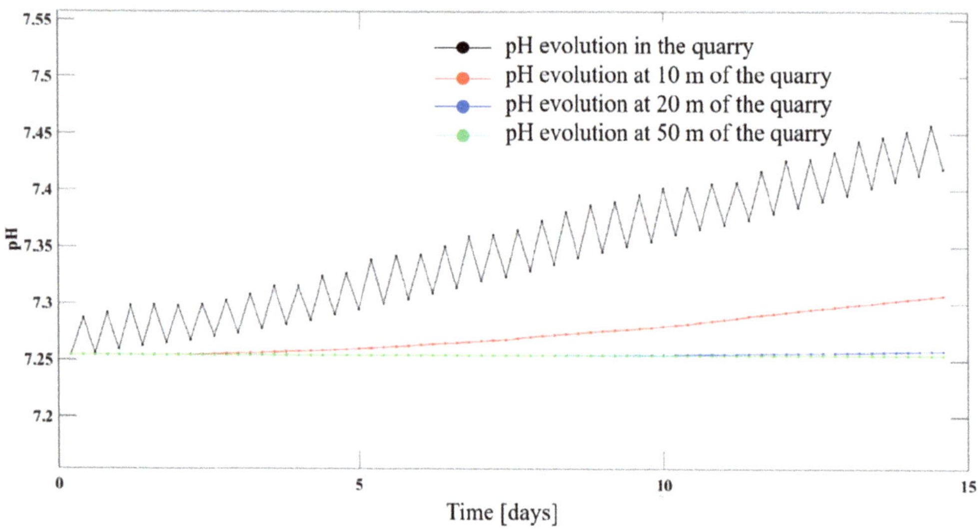

Figure 10. pH evolution in the quarry and at 10, 20, and 50 m around the quarry.

The increase was lower than in the quarry because of the quantity of water present in the aquifer and the buffering effect of the chalk rock. In addition, the propagation of hydrochemical stresses induced by the pumping–discharge cycles did not lead to the dissolution of calcite within the chalk aquifer since the acidification phenomenon was not intensive enough.

5. Conclusions

The numerical models developed in this study were realistic and representative since they were based on a real quarry that was considered for PSH. The developed numerical models aimed to assess the interactions between the Obourg quarry and the surrounding aquifer systems. From a feasibility point of view, an important problem related to the case of the Obourg quarry is the possible interference of the PSH activities with nearby drinking water catchments.

Results showed that the fluctuation zone of the piezometric head did not reach these catchments after 14 days of pumping–discharge cycles. Concerning the impacts on the water quality, PSH activities tended to soften the water. The precipitation of minerals containing major chemical elements (MnO_2, $FeO(OH)$, $CaCO_3$) in the upper reservoir reduced its hardness, causing a slight increase in pH. These hydrochemical changes were strongly attenuated in the aquifer and were only visible in the first few meters around the quarry. Concerning the specific case of the Obourg quarry and its local context, simulated results showed that the hydrochemical changes in the groundwater around the quarry should not influence the hydrochemical characteristics of the groundwater abstracted in the nearby drinking water pumping wells.

PSH activities thus imply an environmental impact around the quarry through the propagation of the piezometric head fluctuations within the chalk aquifer and the modification of the hydrochemical characteristics of the groundwater, which can reduce its quality. Overall, the system behavior and its consequences are strongly dependent on the hydraulic properties of the surrounding medium, and therefore, on the geological and hydrogeological contexts in which a quarry is located. The concentrations of chemical elements and the related equilibria were here considered with relatively constraining assumptions to evaluate the evolution of the hydrochemical conditions in the most problematic case. Achieved results thus correspond to some strong hypotheses. The impact assessment on hydrochemistry was thus performed by considering a certain degree of safety in relation, which is relevant considering the presence of drinking water catchments located near the quarry. The quarry, however, is located in the capture zone of the abstraction stations. In this context, it is important to make sure that the PSH equipment does not emit contaminants, especially regarding any lubricants that may be used.

An equivalent porous media approach was here considered to develop the models. This hypothesis did not consider possible heterogeneities within hydrogeological units. At the scale of the modelled area, this heterogeneity was limited, but it could have an impact on groundwater flow, hydrochemical impact distribution, and related influence zones. For sites with strong heterogeneity, as for example in the case of the karstified limestones, the hypothesis of a homogeneous medium becomes unrealistic. In those cases, the consideration of more complex models would be needed. In the framework of the Obourg quarry model, the regional gradient was neglected. The results of Poulain et al. (2018) nevertheless show that the presence of a regional gradient only has a very slight influence on piezometric head fluctuations and the distance of influence around the quarry. The area around the quarry, where hydrochemical conditions are influenced, is probably not strongly influenced by a realistic hydraulic gradient, although this should be formally evaluated. Finally, the presence of bacteria in the reservoir could also promote chemical reactions during the precipitation of minerals [34]. It would be interesting to assess the impact of organic matter and the bacterial ecosystem present in the groundwater on the hydrogeochemical behavior of the groundwater during the pumping–discharge cycles.

In conclusion, this study highlighted the importance of considering both the hydrodynamic and the hydrochemical aspects of PHS plants. The following points must be considered:

- Quarries cannot be considered as simple impervious reservoirs. They are in strong interaction with the surrounding aquifers. During PSH operations, water exchanges between the quarry and adjacent aquifers occur. The magnitude of the water exchanges depends on the hydrogeological context. This interaction has an impact on the fluctuation of the hydraulic head in the reservoir, on the difference in level between the upper and lower reservoirs, and on the efficiency of the system [8].
- The pump–discharge cycles in the reservoirs generate rapid and periodic fluctuations in the water level, which are propagated into the surrounding rock media. This can have an impact on other possible activities (e.g., groundwater abstraction station) in the vicinity. These load/discharge cycles in the rock media can also create potentially significant stability problems. These stability problems may, for example, occur at quarry walls, or in areas of altered rock in the form of localized collapses. These kinds of collapses are, for example, regularly observed around some limestone quarries where the groundwater table is lowered [35].
- Finally, it is also necessary to evaluate the hydrogeochemical behavior of the system during the PSH activities. From a hydrochemical perspective, carrying out pumping–discharge induces the aeration of the water. This aeration phenomenon can destabilize chemical balances between groundwater and minerals present in the rock. These cycles can therefore influence the chemical composition of the quarry water as well as that of the adjacent aquifer. It is important to study these possible hydrochemical modifications, both from the point of view of the exploited drinking water reservoir and of the water–rock interactions themselves.

This kind of numerical modelling can be developed for any PSH project, considering in each situation the characteristics of the quarry, the geology, hydrogeology and lithology, and the existing mineral species. For a PSH project using a quarry as a lower reservoir, a piezometric network system around the quarry would allow the control of the chemical evolution of the groundwater and would ensure a rapid reaction in case of contamination or unexpected hydrochemical changes.

Author Contributions: Conceptualization, A.P. and P.G.; methodology, A.P., E.P., and P.G.; software, A.P., P.G., and E.P.; validation, AP., E.P., and P.G.; formal analysis, A.P.; investigation, A.P., E.P. and P.G.; resources, A.P., E.P., and P.G.; data curation, A.P., E.P., and P.G.; writing—original draft preparation, A.P.; writing—review and editing, A.P., E.P., and P.G.; visualization, A.P., E.P., and P.G.; supervision, E.P. and P.G.; project administration, P.G.; funding acquisition, P.G. All authors have read and agreed to the published version of the manuscript.

Funding: This research was supported by the Public Service of Wallonia, Department of Energy and Sustainable Building in the framework of the SMARTWATER project. IDAEA-CSIC is a Centre of Excellence Severo Ochoa (Spanish Ministry of Science and Innovation, Project CEX2018-000794-S). E.P. was also funded by the Barcelona City Council through the Award for Scientific Research into Urban Challenges in the City of Barcelona 2020 (20S08708).

Institutional Review Board Statement: Not applicable.

Informed Consent Statement: Not applicable.

Data Availability Statement: All analyzed data in this study has been included in the manuscript.

Conflicts of Interest: The authors declare no conflict of interest. The funders had no role in the design of the study; in the collection, analyses, or interpretation of data; in the writing of the manuscript; or in the decision to publish the results.

References

1. Menendez, J.; Ordóñez, A.; Fernández-Oro, J.M.; Loredo, J.; Díaz-Aguado, M.B. Feasibility analysis of using mine water from abandoned coal mines in Spain for heating and cooling of buildings. *Renew. Energy* **2020**, *146*, 1166–1176. [CrossRef]
2. Rehman, S.; Al-Hadhrami, L.M.; Alam, M. Pumped hydro energy storage system: A technological review. *Renew. Sustain. Energy Rev.* **2015**, *44*, 586–598. [CrossRef]
3. Uddin, P.N.; Asce, M. Preliminary design of an underground reservoir for pumped storage. *Geotech. Geol. Eng.* **2003**, *21*, 331–355. [CrossRef]
4. Erpicum, S.; Archambeau, P.; Dewals, B.; Pirotton, M.; Pujades, E.; Orban, P.; Dassargues, A.; Cerfontaine, B.; Charlier, R.; Poulain, A.; et al. Underground Pumped Hydro-Energy Storage in Wallonia (Belgium) Using Old Mines—Potential and Challenges. In Proceedings of the HYDRO2017 Conference, Sevilla, Spain, 9 October 2017.
5. Madlener, R.; Specht, J. *An Exploratory Economic Analysis of Underground Pumped-Storage Hydro Power Plants in Abandoned Coal Mines*; Social Science Research Network: Rochester, NY, USA, 2013.
6. Toubeau, J.-F.; Bottieau, J.; De Greve, Z.; Vallee, F.; Bruninx, K. Data-Driven Scheduling of Energy Storage in Day-Ahead Energy and Reserve Markets with Probabilistic Guarantees on Real-Time Delivery. *IEEE Trans. Power Syst.* **2020**, *1*. [CrossRef]
7. Poulain, A.; de Dreuzy, J.-R.; Goderniaux, P. Pump Hydro Energy Storage systems (PHES) in groundwater flooded quarries. *J. Hydrol.* **2018**, *559*, 1002–1012. [CrossRef]
8. Pujades, E.; Orban, P.; Bodeux, S.; Archambeau, P.; Erpicum, S.; Dassargues, A. Underground pumped storage hydropower plants using open pit mines: How do groundwater exchanges influence the efficiency? *Appl. Energy* **2017**, *190*, 135–146. [CrossRef]
9. Pujades, E.; Jurado, A.; Orban, P.; Ayora, C.; Poulain, A.; Goderniaux, P.; Brouyère, S.; Dassargues, A. Hydrochemical changes induced by underground pumped storage hydropower and their associated impacts. *J. Hydrol.* **2018**, *563*, 927–941. [CrossRef]
10. Xuan, P.T.; Van Pho, N.; Gas'Kova, O.L.; Bortnikova, S.B. Geochemistry of surface waters in the vicinity of open pit mines at the Cay Cham deposit, Thai Nguyen province, northern Vietnam. *Geochem. Int.* **2013**, *51*, 931–938. [CrossRef]
11. Younger, P.L.; Banwart, S.A.; Hedin, R.S. *Mine Water: Hydrology, Pollution, Remediation*; Environmental Pollution; Springer: Dordrecht, The Netherlands, 2002; ISBN 978-1-4020-0137-6.
12. España, J.S.; Pamo, E.L.; Pastor, E.S.; Andrés, J.R.; Rubí, J.A.M. The Removal of Dissolved Metals by Hydroxysulphate Precipitates during Oxidation and Neutralization of Acid Mine Waters, Iberian Pyrite Belt. *Aquat. Geochem.* **2006**, *12*, 269–298. [CrossRef]
13. Severson, M.J. *Preliminary Evaluation of Establishing an Underground Taconite Mine, to Be Used Later as a Lower Reservoir in a Pumped Hydro Energy Storage Facility, on the Mesabi Iron Range, Minnesota*; University of Minnesota Duluth: Duluth, MI, USA, 2011.
14. Habils, F.; Roland, S.; Rorive, A. *Carte Hydrogéologique de Wallonie Au 1/25 000ème (Avec Notice Explicative)*; Planche Jurbise—Obourg N° 45/3-4; Service Public de Wallonie: Namur, Belgium, 2017.
15. Hallet, V. *Étude de la Contamination de la Nappe Aquifère de Hesbaye par les Nitrates: Hydrogéologie, Hydrochimie et Modélisation Mathématique des Écoulements et du Transport en Milieu Saturé (Contamination of the Hesbaye Aquifer by Nitrates: Hydrogeology, Hydrochemistry and Mathematical Modeling)*; University of Liège: Liège, Belgium, 1998.

16. Brouyère, S. *Etude et Modélisation du Transport et du Piégeage des Solutés en Milieu Souterrain Variablement Saturé. Evaluation des Paramètres Hydrodispersifs par la Réalisation et L'interprétation D'essais de Traçage In Situ*; Université de Liège: Sart Tilman, Belgique, 2001.
17. Dargne Carte Hydrogéologique de Wallonie [WWW Document]. Carte Hydrogéologique Wallonie. 2017. Available online: http://Environnement.Wallonie.Be/Cartosig/Cartehydrogeo/Application.Htm# (accessed on 29 January 2018).
18. Gaviglio, P.; Bekri, S.; Vandycke, S.; Adler, P.; Schroeder, C.; Bergerat, F.; Darquennes, A.; Coulon, M. Faulting and deformation in chalk. *J. Struct. Geol.* **2009**, *31*, 194–207. [CrossRef]
19. Goderniaux, P.; Beyek, A.; Tchotchom, A.; Poulian, A.; Wattier, M.-L.; Vandycke, S. Study of the heterogeneity of hydraulic properties in a chalk aquifer unit, using sequential pumping and tracing experiments with packer systems. In *Engineering in Chalk: Proceedings of the Chalk 2018 Conference, London, UK, 17–18 September 2018*; ICE Publishing: London, UK, 2018; pp. 675–680. ISBN 978-0-7277-6407-2.
20. Hoffmann, R.; Goderniaux, P.; Jamin, P.; Chatton, E.; De La Bernardie, J.; Labasque, T.; Le Borgne, T.; Dassargues, A. Continuous dissolved gas tracing of fracture-matrix exchanges. *Geophys. Res. Lett.* **2020**, *47*, e2020GL088944. [CrossRef]
21. Marlière, R. *Carte Géologique et Texte Explicatif de La Feuille Jurbise—Obourg*; Service Géologique de Belgique: Mons, Belgium, 1964.
22. Mercado, A. The kinetics of mineral dissolution in carbonate aquifers as a tool for hydrological investigations, II. Hydrogeochemical models. *J. Hydrol.* **1977**, *35*, 365–384. [CrossRef]
23. Edmunds, W.; Cook, J.; Darling, W.; Kinniburgh, D.; Miles, D.; Bath, A.; Morgan-Jones, M.; Andrews, J. Baseline geochemical conditions in the Chalk aquifer, Berkshire, U.K.: A basis for groundwater quality management. *Appl. Geochem.* **1987**, *2*, 251–274. [CrossRef]
24. Edmunds, W.; Shand, P.; Hart, P.; Ward, R. The natural (baseline) quality of groundwater: A UK pilot study. *Sci. Total. Environ.* **2003**, *310*, 25–35. [CrossRef]
25. Gillon, M.; Crançon, P.; Aupiais, J. Modelling the baseline geochemistry of groundwater in a Chalk aquifer considering solid solutions for carbonate phases. *Appl. Geochem.* **2010**, *25*, 1564–1574. [CrossRef]
26. Rorive, A. Determination Des Ressources Souterraines de La Nappe Du Crétacé de La Vallée de La Haine—Convention RW-IDEA, Rapport Final. 1983.
27. Harbaugh, A.W. MODFLOW-2005: The U.S. Geological Survey modular ground-water model—The ground-water flow process. In *Techniques and Methods*; US Geological Survey: Reston, VI, USA, 2005; Volume 6.
28. Parkhurst, D.L.; Kipp, K.L.; Engesgaard, P.; Charlton, S.R. PHAST—A program for simulating ground-water flow, solute transport, and multicomponent geochemical reactions. *U.S. Geol. Surv. Tech. Methods* **2005**, *6*, A8.
29. Kipp, K.L. *HST3D: A Computer Code for Simulation of Heat and Solute Transport in Three-Dimensional Ground-Water Flow Systems*; Water-Resources Investigations Report; U.S. Geological Survey: Reston, VA, USA, 1987; Volume 86-4095.
30. Parkhurst, D.; Appelo, T.; Parkhurst, D.L.; Appelo, C.A.J. *Description of Input and Examples for PHREEQC Version 3—A Computer Program for Speciation, Batch-Reaction, One-Dimensional Transport, and Inverse Geochemical Calculations. US Geological Survey Techniques and Methods*; US Geological Survey: Denver, CO, USA, 2013; Book 6, Chapter A43; 497p.
31. Pujades, E.; Willems, T.; Bodeux, S.; Orban, P.; Dassargues, A. Underground pumped storage hydroelectricity using abandoned works (deep mines or open pits) and the impact on groundwater flow. *Hydrogeol. J.* **2016**, *24*, 1531–1546. [CrossRef]
32. Pujades, E.; Jurado, A.; Orban, P.; Dassargues, A. Parametric assessment of hydrochemical changes associated to underground pumped hydropower storage. *Sci. Total. Environ.* **2019**, *659*, 599–611. [CrossRef] [PubMed]
33. Chimie et Pollutions des Eaux Souterraines ATTEIA Olivier. Available online: https://www.lavoisier.fr/livre/environnement/chimie-et-pollution-des-eaux-souterraines/atteia/descriptif-9782743020095 (accessed on 9 April 2021).
34. Papier, S. *Géomicrobiologie des Dépôts Bactériens Ferrugineux: Influence des Facteurs Biologiques et Environnementaux sur L'oxydation du fer en Milieu Continental*; Université de Mons: Mons, Belgium, 2014.
35. Kaufmann, O.; Quinif, Y. Geohazard map of cover-collapse sinkholes in the 'Tournaisis' area, southern Belgium. *Eng. Geol.* **2002**, *65*, 117–124. [CrossRef]

applied
sciences

Article

Monitoring Scheme for the Detection of Hydrogen Leakage from a Deep Underground Storage. Part 2: Physico-Chemical Impacts of Hydrogen Injection into a Shallow Chalky Aquifer

Philippe Gombert [1,*], Stéphane Lafortune [1], Zbigniew Pokryszka [1], Elodie Lacroix [1,2], Philippe de Donato [2] and Nevila Jozja [3]

[1] Ineris, Parc Technologique Alata, 60550 Verneuil-en-Halatte, France; stephane.lafortune@ineris.fr (S.L.); zbigniew.pokryszka@ineris.fr (Z.P.); elodie.lacroix@ineris.fr (E.L.)

[2] GéoRessources Laboratoire, Université de Lorraine-CNRS, 54500 Vandœuvre-lès-Nancy, France; philippe.de-donato@univ-lorraine.fr

[3] École Polytechnique, Laboratoire CETRAHE, Université d'Orléans, 8 Rue Léonard de Vinci, 45100 Orléans, France; nevila.jozja@univ-orleans.fr

* Correspondence: philippe.gombert@ineris.fr; Tel.: +33-(0)-344-556-234

Featured Application: Implementation of a physicochemical and hydrogeochemical monitoring in shallow aquifers above future underground hydrogen storage sites in salt caverns.

Abstract: This paper presents the results of an experiment to simulate a sudden and brief hydrogen leak from a potential deep geological storage site. A 5 m^3 volume of groundwater was extracted, saturated with hydrogen, and then reinjected into the aquifer. Saturating the water with hydrogen caused a decrease in the oxidation-reduction potential, the dissolved gas content (especially O_2 and CO_2), the electrical conductivity, and the concentration of alkaline earth bicarbonate ions and a slight increase in pH. These changes are observed until 20 m downstream of the injection well, while the more distant piezometers (from 30 to 60 m) are not significantly affected. During this experiment, no indicators of the development of chemical or biochemical reactions are observed, because of the rapid transfer of the dissolved hydrogen plume through the aquifer and its significant dilution beyond 10 m downstream of the injection well. Here, hydrogen behaved as a conservative element, reacting very slightly or not at all. However, this experiment demonstrates the existence of direct and indirect impacts of the presence of hydrogen in an aquifer. This experiment also highlights the need to adapt the monitoring of future underground hydrogen storage sites.

Keywords: hydrogen; underground storage; leakage; environmental impact; monitoring

Citation: Gombert, P.; Lafortune, S.; Pokryszka, Z.; Lacroix, E.; de Donato, P.; Jozja, N. Monitoring Scheme for the Detection of Hydrogen Leakage from a Deep Underground Storage. Part 2: Physico-Chemical Impacts of Hydrogen Injection into a Shallow Chalky Aquifer. *Appl. Sci.* **2021**, *11*, 2686. https://doi.org/10.3390/app11062686

Academic Editors: Jorge Loredo and Javier Menéndez

Received: 23 February 2021
Accepted: 15 March 2021
Published: 17 March 2021

Publisher's Note: MDPI stays neutral with regard to jurisdictional claims in published maps and institutional affiliations.

1. Introduction

1.1. General Information on Underground Hydrogen Storage

In 2015, France promulgated the Law on Energy Transition for Green Growth in order to contribute more effectively to the fight against climate change and the preservation of the environment, as well as to strengthen its energy independence [1]. This law aims in particular to develop renewable energy sources, some of which are of a fluctuating or intermittent nature and thereby require significant storage capacity. The underground environment is well suited to large-scale storage, and France already has 78 salt caverns with a capacity to store liquid or liquefied hydrocarbons as well as natural gas [2]. The gradual phasing out of fossil fuels gives rise to the hypothesis that future underground storages of hydrogen (H_2) will need to be developed, as already in the United Kingdom or the United States [3]. This technology is likely to be of interest to several European countries: Germany, the Netherlands, Denmark, Poland, France, etc. [4].

This is why the French Scientific Interest Group GEODENERGIES chose to fund a research project in 2017 on the risks and opportunities of the geological storage of hydrogen in salt caverns, entitled ROSTOCK-H. Among other things, this project focuses on the risks associated with possible hydrogen gas leaks from deep storage. In the event of such leakage, the hydrogen would migrate to the surface and encounter at least one aquifer where the gas would dissolve in the water until saturation [5]. However, considering the very low solubility of this gas (in the order of $1.8~mg \cdot L^{-1}$ at saturation under surface conditions), there is a risk that the leak flow rate may exceed the dissolution potential for hydrogen in water. In this case, part of the gas would continue its migration to the surface in gaseous form until it encounters an impermeable formation or a void where it can accumulate (mine, underground networks, cellar, underground car park, tunnel, etc.): hydrogen may then cause asphyxiation or, due to its very weak lower flammability limit (around 4%), explosion or fire [2].

To study the risks associated with this new storage technology, particularly in the event of a leak towards the surface, we carried out an experiment to simulate a hydrogen leakage from a possible deep geological storage. To do this, we injected hydrogen into the chalk aquifer, which is a major drinking water resource in the Paris region. Several monitoring devices were set up directly in the aquifer up to 60 m downstream from the injection point, in order to monitor the evolution of the dissolved hydrogen plume and the associated physico-chemical and hydrochemical phenomena. Under the proposed experimental conditions, and due to the brevity of the injection, we did not expect any biochemical reactions that require long time scales to occur. Moreover, additional studies [6–9] pointed out that a sudden injection of hydrogen saturated water into an aquifer increases the dissolved hydrogen content of the water, and consequently reduces its oxidation-reduction potential, as well as the concentration of other dissolved gases (O_2, CO_2, N_2, etc.). The injection of hydrogen was preceded by the injection of tracers (helium and hydrogeological tracers) in order to facilitate the monitoring of its progress underground. This experiment was conducted at the Catenoy (Oise) experimental site in a surface aquifer representative of the carbonated hydrogeological context of the Paris Basin. It followed characterization work for this site, previously carried out during CO_2 leak simulations [10,11], as well as the realization of a baseline and a preliminary injection of helium [6].

1.2. Hydrogen Reactivity in a Natural Environment

The impacts expected from a hydrogen leak in the underground environment are linked to the fact that it is a strongly reducing gas acting as a potential electron donor for numerous chemical species: metal sulfides, sulfates, carbonates, oxides (in particular of iron and magnesium), nitrates, ferrous ions, and gases (CO and CO_2) [7,8]. The resulting oxidation-reduction reactions can thus modify the chemical composition of water or the mineral composition of aquifer rocks [6]. However, most reactions that occur in the presence of hydrogen require—at least in the laboratory—high temperatures or pressures or the presence of catalysts (Table 1). Under ambient conditions, hydrogen-consuming oxidation-reduction reactions have slow kinetics because hydrogen is not a polar molecule and the H-H bond is difficult to break owing to its elevated binding energy ($436~kJ \cdot mol^{-1}$). However, Truche et al. [12] showed that the reduction of pyrite by hydrogen could have significant kinetics at low pressure and temperature in the presence of catalysts, in this case clay minerals. According to these authors, catalysts could also be bacteria, other mineral surfaces, or certain metals (iron, carbon steel, stainless steel, copper, nickel, platinum, and palladium). In the case of shallow aquifers, frequently used for water production (wells and boreholes for drinking, agricultural, industrial, or mineral water) or potentially crossed by other types of underground structures (wells, geotechnical foundations, etc.), the presence of metal parts made of iron, steel, or stainless steel could therefore play this catalytic role locally (Table 1).

Table 1. Some examples of abiotic reductions due to H_2 under experimental laboratory conditions.

Impacted Species	H_2 Partial Pressure (bar)	Temperature (°C)	pH	Catalyst	Resulting Species	Reference
SO_4^{2-}	4–16	250–300	≤ 5.6	None	H_2S	[13]
FeS_2 *	8–18	90–180	6.9–8.7	None	FeS_{1+x}, H_2S	[14]
FeS_2 *	3–30	90–250	7.8–9.8	None	FeS_{1+x}, H_2S	[12]
NO_3^-	0.1	22	7.0–8.7	Fe	NO_2^-	[15]
NO_3^-	0.05	200	~6	Fe	NH_4^+	[16]
NO_3^-	7.5	90–180	4–9	Carbon steel	NH_4^+	[17]
NO_3^-	0.2–7.5	90–150	4–9	Stainless steel	NH_4+	[18]
NO_3^-	0.1–0.5	7–25	5–11	Pd, Cu	N_2	[19,20]

* Pyrite transformed into pyrrhotite (FeS_{1+x} with $0 < x < 0.125$).

In addition, the natural leak analogs, which are the hydrogen emission sites, also show that the surface and subsurface environments affected by these emissions can be significantly altered, even under ambient temperature and pressure: e.g., decrease in sulfates, decrease in oxidation-reduction potential, and increase in pH [9]. However, these are usually biogeochemical reactions that also have slow kinetics [12,21,22]. As such, the laboratory experiment carried out by Berta et al. [22] under ambient conditions lasted 180 days: it demonstrated a biochemical reduction of sulfates and carbon dioxide, a decrease in the calcium concentration, and an increase in the silica concentration concurrent with an increase in pH.

1.3. Risk of Contamination of Drinking Water Aquifers

Monitoring the aquifers would therefore be a way of detecting an ongoing hydrogen leak, for alerting purposes, especially since this gas is not generally present in groundwater. In the hydrogeological context of the Paris Basin, two aquifers could fulfill the role of barrier and surveillance zone: the deep Albian-Neocomian aquifer and the shallow Cretaceous chalk aquifer. The latter contains a generally unconfined water table which is used to supply drinking water to a part of Paris and almost all of the neighboring towns. When it is close to the surface, its water is oxygenated, thus under oxidizing conditions, and often enriched with nitrates from anthropogenic surface activities, as well as, locally, with sulfates from the overlying Tertiary formations. The arrival of hydrogen in such an aquifer has the immediate effect of increasing the dissolved hydrogen content of the water and, as a result, reducing its oxidation-reduction potential and possibly its content in natural dissolved gases (mainly N_2, O_2, and CO_2). The hydrogeochemical impacts expected in the aquifer zone affected by the presence of hydrogen could be as follows [12,13,15–17,22,23]:

1. Reduction of nitrates (NO_3^-) to nitrites (NO_2^-), or even to ammonium (NH_4^+), and then to gaseous nitrogen (N_2)
2. Reduction of sulfates (SO_4^{2-}) to sulfides (SO_3^-), or even to hydrogen sulfide (H_2S)
3. Reduction of iron III to iron II
4. Dissolution of metallic trace elements potentially present in the aquifer rock following the drop in the redox potential

The consequences of a possible reduction in ion oxides such as NO_3^- can be significant because the standards for drinking water destined for human consumption stipulate $50 \text{ mg} \cdot L^{-1}$ for nitrates against $0.50 \text{ mg} \cdot L^{-1}$ for nitrites and $0.10 \text{ mg} \cdot L^{-1}$ for ammonium [24]. Thus, allowing for a mean concentration of nitrates of $33 \text{ mg} \cdot L^{-1}$ in the groundwater at the Catenoy experimental site, during the baseline [6], it would be sufficient to reduce only 2% of these ions into nitrites or 0.5% into ammonium to render the water unfit for human consumption in the first case or to trigger health warning measures (information, reinforced surveillance, treatment) in the second case.

However, the literature shows that, under normal pressure and temperature conditions, the reduction of these nitrates and sulfates cannot take place except in the presence of a catalyst such as iron, nickel, copper, or platinum. Nevertheless, the frequent use of iron and stainless steel, which contains up to 20% of nickel, in the composition of the metal tubing of underground structures will likely bring some of these catalysts into contact with the groundwater. This paper describes a specific monitoring scheme which has been applied on a pilot site allowing to follow the impacts of H_2 injection into a shallow chalky aquifer. For this, many chemical-physical parameters are monitored as well as some dissolved gases such as H_2 and O_2. The temporal evolution of all these parameters is compared in order to better understand the effect of a sudden appearance of H_2 in an aquifer and assess our ability to detect H_2 leaks.

2. Material and Methods

2.1. Overview of the Site Context and Equipment

The experimental site is located at Catenoy (Oise), about 50 km north of Paris [6]. It is situated in the unconfined chalk aquifer of the Paris Basin. The geology corresponds to a few meters of Quaternary deposits and Tertiary formations, lying over a hundred meters of Senonian chalk that is only visible in the thalwegs (see also Figure 1). Under the site, the underground geology of the first 25 m is: 3 m of colluvium, 4 m of Thanetian sands, and 18 m of chalk. The chalk encloses an aquifer with a static level at a depth of 13 m, which flows in the WSW–ENE direction. The storage coefficient is from 1.1×10^{-2} to 6.5×10^{-2} and the hydraulic conductivity is from 6.4×10^{-4} to 1.4×10^{-3} m·s^{-1}.

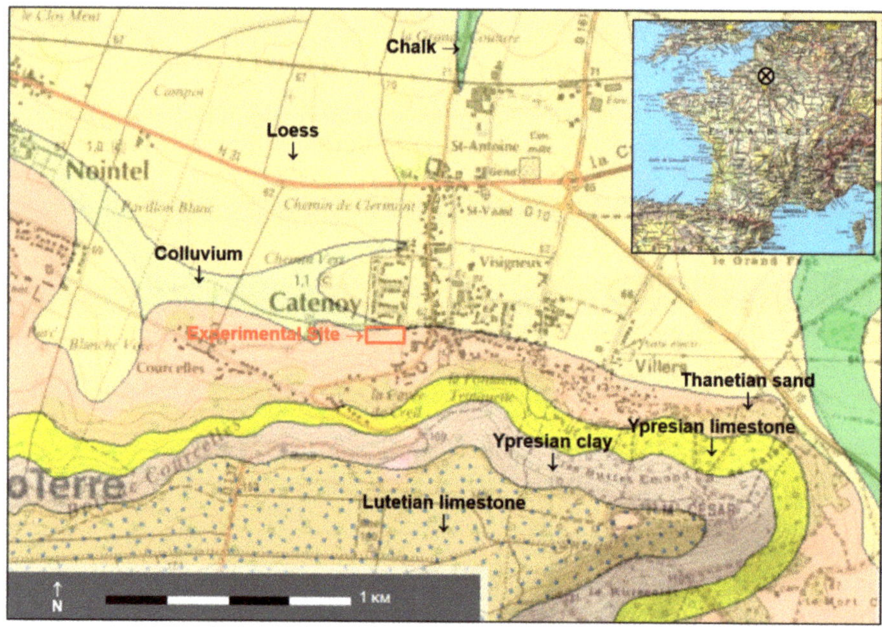

Figure 1. Location and geological context of the experimental site.

The experimental site has eight piezometers which are 25 m deep and aligned in the direction of the aquifer flow (Figure 2), over a distance of 80 m. The PZ1 piezometer, which functions as a control, is located 20 m upstream of the PZ2 injection well. The PZ3, PZ4, PZ5, and PZ6 piezometers are, respectively, located 10, 20, 30, and 60 m downstream of the injection well. There are also two close monitoring piezometers: the PZ2BIS located at 5 m, which is the main monitoring piezometer for this study, and the PZ2TER located at 7.5 m, which contains Infrared and Raman spectrometers (the specific results from these are the subject of another article in preparation). The site is also equipped with a technical room and a meteorological station. The data are transmitted in real time to the dedicated e.cenaris cloud monitoring system [25,26].

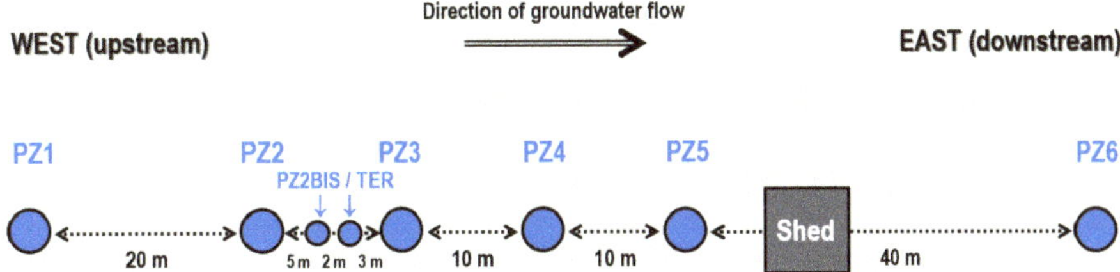

Figure 2. Detailed plan view of the experimental site.

The hydrodynamic characteristics of the aquifer were determined during a pumping test carried out previously [10]. Depending on the piezometer considered, the porosity varies from 1.1×10^{-2} to 6.5×10^{-2} and the hydraulic conductivity from 6.4×10^{-4} to 1.4×10^{-3} m·s^{-1}. The groundwater flow velocity, measured during a tracing test, varies from 3 m·day^{-1} at PZ3 and PZ5 to 10 m·day^{-1} at PZ4, which is situated in a preferential flow path (probably a fissured zone). This highlights the dual porosity of the aquifer studied, which may be reflected hydrogeologically by contrasts in flow velocity and variations in transit times from one piezometer to another.

2.2. Overview of the Monitoring Protocol

The monitoring protocol for the saturated zone was detailed by Lafortune et al. [6] and is based on the preliminary injection of helium, because hydrogen and helium have a comparable physical behavior, in particular a very low solubility and a high diffusion coefficient in water.

As a reminder, the monitoring protocol includes (Figure 3):

1. Two physicochemical probes measure temperature, pH, electrical conductivity, oxidation-reduction potential, and dissolved oxygen. One is installed in the PZ2bis while the other is mobile in order to take measurements at all the piezometers. It should be noted that this last probe had to be replaced on 19 November 2019 at 14:00, which is visible on the graphs as a deviation due to the new calibration (just before the injection of dissolved hydrogen).
2. Two field fluorometers allow on site measure of the water fluorescence. As before, one is installed in the PZ2bis, while the other is mobile in order to take measurements in all the piezometers by groundwater sampling.
3. Six submerged electric pumps are installed at a depth of 16 m at the PZ1, PZ2BIS, PZ3, PZ4, PZ5 and PZ6 piezometers. They are used to regularly sample the groundwater in order to perform laboratory analyses of tracers, dissolved gases (CH_4, He, H_2, and H_2S), major elements (Ca^{2+}, Mg^{2+}, Na^+, K^+, HCO_3^-, Cl^-, SO_4^{2-}, and NO_3^-, with detection limits of 0.01–0.05 mg·L^{-1}), and minor elements (SO_3^-, S^{2-}, NO_2^-, and NH_4^+, with detection limits of 0.01–0.02 mg·L^{-1}).
4. Raman and Infrared (IR) spectrometers installed at the PZ2TER [25,27] make it possible to analyze the concentration in the groundwater of mononuclear diatomic molecules (H_2, O_2, and N_2 for Raman) and polar molecules (CO_2 and CH_4 for Raman and IR). It should be noted that the data acquired by these devices are not presented here but are the subject of another specific publication to come. Here, we selected Raman and IR technologies because they were already used in other studies, which ensures they suit our needs. New sensing technologies using optical fiber grating platforms are presently developed and tested on lab benches and would be promising for field studies in the close future [28].

5. There are specific analyzers for measuring the gas concentration in the piezometer head spaces and in the gas mixture released from the water extracted from the aquifer. A DRÄGER multigas analyzer equipped with a catalytic cell (resolution of 0.1% vol.) and a portable Biogas analyzer equipped with an electrochemical cell (resolution of 1 ppm) are used for hydrogen, while an ALCATEL ASM 122D transportable mass spectrometer is used for helium.

6. Device for extracting and degassing water by mechanical agitation serves to establish the dissolved gas concentration in the groundwater, in conjunction with the gas analyzers detailed above.

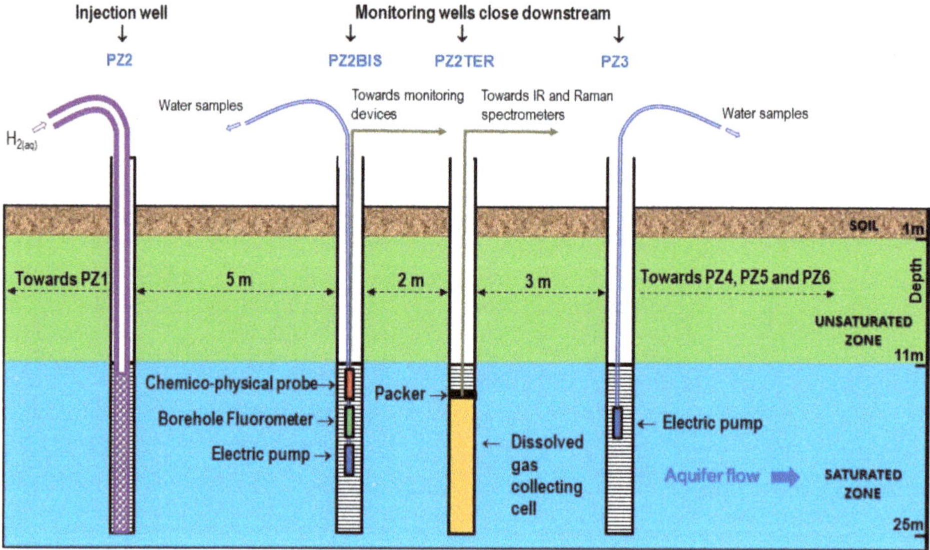

Figure 3. Various devices installed at the injection well and the near downstream monitoring wells.

The migration of the plumes injected into the aquifer is monitored automatically at PZ2BIS and at PZ2TER by means of devices installed in situ and manually at the other piezometers by sampling the water. All measuring devices or probes are installed between 15 and 16 m deep, i.e., in the most productive zone of the aquifer under study [10].

During the first week, the monitoring was relatively frequent with 4–15 samples per day depending on the piezometer. During the following weeks, a follow-up with a more spaced measurement frequency was set up, at a rate of 1–3 samples per piezometer and per week. In total, this monitoring made it possible to take 130 water samples to analyze the tracers and chemical elements and to conduct 104 physicochemical measurements. In addition, 127 water samples were taken to measure in situ the helium (tracer gas) and hydrogen contents using the method of partial degassing by mechanical agitation.

2.3. Experimental Protocol

Figure 4 shows a view of the experimental site (Catenoy, France). The experiment consisted of extracting 5 m^3 of groundwater from PZ2, saturating it with hydrogen gas in a tank (Figure 5), and then injecting this hydrogenated water into the aquifer through the same piezometer. Another 1 m^3 tank contained groundwater with tracers to monitor the propagation of the injected plume.

Figure 4. View of the experimental site during the experiment.

Figure 5. View of the injection device.

The tracers were selected during the preliminary test carried out in April 2019 [6]. They consisted of a neutral gas (helium) and two tracers (uranine and lithium chloride). These tracers were not added to the 5 m^3 tank in order to avoid affecting the dissolution of hydrogen, a gas that is very poorly soluble naturally, with a solubility at saturation of around 1.8 mg·L^{-1} under surface pressure and temperature conditions (compared with that of other gases generally present in groundwater: 11 mg·L^{-1} for dioxygen, 24 mg·L^{-1} for nitrogen, and 2500 mg·L^{-1} for carbon dioxide).

The two tanks were filled on 18 November 2019 from 10:00 to 15:30 using groundwater. The bubbling of helium gas in the first 1 m^3 tank began immediately after filling it and continued overnight until the next day at 14:00, which made it possible to achieve complete saturation of the water with helium. The tracers were then dissolved in the water of this tank at the concentration of 10 mg·L^{-1} each. These tracers were lithium in the form of LiCl (a conservative but a colorless tracer, only detectable by water sampling and laboratory analyses a posteriori) and uranine (a less conservative tracer but colored and detectable in situ, in real time, by fluorimetry). The water from the first tank was injected by gravity into PZ2 on 19 November 2019 from 14:00 to 14:20 at a rate of 3 m^3·h^{-1}.

The second tank, with a volume of 5 m^3, contained the hydrogenated water. This is pure hydrogen from two compressed gas cylinders (50 L and 200 bars each). The bubbling of hydrogen at a flow rate of approximately 20 L·min^{-1} started immediately after filling the tank and continued until 21:30. It was then interrupted in the evening, for safety reasons, to resume the next morning at 06:30 until 15:10. This made it possible to reach a concentration of dissolved hydrogen of 1.76 mg·L^{-1}, or 95% of water saturation by hydrogen under the average temperature and hydrostatic pressure conditions within the tank (Figure 6). The total quantity of hydrogen dissolved in this tank was approximately 9 g or 100 L under Standard Temperature and Pressure (STP). H$_2$-saturated water was injected below the water table, i.e., a slight undersaturation with respect to hydrostatic conditions to prevent or limit H$_2$ degassing.

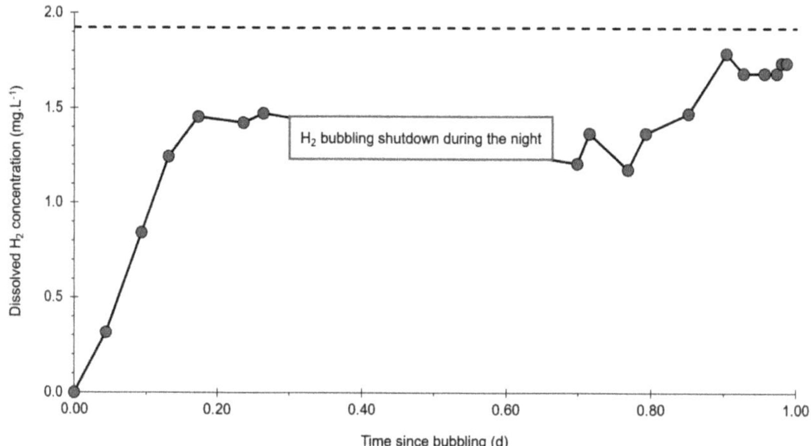

Figure 6. Evolution of the dissolved hydrogen concentration of the water in the second tank during bubbling (measured by partial degassing after mechanical agitation; the dashed line corresponds to saturation).

The physicochemical parameters of the water were monitored during this operation. During the saturation of the water with hydrogen, the oxidation-reduction potential increased from +148 mV to −224 mV and the concentration of dissolved oxygen dropped from 7.21 mg·L^{-1} to 0.74 mg·L^{-1} (Figure 7a). At the same time, the pH increased from 6.95 to 7.51, while the electrical conductivity remained stable around 552 µS·cm^{-1}. The temperature of the water in the tank regularly decreased from 11.5 °C to less than 9 °C, under the effect of nocturnal cooling. The water from this tank was then injected by gravity into PZ2 on 19 November 2019 from 14:50 to 17:20, i.e., 30 min after completing injection of the first tank. The average injection flow rate was 2 m^3·h^{-1}.

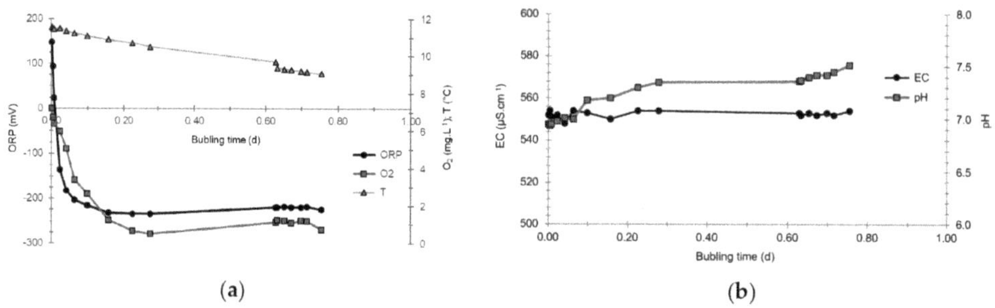

Figure 7. Evolution of the physicochemical parameters of the water in the tank during hydrogen bubbling: (**a**) oxidation-reduction potential (ORP), temperature (T), and dissolved O$_2$; and (**b**) electrical conductivity (EC) and pH.

3. Results

3.1. Aquifer Piezometry

Long-term piezometric monitoring was set up more than a year before injection [6]. During the 2019–2020 hydrological cycle, the average water table of the site reached its maximum in mid-April 2019 with 47.67 m ASL and its minimum at the beginning of October 2019 with 46.80 ASL (Figure 8a). Since that date, i.e., 49 days before the experiment, the piezometric level rose regularly as a result of the autumn recharge.

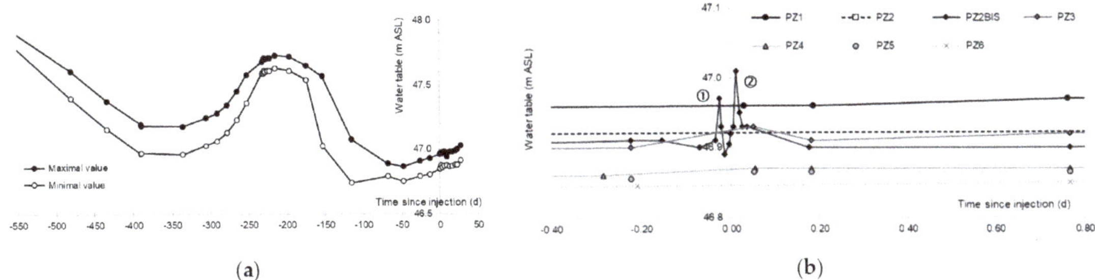

Figure 8. Piezometric monitoring of the experimental site: (**a**) piezometry from the start of monitoring; and (**b**) piezometry before and after the injections.

During the injection, the average piezometric level of the site remained stable at 46.90 m ASL, with the exception of PZ2BIS (+0.07 m) and PZ3 (+0.03 m), which were affected by the injections (Figure 8a). This was probably also the case for PZ2TER, located between the two previous piezometers, but it was not possible to measure its water table due to the presence of a packer.

At PZ2BIS, a first piezometric peak of +0.07 m occurred 0.25 h after the start of the first injection (see circled 1 in Figure 8b), and then the water table returned to its original level just before a new peak of +0.11 m, which occurred 0.33 h after the start of the second injection (see circled 2 in Figure 8b). The water table returned to its initial state 2 h after the end of the second injection. This piezometer was therefore influenced by the average overpressures of 1.4 bar induced by the successive injections and, to a lesser extent, same for the PZ2TER and the PZ3.

3.2. Tracing of the Injected Water

PZ2 remained sealed for 24 h after the injections to limit the risk of hydrogen degassing. When it was reopened, despite the leaching induced by the injection of 5 m^3 of hydrogenated water that followed the injection of 1 m^3 of tracer-holding water, the tracer concentrations still reached 208 µg·L^{-1} for lithium, 117 µg·L^{-1} for uranine, and 3.9 µg·L^{-1} for helium (Figure 9). Depending on the tracer, this represents 0.3–2% of the concentration injected the day before. Since the injection induced an overpressure, this phenomenon is probably due to the retention of part of the tracers within the chalk's porous matrix (which is of micrometric dimension), and then to their slow release when the natural flow of the aquifer resumed. This phenomenon has previously been demonstrated at other sites where the chalk shows dual porosity [29,30]. Thus, in the injection well, the tracer concentration remained higher than the background noise for 12.9 days for lithium and helium and until the end of the monitoring (t ≥ 49.9 days) for uranine, on which date the analyses still showed the presence of 0.29 µg·L^{-1} of this tracer. This retention phenomenon therefore generated a decreasing but clearly significant background noise for about two weeks after injection.

At PZ2BIS, 5 m downstream, three successive tracer plumes were observed (Figure 10b):

1. A first plume caused a significant peak (see circled 1 in Figure 10b). It started at 14:30, i.e., 0.5 h after injection of the tracers, and peaked at 14:45 at a concentration of 1678 µg·L^{-1} for lithium and 1188 µg·L^{-1} for uranine (a signal which saturated the fluorimeters). Due to an initially inadequate measurement interval scheme, no helium was detected in this first plume that corresponds to the passage of water from the tracer tank. During this peak, which lasted approximately 1 h, the tracer concentration reached 17% of the lithium injection concentration and 12% of the uranine injection concentration, i.e., a dilution factor of 5.96 and 8.42, respectively.

2. A second more intense and longer plume lasted about 3.5 h, synchronous with the injection of the hydrogenated water (see circled 2 in Figure 10b). It peaked at 15:00,

i.e., approximately 10 min after the start of injection, at a concentration of 3913 $\mu g \cdot L^{-1}$ for lithium, 2667 $\mu g \cdot L^{-1}$ for uranine (again inducing saturation of the fluorimeters), and 160 $\mu g \cdot L^{-1}$ for helium. The tracer dilution factors reached 2.56, 3.75, and 9.28, respectively. This plume is interpreted as the result of leaching of the tracers contained in the porous matrix of the aquifer rock induced by the arrival of water at a slight overpressure from the second tank.

3. In the third, weaker and more spread-out plume (see ③ in Figure 10b; note the logarithmic axes), uranine was detected by the fluorimeter. It peaked at 189 $\mu g \cdot L^{-1}$ on 20 November 2019 at 02:45 (t = 0.531 days) and decreased very slowly, finally lasting about a month; 0.24 $\mu g \cdot L^{-1}$ of uranine remained at t = 26.8 days, lithium and helium having disappeared by t = 12.9 days. When the concentration peak passed through, the dilution factor reached 52.9 for uranine, the only tracer measured using the recording fluorimeter (because the peak occurred in the middle of the night), which must have corresponded to a dilution factor of about 36 for lithium and about 130 for helium.

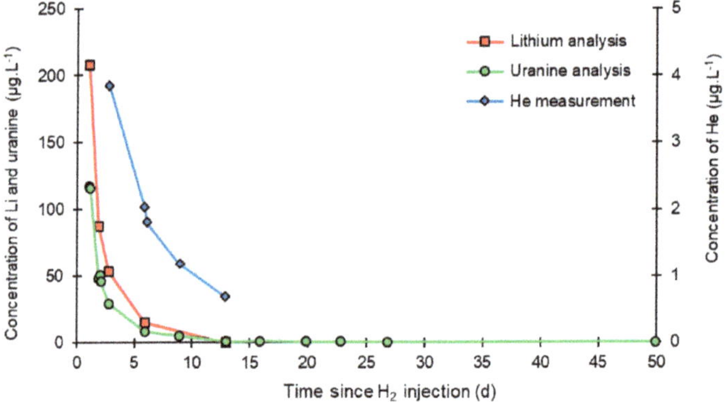

Figure 9. Evolution of the tracers at PZ2 (injection well).

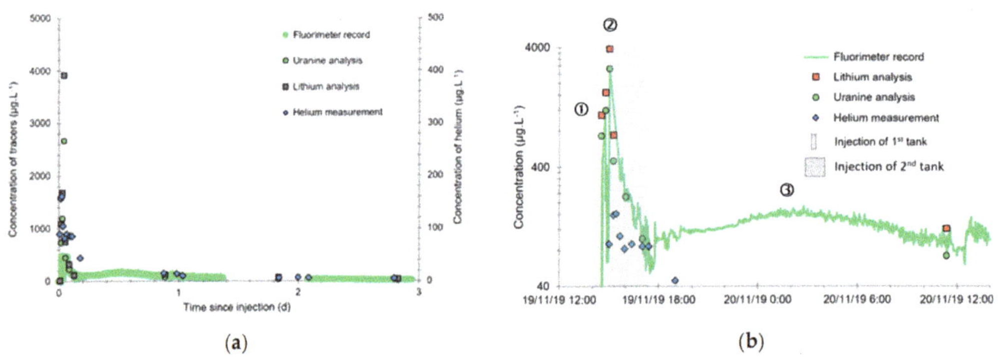

(a)

(b)

Figure 10. Breakthrough curves for the tracers at PZ2BIS (5 m downstream): (a) during the first three days of monitoring; and (b) detail of the first day of monitoring.

As regards the other downstream piezometers, a peak of tracers is clearly noticeable at PZ3 (10 m downstream) and PZ4 (20 m downstream) at t = 2 days with respective concentrations of 9.75 and 9.08 $\mu g \cdot L^{-1}$ for uranine (Figure 11a). These peaks seem to be shifted to t = 6 days for lithium with respective concentrations of 15 and 10 $\mu g \cdot L^{-1}$

(see circled 1 in Figure 11b), and for helium with respective concentrations of 1 and 2 µg·L^{-1} (Figure 11c). At PZ4, a second lithium peak is also observed at t = 20 days with a maximum concentration of 24 µg·L^{-1} (see circled 2 in Figure 11b). This distinct behavior of piezometers PZ3 and PZ4 has been observed previously. It means that the tracers arrive as quickly and at the same concentrations at PZ3 as at PZ4, which is twice as far from the injection well, with respective speeds of 5 and 9.5 m·day^{-1}. It is assumed that the groundwater takes a dual flow path, which is more marked at PZ4 than at PZ3, with a rapid circulation within a more permeable (probably fissured) zone, and a slow circulation within the aquifer's porous matrix.

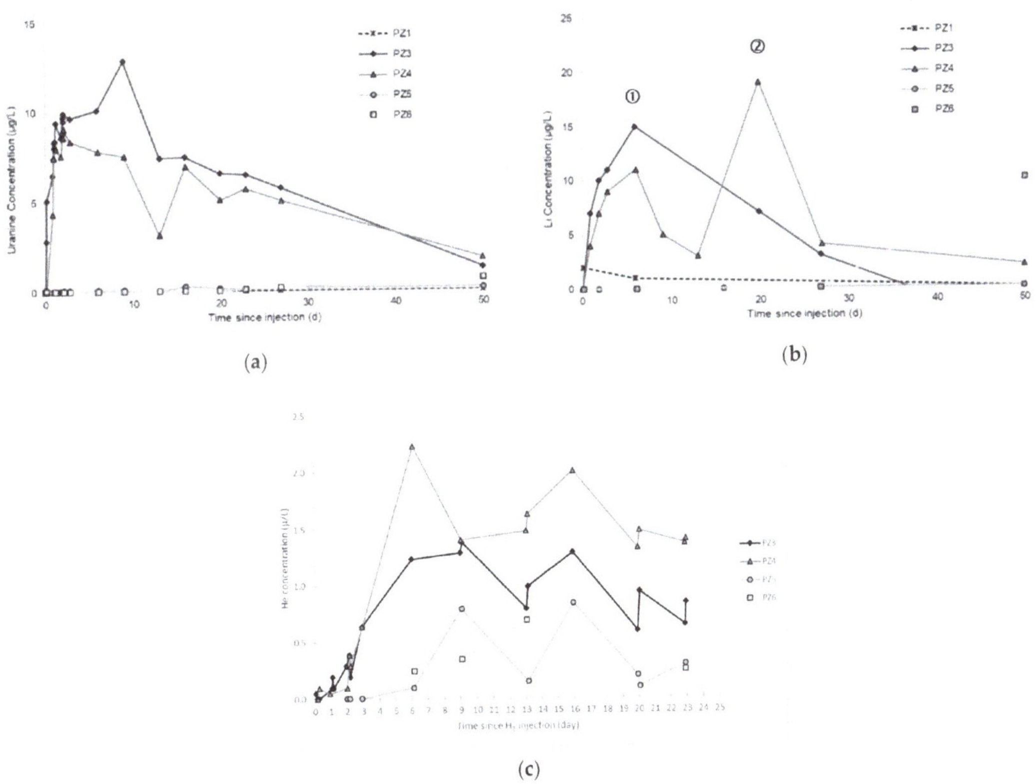

Figure 11. Breakthrough curves for the hydrogeological tracers at the other downstream piezometers: (**a**) uranine; (**b**) lthium; and (**c**) helium.

Regarding the piezometers located far downstream, namely PZ5 at 30 m and PZ6 at 60 m, traces of each of the two tracers are observed there approximately two weeks after injection (about 0.6 µg·L^{-1} for uranine, about 10 µg·L^{-1} for lithium), and helium is detected six days after injection with a concentration lower than 1 µg·L^{-1} (Figure 11a). Although the peak concentration of tracers has not yet been reached when the monitoring was stopped, it can be estimated that the corresponding transfer rates here are less than 1–2 m·day^{-1}. The dilution ratio is from 0.01% to 0.1%, which means that the impact of the hydrogen injected into the aquifer cannot be measured at these distances, under the experimental conditions created. Note that this result is in line with that obtained by PHREEQC modeling during the previous helium injection test [6].

3.3. Dissolved Hydrogen

In the water samples obtained by pumping, the dissolved hydrogen is extracted by the method of partial degassing by mechanical agitation, after which its content is measured with a portable Biogas analyzer with a detection threshold of 0.5 $\mu g \cdot L^{-1}$. The dissolved CH_4 and H_2S are also measured by the same method, with respective detection limits of approximately 1 and 0.6 $\mu g \cdot L^{-1}$. In parallel, the gases dissolved in the water of the PZ2TER are analyzed by Raman and Infrared spectrometry [25].

The results obtained are as follows (Figure 12):

1. No trace of dissolved CH_4 or H_2S is detected in any piezometer using the method of partial degassing by mechanical agitation.
2. No trace of dissolved H_2 is detected at PZ1 located 20 m upstream of the injection well or at PZ5 and PZ6, respectively, located 30 and 60 m downstream of the injection well.
3. At the injection well (PZ2), the dissolved H_2 concentration, which is 1.76 $mg \cdot L^{-1}$ at the time of injection, is still 0.084 $mg \cdot L^{-1}$ when the well was reopened, i.e., 1.8 days after the start of the injection. This residual concentration represents 5% of the concentration of the injected water.
4. At the main monitoring piezometer (PZ2BIS), located 5 m downstream, the concentration peak passed 2 h after the injection started (i.e., 0.08 days) with a value of 0.63 $mg \cdot L^{-1}$. This corresponds to a theoretical transfer velocity of 60 $m \cdot day^{-1}$ what is not a representative value of the mean transfer velocity of the water in the aquifer (which is approximately 3–10 $m \cdot day^{-1}$ according to Gombert et al. [7]) because it is strongly influenced by the injection.
5. At the monitoring piezometer PZ2TER located 7 m downstream, the dissolved H_2 concentration peak detected by Raman spectrometry passed through 9.7 h after the start of injection (i.e., 0.40 days) with a value of 0.17 $mg \cdot L^{-1}$ [25]. This corresponds to a theoretical transfer velocity of 17 $m \cdot day^{-1}$; value still strongly influenced by the injection.
6. At the PZ3 piezometer, located 10 m downstream, the first traces of dissolved H_2 appeared at 1.05 days and the concentration reached its maximum of 1.7 $\mu g \cdot L^{-1}$ after 2.02 days, which corresponds to a transfer velocity of 5 $m \cdot day^{-1}$.
7. At the PZ4 piezometer, located 20 m downstream, the first traces of dissolved H_2 appeared at 1.2 days and the concentration peaked a first time at 1.5 $\mu g \cdot L^{-1}$ after two days and a second time at 1.74 $\mu g \cdot L^{-1}$ after 2.8 days. This corresponds to transfer speeds of around 10 and 7 $m \cdot day^{-1}$, respectively.

A detailed analysis of the evolution of the dissolved hydrogen concentration at the main monitoring piezometer PZ2BIS during the injection day shows that the first peak appears only 38 min after the start of the hydrogenated water injection at 0.36 $mg \cdot L^{-1}$ (see circled 1 in Figure 13), following which a second more significant peak reaching 0.63 $mg \cdot L^{-1}$ occurred after 2 h (see circled 2 in Figure 13). These two peaks occurred during the injection of hydrogenated water, which lasted 2.5 h. Thus, at this close distance to the injection well, PZ2BIS appears to be strongly disturbed by the experimental conditions and particularly the overpressure induced. Similar to the tracers, the theoretical transfer speed it provides is not representative of the natural flow of the aquifer water, but instead of a flow disturbed by the successive injections.

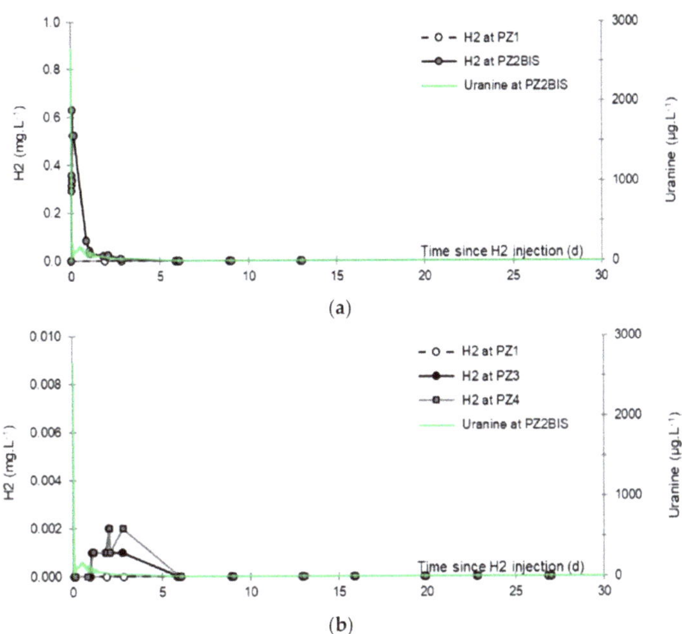

(a)

(b)

Figure 12. Comparative evolution of dissolved H_2 and tracer concentrations at the different piezometers: (**a**) PZ1 (20 m upstream) and PZ2BIS (5 m downstream); and (**b**) PZ1 and PZ3 (10 m downstream) and PZ4 (20 m downstream).

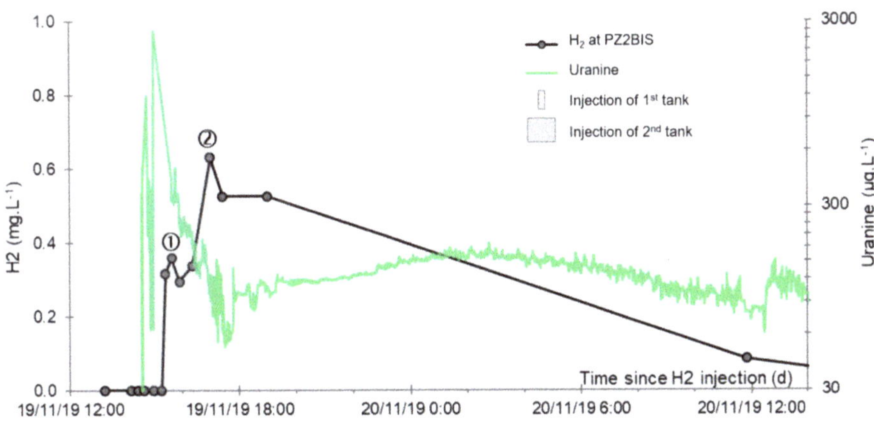

Figure 13. Detailed evolution of the dissolved H_2 and tracer concentration at PZ2BIS (5 m downstream) during the day of injection.

3.4. Oxidation-Reduction Potential

At the PZ2 injection well, the baseline showed an average oxidation-reduction potential of +192 mV prior to the injections [6]. When the measurements are resumed with the reopening of the well the day after the injections (t = 1.01 days), the oxidation-reduction potential is still low, with a value of +94 mV and it did not regain its initial value until t = 2.8 days onwards (Figure 14a).

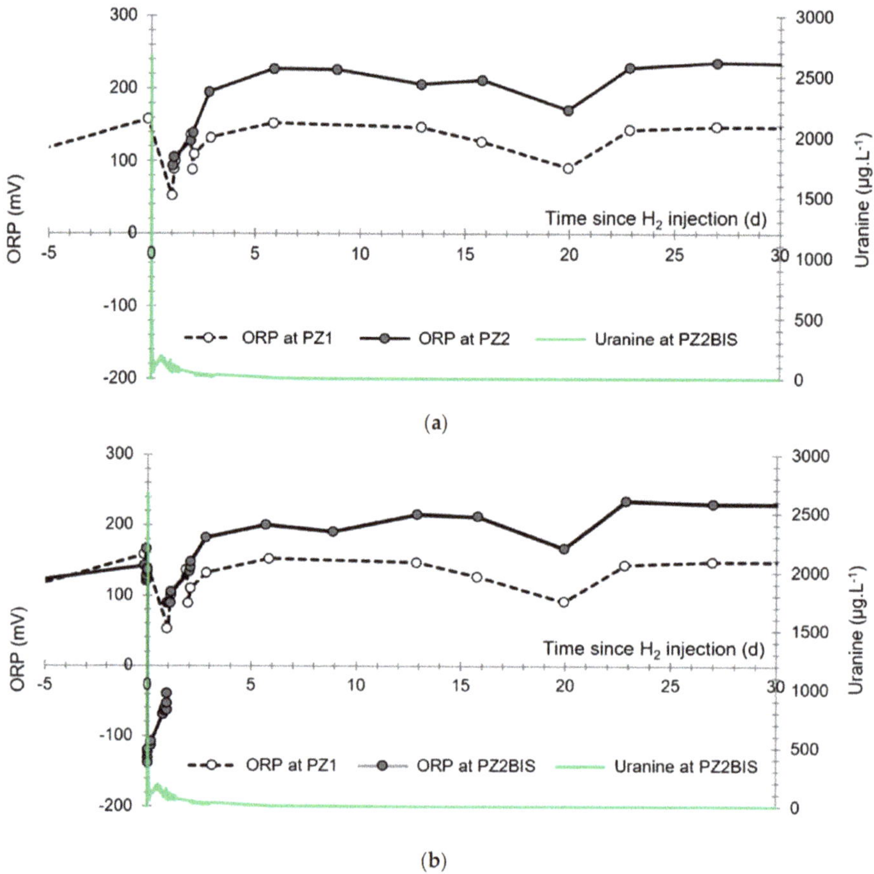

Figure 14. Comparative evolution of the oxidation-reduction potential and the tracer concentration at PZ1 (20 m upstream), PZ2 (injection well), and PZ2BIS (5 m downstream): (**a**) PZ1 and PZ2; and (**b**) PZ1 and PZ2BIS. The gap in the redox potential values corresponds to the change of probe.

At PZ2BIS, the oxidation-reduction potential decreased from +154 mV prior to the injections to a minimum of −139 mV during the injection (Figure 14b).Despite the shift in the values due to a difference in the calibration of the measuring probes, it is noted that the water remained reductive for at least 1.8 days and did not return to its normal value until after 2.8 days.

Again, at PZ2BIS, a detailed examination of the first day of the experiment shows the existence of two successive minima (see circled 1 and 2 in Figure 15):

1. The first is not very marked (−22%) and reached +120 mV at 14:15, i.e., 15 min after injection from the first tank containing helium and hydrological tracers. It corresponds to the passage of less oxidizing water, probably because of its deoxygenation induced by the introduction of dissolved helium.
2. The second is very clear (−190%) and it reached −139 mV at 15:45, i.e., 55 min after injection from the second tank containing dissolved hydrogen. This drop is in fact synchronous with the second tracer peak.

Due to the persistence of reducing conditions in the aquifer, in particular because of the release of the hydrogenated water stock present in the chalk's porous matrix in the immediate surroundings of the injection well, the oxidation-reduction potential remained at low values for the first day after the injection at PZ2BIS. It then increased in regular fashion at a mean speed of +59 mV·day^{-1} as a result of three mechanisms acting in unknown proportions: (i) dilution of the plume injected into the aquifer; (ii) partial degassing of the hydrogen; and (iii) potential chemical or biochemical reaction between the hydrogen and certain elements present in the water or the aquifer rock.

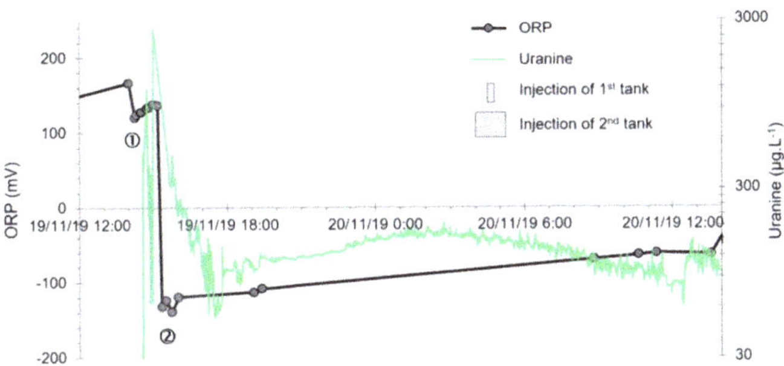

Figure 15. Detailed evolution of the oxidation-reduction potential and the tracer concentration at PZ2BIS (5 m downstream) during the day dissolved hydrogen was injected.

For the piezometers located further downstream, we consistently noted the existence of fluctuations in the oxidation-reduction potential during the first two days (Figure 16). These are in fact artifacts arising from the increased frequency of measurements, which demonstrate intraday fluctuations that were not visible during the looser monitoring on the other days. Apart from this, there is no evidence of an impact of the injection of hydrogenated water on the oxidation-reduction potential once the distance downstream of the injection well reaches 10 m.

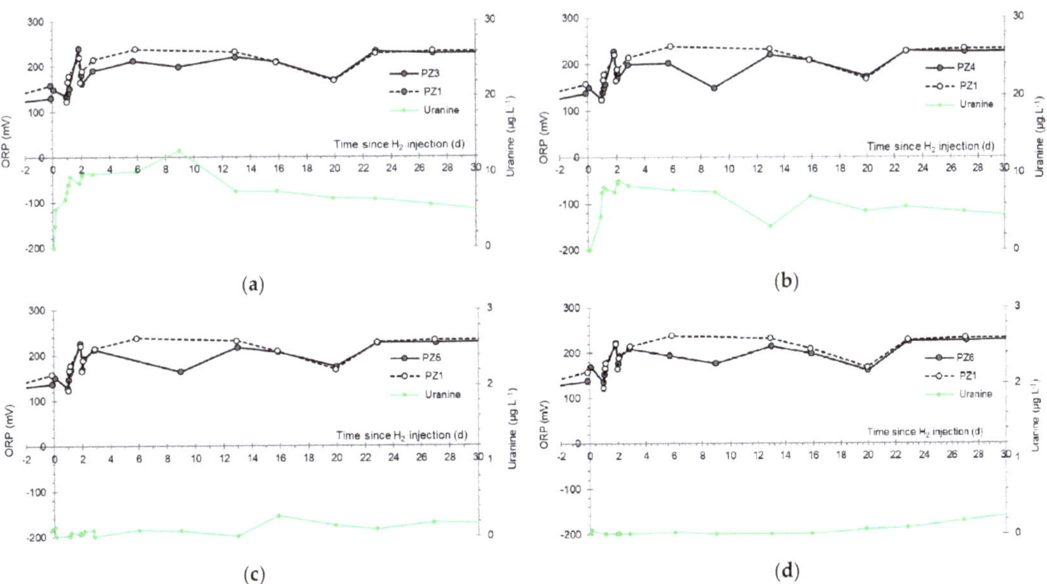

Figure 16. Comparative evolution of the oxidation-reduction potential and the tracer concentration at the other downstream piezometers: (**a**) PZ3 (10 m); (**b**) PZ4 (20 m); (**c**) PZ5 (30 m); and (**d**) PZ6 (60 m).

3.5. Dissolved Oxygen

At the PZ2 injection well, the mean concentration of dissolved oxygen during baselining equaled 6.30 mg·L^{-1} [6]. The day after the injections, the first measurement taken at 6.27 mg·L^{-1} has the same order (Figure 17a) of magnitude. This shows that the impact of the injection on this parameter is already no longer visible at this point.

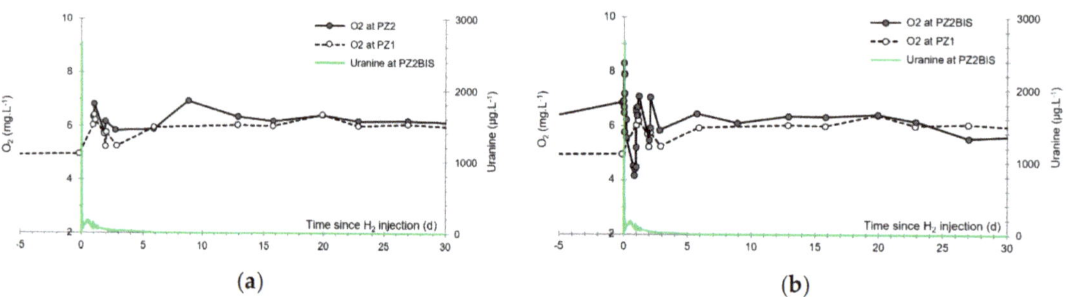

Figure 17. Comparative evolution of the dissolved oxygen and the tracer concentrations at the main piezometers: (**a**) PZ1 (20 m upstream) and PZ2 (injection well); and (**b**) PZ1 and PZ2BIS (5 m downstream).

In contrast, at PZ2BIS, the mean concentration dropped from 6.24 mg·L^{-1} prior to the injections to a minimum of 4.17 mg·L^{-1} at t = 0.82 days (Figure 17b). As with the oxidation-reduction potential, there are also significant fluctuations in the dissolved oxygen concentration during the first two days following the injections, probably also linked to the increase in the frequency of measurements and the disturbances induced by operations at the piezometers.

A detailed inspection of the first day of the experiment clearly shows the existence of three successive minima on 19 November 2019 at 14:45 with 5.13 mg·L^{-1}, on 19 November 2019 at 15:45 with 5.77 mg·L^{-1}, and on 20 November 2019 at 10:35 with 4.17 mg·L^{-1} (Figure 18):

1. The first one, synchronous with the first tracer peak, corresponds to the passage of water from the first tracer tank (see circled 1 in Figure 18). This water is deoxygenated due to the bubbling of helium.
2. The second one, synchronous with the second tracer peak, corresponds to the rapid passage of water from the second tank. This water is deoxygenated due to the bubbling of hydrogen (see circled 2 in Figure 18).
3. The third one, synchronous with the third tracer peak, corresponds to the slow passage of water from the second tank (see circled 3 in Figure 18). It signals the arrival of the main plume of hydrogenated water, which circulated more slowly within the aquifer and still contained under-oxygenated water for about a day after the injections.

Regarding the other piezometers located further downstream, fluctuations are systematically noted during the first two days of monitoring with dissolved oxygen concentrations below 6 mg·L^{-1}: this is also considered as an artifact associated with the higher frequency of measurements (Figure 19). Apart from these fluctuations, as the distance downstream of the injection well approaches 10 m, there is no evidence of an impact due to the injections on the dissolved oxygen concentration in the water.

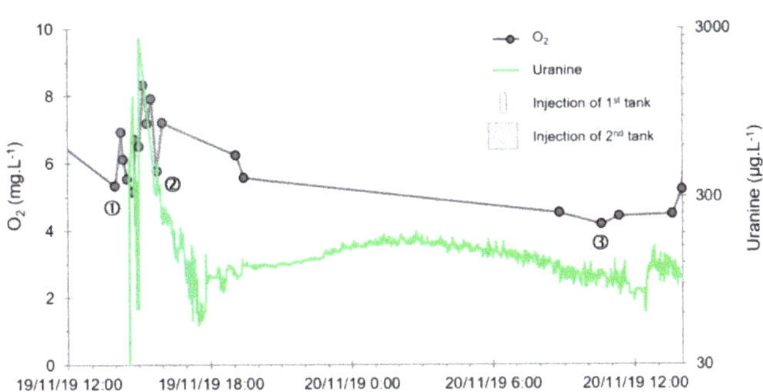

Figure 18. Detailed evolution of the dissolved oxygen and tracer concentrations at PZ2BIS (5 m down-stream) during the first day of dissolved hydrogen injection.

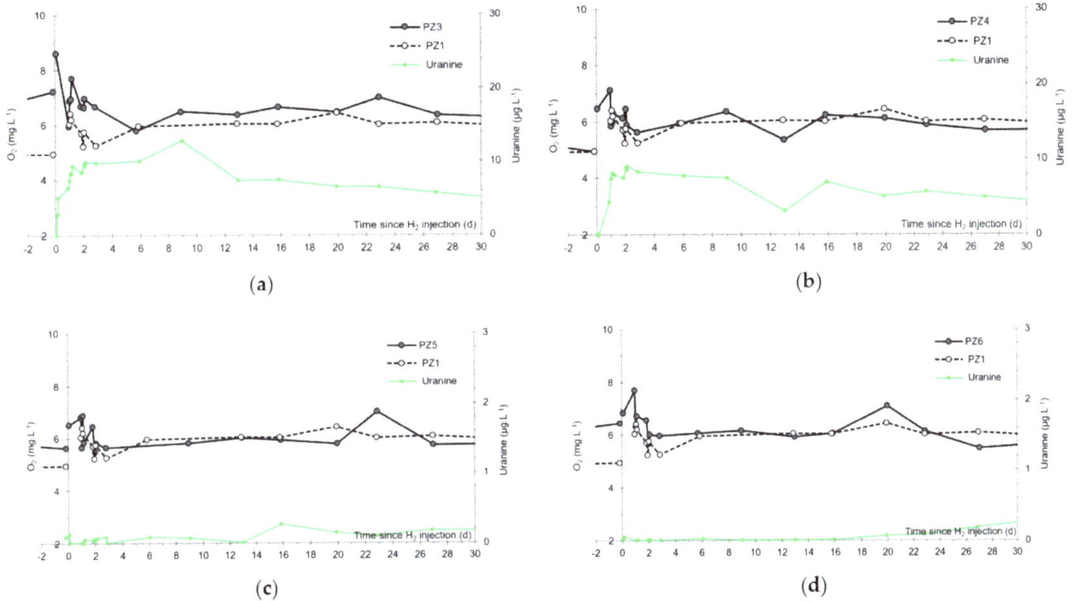

Figure 19. Comparative evolution of the dissolved oxygen during the dissolved hydrogen injection experiment at PZ1 (20 m upstream piezometer) and at the downstream piezometers: (**a**) PZ3 (10 m); (**b**) PZ4 (20 m); (**c**) PZ5 (30 m); and (**d**) PZ6 (60 m).

3.6. Other Physicochemical Parameters

Regarding the electrical conductivity, we note the existence of natural variations within the aquifer, recorded at PZ1 upstream of the injection well: at this piezometer, before injection and up to six days after, the average conductivity is 550 μS·cm^{-1}; it then decreases to 492 μS·cm^{-1} after this date. This fluctuation, which is observed upstream of the area affected by the injections, therefore affects the entire aquifer with successive repercussions on all piezometers: at PZ2BIS, the conductivity thus dropped from 558 to 495 μS·cm^{-1} (Figure 20a). On the other hand, a detailed analysis of the first day following the injection again shows the same three successive decreases in concentration already encountered at this piezometer with the previous physicochemical parameters (Figure 20b).

The minimal values are reached on the day of injection at 14:30 with 490 $\mu S \cdot cm^{-1}$, at 15:20 with 564 $\mu S \cdot cm^{-1}$, and at 16:00 with 502 $\mu S \cdot cm^{-1}$. As before, this corresponds to the rapid passage of the plumes of water from the two tanks, as well as the slow release into the aquifer of water that has remained trapped in the porous matrix around the injection well.

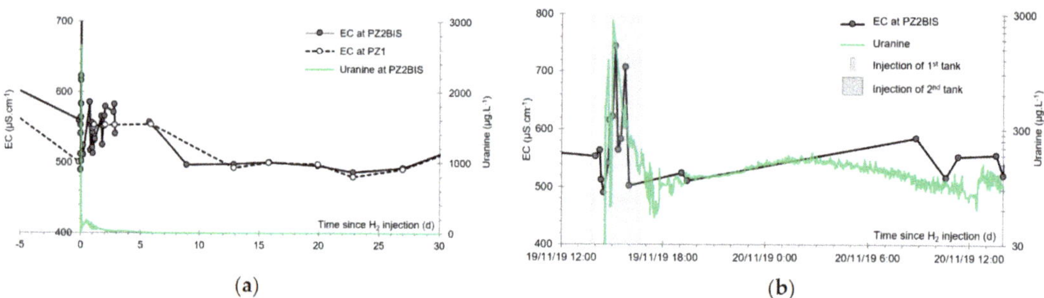

(a)   (b)

Figure 20. Comparative evolution of the electrical conductivity (EC) and the tracer concentration at PZ1 (20 m upstream) and PZ2BIS (5 m downstream): (**a**) over a period of one month; and (**b**) during the first 24 h.

We show below that this correspond mainly to a decrease in the cumulative concentration of the dominant ions (Ca^{2+}, Mg^{2+}, and HCO_3) of this bicarbonate-calcic groundwater: this is interpreted as a consequence of the precipitation of $CaCO_3$ and $MgCO_3$ from the degassing of dissolved CO_2 following the saturation of the water with He or H_2. On the other hand, these decreases are immediately followed by a peak in conductivity reaching 744 $\mu S \cdot cm^{-1}$ for the first, 707 $\mu S \cdot cm^{-1}$ for the second, and, later ($t = 2.80$ days), 582 $\mu S \cdot cm^{-1}$ for the third. These are the highest values recorded since the implementation of baseline monitoring over more than a year ago. These cycles of decrease and increase should therefore be related to the injections conducted: it seems that the observed increases can result from the dissolution of the carbonaceous aquifer rock by the injected water, which became aggressive to calcite, following the degassing of CO_2 induced by the bubbling with other gases.

Regarding pH, we note that it is already varying cyclically from 6.8 to 7.4 at PZ1 (upstream) before the injections. As is the case for the conductivity, we assumed that these are natural groundwater fluctuations since they are measured upstream of the injection well, and they affect the entire experimental site up to the most downstream piezometers: at PZ2BIS, the pH thus varies from 7.0 to 7.7 (Figure 21a). On the other hand, the detailed analysis of the first day following the injection shows, at this piezometer, a weak but sudden increase in pH from 7.2 to 7.4 when the first two tracer peaks passed through (Figure 21b). This behavior is to be linked to the degassing of dissolved CO_2, following the dissolution of He and H_2. The pH then decreased to 6.9 after $t = 2.88$ days, i.e., for the duration of the passage of the third plume of water, while at the same time the water at PZ1 showed an increase of natural origin. One could possibly interpret this increase as an impact of the slow arrival of hydrogenated water with the main aquifer flow.

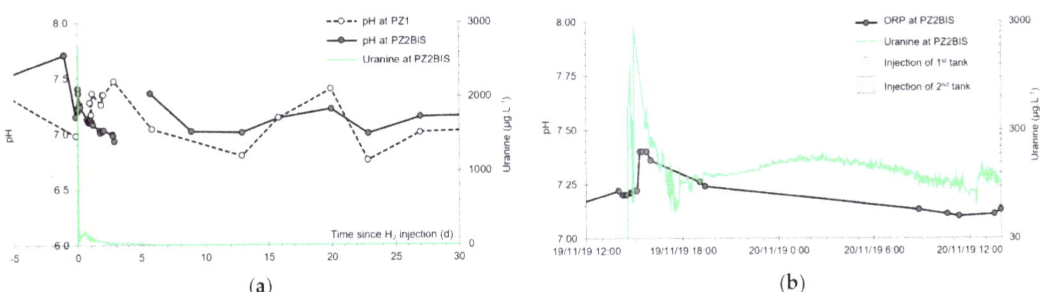

Figure 21. Comparative evolution of the pH and the tracer concentration at PZ2BIS (5 m downstream): (**a**) over a period of one month; and (**b**) during the first 24 h.

3.7. Dominant Ions (Ca^{2+}, Mg^{2+} and HCO_3^-)

The saturation of the water in the first tank with helium and then that of the second tank with hydrogen caused the degassing of the natural gases initially present, mainly an admixture of CO_2, N_2, and O_2. The degassing of CO_2 notably upsets the calco-carbonic equilibrium in the water, which seems to have led to the precipitation of $CaCO_3$ at the bottom of the tanks and $MgCO_3$ to a lesser extent, totaling a probable loss of around 29 mg·L^{-1} of dissolved elements. The water injected into the aquifer is therefore undersaturated with calcite and dolomite, which explains the observed decrease in the concentrations of Ca^{2+}, Mg^{2+}, and HCO_3^- in the samples taken during the first day (Figure 22). The cumulative concentration of these three ions decreased from 401 mg·L^{-1} before injection to a minimum of 372 mg·L^{-1} during injection, which corresponds to a 9.1% drop in their molar concentration shortly after the passage of the injected plumes. A maximum of 427 mg·L^{-1} is then observed at t = 2.80 days. This is interpreted as the dissolution of the carbonaceous aquifer rock by the injected water, which became more aggressive with respect to calcite and dolomite. The concentrations then reached their normal baseline values between the third and sixth days, with a cumulative concentration of 405 mg·L^{-1} until the end of monitoring.

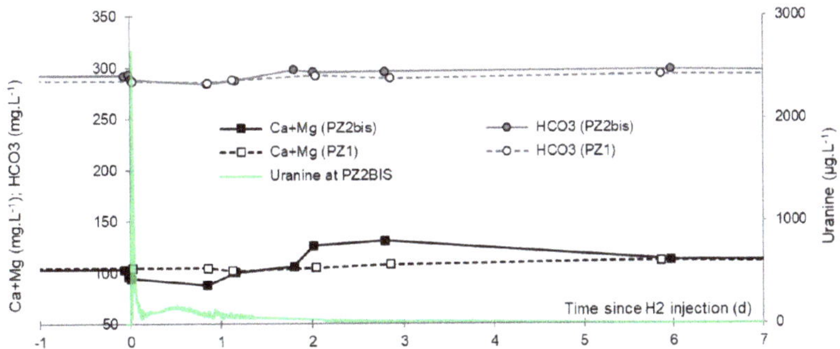

Figure 22. Comparative evolution of the concentrations of dominant ions and tracer at PZ1 (20 m upstream) and PZ2BIS (5 m downstream) during the week following the injection of dissolved hydrogen.

These cycles of variations in concentrations does not appear at any of the piezometers located further downstream (Figure 23). This shows that the aquifer has already returned to its natural calco-carbonic equilibrium at a distance of 10 m downstream from the injection well.

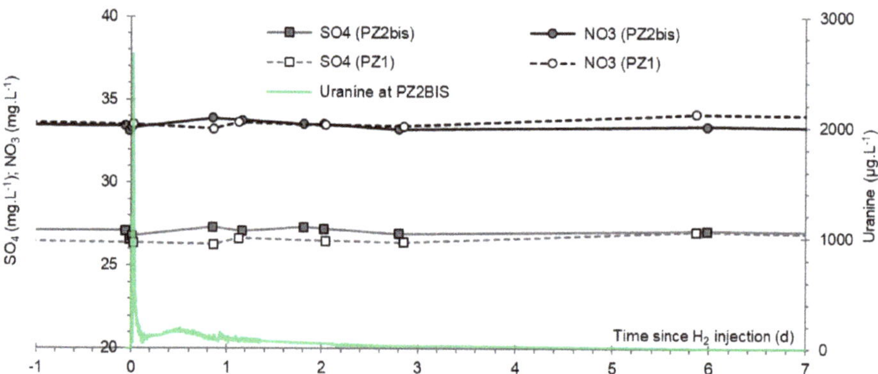

Figure 26. Comparative evolution of the concentrations of sulfates, nitrates, and tracer at PZ1 (20 m upstream) and PZ2BIS (5 m downstream) during the week following the injection of dissolved hydrogen.

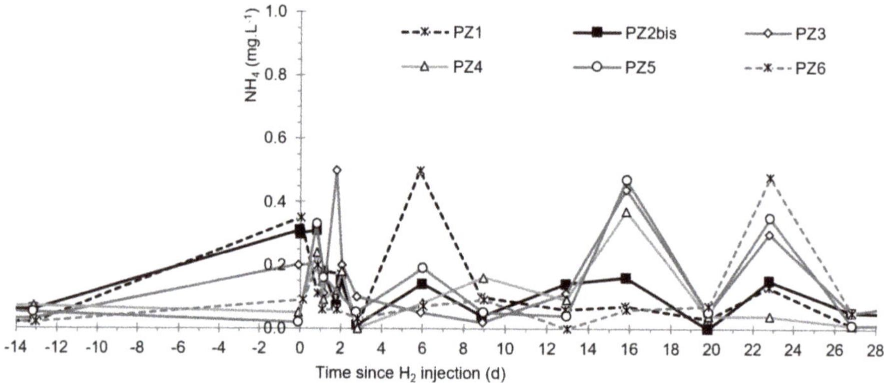

Figure 27. Comparative evolution of the ammonium ion concentrations at all piezometers before and during the dissolved hydrogen injection experiment.

4. Discussion

Panfilov [3] pointed out in 2015 that few articles have been published on the scientific aspects of hydrogen behavior in geological structures. Since then, studies on the impact of hydrogen leaks in an aquifer are still very rare. These include the experimental laboratory work of Berta et al. [22] and an ongoing experiment by the same team at the University of Kiel (D) on a shallow aquifer, the results of which have not yet been published. The results obtained by Berta et al. [22] are however not extrapolatable to our case: they concerned a long contact period (six months) between hydrogen under pressure (2–15 bars) and a reconstituted aquifer medium. Under these specific conditions, the reduction of sulfates, the production of acetate, the precipitation of calcite, and, consequently, an increase in pH and a decrease in electrical conductivity were observed. These parameters were considered as potential targets of a monitoring network covering a hydrogen gas storage site. However, it has been shown that these modifications resulted from the development of a hydrogenotrophic microbial community including sulfate-reducing bacteria.

In our experiment, several physicochemical and hydrogeochemical phenomena occurred in the hydrogen saturation tank, and then, following the injection of the hydrogenated water, at the piezometers near downstream to the injection well. Some of these phenomena are the result of the injection of hydrogenated water (drop in the redox potential dissolved hydrogen in water), but some others are the results of the preliminary dissolution of helium and tracers (mainly, deoxygenation of water and degassing of CO_2). At PZ2BIS, the main monitoring piezometer located 5 m downstream to the injection well, such variations are synchronous with the passage of the first uranine peak (fluorescent tracer), which signaled the arrival of the hydrogen-free (helium-saturated) water from the first tank. Once injected into the aquifer, this first water therefore caused the same phenomena as those observed in the tank, but at a lower intensity as a result of its dilution: A drop in the oxidation-reduction potential, the dissolved O_2 concentration, the electrical conductivity and the concentration of dominant ions (Ca^{2+}, Mg^{2+}, HCO_3^-). In addition, the water from this tank also contained an excess of Cl^- ions, linked to the presence of LiCl as an ionic tracer: it thus caused an increase of Cl^- ions in the aquifer. Since this ion is particularly conservative, it makes it possible to calculate that the dilution factor of the first injected plume must have reached 60% during the passage of the tracer peak at this piezometer.

The dissolution of hydrogen in the second tank also resulted in the same phenomena related to the degassing of dissolved O_2 and CO_2: the dissolved O_2 concentration dropped and the pH increased as a result of the degassing of dissolved CO_2. This must have also induced the precipitation of some of the dominant ions (Ca^{2+}, Mg^{2+} and HCO_3^-) in the water of the tank. The presence of dissolved hydrogen at a concentration close to saturation, additionally caused a significant drop in the oxidation-reduction potential, which made the water in this tank highly reducing: this is the only real impact directly linked to the saturation of water with hydrogen. Shortly after its injection into the aquifer, the oxidation-reduction potential of the groundwater thus fell at PZ2BIS, making reductive the initially oxidizing groundwater. This drop coincides with the passage of the second tracer peak induced by the injection of water from the second tank. Compared to the baseline state, the concentration of dissolved O_2 also decreased at that time, as well as those of the dominant ions (Ca^{2+}, Mg^{2+}, and HCO_3^-), while the pH increased slightly. In the piezometers located further downstream, i.e., at a distance of 10 m or more from the injection well, no impact of this type is measured in the aquifer: only the tracers show a weak presence which clearly reflects the passage of the plumes of water injected from the tanks. In addition, no piezometer showed significant variations, other than natural ones, in the concentrations of the oxidized and potentially reactive NO_3^- and SO_4^{2-} ions or in any of their expected metabolites (NO_2^-, NH_4^+, SO_3^-, and S^{2-}). It should be specifically noted that no element likely to play a catalytic role (especially metals) came into contact with the groundwater on the site.

The behavior of the aquifer during the passage of the third tracer plume is also interesting. This plume corresponds to the slow release of the water stock that has remained trapped in the poorly permeable zone at the base of the injection well PZ2 or in the porous matrix of its immediate surroundings. Thus, at PZ2BIS, the oxidation-reduction potential remained low throughout the first day following injection, although it regularly increased at an average rate of +59 mV·day^{-1} due to three possible mechanisms: (i) the dilution of the plume injected into the water table; (ii) the presumed degassing of the hydrogen; and (iii) the potential chemical or biochemical reaction of the hydrogen with some elements present in the aquifer rock or in the groundwater. Because the hydrogen is injected under conditions of undersaturation with respect to the aquifer (hydrostatic pressure > saturation pressure of the 5 m^3 tank), and no impact of the injection of hydrogen on the concentration of nitrates and sulfates could be demonstrated, the dilution of the injected plume seems to be the principal mechanism. Again, at PZ2BIS, the dissolved O_2 concentration of the aquifer remained low until at least 2.80 days after the start of the injections. During this same period, the pH and the concentration of dominant ions dropped below their initial values. Regarding the conductivity, the decreases observed during the passage of each injected plume of water are followed by brief increases before returning to the initial values. The decreases correspond to the passage of undersaturated water due to the precipitation of calcite and dolomite within the tanks, linked to the degassing of CO_2 due to the bubbling of He or H_2. The observed increases however could represent a brief renewal of the carbonaceous aquifer rock dissolution following the passage of more aggressive water due to its lower concentration of dominant ions (Ca^{2+}, Mg^{2+}, and HCO_3^-).

According to lithium breakthrough curves and the corresponding calculated dilution factors, the theoretical hydrogen concentration should have reached a maximum of around 0.70 mg·L^{-1} at PZ2BIS and 0.002 mg·L^{-1} at PZ3. The values observed of 0.63 and 0.002 mg·L^{-1}, respectively, correspond well with this hypothesis. During this experiment, the hydrogen therefore behaved mainly as a conservative tracer (Figure 28). It can also be estimated that, at the more distant piezometers, the maximum concentration of dissolved H_2 must have been lower than the analytical detection threshold, which meant that this element could not be detected, nor could its possible impact on the aquifer.

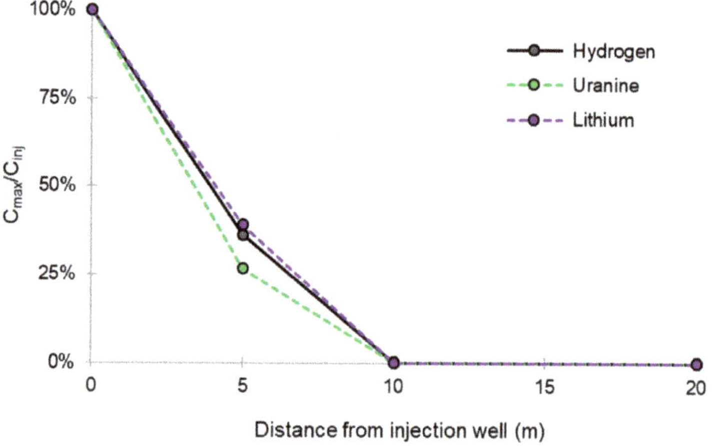

Figure 28. Comparative evolution of the ratio of maximum concentrations (C_{max}) to the injection concentration (C_{inj}) of tracers and hydrogen up to 20 m downstream of the injection well.

5. Conclusions

The hydrogen leak simulation experiment consisted of extracting water from the aquifer, saturating it with hydrogen gas in a tank, and then reinjecting it into the aquifer. The saturation of the water with gaseous hydrogen initially caused several physicochemical and hydrochemical phenomena within the tank and, consequently, in the aquifer:

- A sharp decrease of the oxidation-reduction potential
- The almost total disappearance of dissolved O_2 and CO_2
- A slight increase of the pH that induced the precipitation of alkaline earth bicarbonated ions and, accordingly, the decrease of electrical conductivity

These variations are primarily observed at the main PZ2BIS monitoring piezometer located 5 m downstream of the injection well and, to a lesser extent, at the PZ3 piezometer located 10 m downstream of the injection well. The other piezometers, located between 20 and 60 m downstream of the injection well, are not significantly affected. Our monitoring results show that, under the experimental conditions, the impact is only significantly measurable up to 10–20 m downstream of the injection well. This demonstrates the utility of closely monitoring the immediate surroundings of a future hydrogen injection well. This surveillance must be applied not only to the groundwater aquifer, but also to all major aquifers located deeper.

These results, however, are only valid under the experimental conditions of this test, which consisted of injecting a limited amount of dissolved hydrogen (9 g or 100 L STP) for a short time to simulate a sudden leak. Thus, in the case of a larger and/or longer leak, it is likely that the physicochemical and hydrogeochemical impacts would be greater across both space and time. In addition, bacterial growth could take place and induce biochemical reactions that may consume some dissolved species (sulfates, hydrogen), as observed by Berta et al. [22].

It is also shown that, during this experiment, the rapid transfer of hydrogen through the aquifer and its significant dilution beyond 10 m downstream of the injection well did not allow the development of significant chemical or biochemical reactions: hydrogen behaved here as a predominantly conservative element and is not (or not very) reactive.

This experiment is a first test of the impact of a hydrogen leak in a shallow unconfined aquifer. It made it possible to show that there are direct and indirect impacts of the arrival of dissolved hydrogen even in low amounts in an aquifer and, therefore, to recommend the implementation of a monitoring of future underground hydrogen storage sites that takes all of these physicochemical and hydrogeochemical parameters into account: concentrations of dissolved H_2, O_2, and CO_2, pH, electrical conductivity, and oxidation-reduction potential. In the case of long-term leakage, the concentrations of sulfates, nitrates, and bicarbonates must also be monitored.

This short time experiment should be considered as a first test intended to fit the injection and monitoring protocols for the effective monitoring of the impacts of a hydrogen leak in an aquifer under geological storage conditions. To be more representative of a real hydrogen leak, it should be supplemented in the future by a continuous leak simulation, lasting several days, the monitoring of which should focus on the relevant parameters previously identified.

Author Contributions: Conceptualization and methodology, P.G., S.L., Z.P. and E.L.; validation, P.G., S.L., Z.P., E.L., P.d.D. and N.J.; data curation, P.G. and S.L; writing—original draft preparation, P.G. and S.L; writing—review and editing, P.G., S.L., Z.P., E.L., P.d.D. and N.J.; visualization, P.G. and S.L.; project administration, P.G.; supervision and funding acquisition, P.G. and S.L. All authors have read and agreed to the published version of the manuscript.

Funding: This research was funded by the French Scientific Interest Group GEODENERGIES in the framework of the ROSTOCK-H project (Risks and Opportunities of the Geological Storage of Hydrogen in Salt Caverns in France and Europe).

Institutional Review Board Statement: Not applicable.

Conflicts of Interest: The authors declare no conflict of interest.

References

1. Légifrance. Loi n° 2015-992 du 17 août 2015 Relative à la Transition Energétique Pour la Croissance Verte. Available online: https://www.legifrance.gouv.fr/eli/loi/2015/8/17/DEVX1413992L/jo/texte (accessed on 25 June 2020).
2. Ineris. Le Stockage Souterrain Dans le Contexte de la Transition Energétique. Maîtrise des Risques et Impacts. *Ineris Références.* 2016. Available online: https://www.ineris.fr/sites/ineris.fr/files/contribution/Documents/ineris-dossier-ref-stockage-souterrain.pdf (accessed on 25 June 2020).
3. Panfilov, M. Underground and pipeline hydrogen storage. In *Compendium of Hydrogen Energy*; Gupta, R.B., Ed.; Elsevier: Amsterdam, The Netherlands, 2015; pp. 92–116. [CrossRef]
4. Caglayan, D.G.; Weber, N.; Heinrichs, H.U.; Linssen, J.; Robinius, M.; Kukla, P.A.; Stolten, D. Technical potential of salt caverns for hydrogen storage in Europe. *Int. J. Hydrog. Energy* **2020**, *45*, 6793–6805. [CrossRef]
5. Lions, J.; Devau, N.; de Lary, L.; Dupraz, S.; Parmentier, M.; Gombert, P.; Dictor, M.-C. Potential impacts of leakage from CO_2 geological storage on geochemical processes controlling fresh groundwater quality: A review. *Int. J. Greenh. Gas Control* **2014**, *22*, 165–175. [CrossRef]
6. Lafortune, S.; Gombert, P.; Pokryszka, Z.; Lacroix, E.; de Donato, P.; Jozja, N. Monitoring scheme for the detection of hydrogen leakage from a deep underground storage. Part 1: On site validation of an experimental protocol via the combined injection of helium and tracers into an aquifer. *Appl. Sci.* **2020**, *10*, 6058. [CrossRef]
7. Foh, S.; Novil, M.; Rockar, E.; Randolph, P. *Underground Hydrogen Storage*; Final Report, [Salt Caverns, Excavated Caverns, Aquifers and Depleted Fields]; 1979. Available online: https://www.osti.gov/servlets/purl/6536941-eQcCso/ (accessed on 22 February 2021).
8. Lord, A.S. *Overview of Geologic Storage of Natural Gas with an Emphasis on Assessing the Feasibility of Storing Hydrogen*; Sandia National Laboratories: Albuquerque, NM, USA, 2009. [CrossRef]

9. Lassin, A.; Dymitrowska, M.; Azaroual, M. Hydrogen solubility in pore water of partially saturated argillites: Application to Callovo-Oxfordian clayrock in the context of a nuclear waste geological disposal. *Phys. Chem. Earth* **2011**. [CrossRef]

10. Gombert, P.; Pokryszka, Z.; Lafortune, S.; Lions, J.; Grellier, S.; Prevot, F.; Squarcioni, P. Selection, instrumentation and characterization of a pilot site for CO_2 leakage experimentation in a superficial aquifer. *Energy Procedia* **2014**. [CrossRef]

11. Gal, F.; Lions, J.; Pokryszka, Z.; Gombert, P.; Grellier, S.; Prévot, F.; Squarcioni, P. CO_2 leakage in a shallow aquifer—Observed changes in case of small release. *Energy Procedia* **2014**. [CrossRef]

12. Truche, L.; Jodin-Caumon, M.C.; Lerouge, C.; Berger, G.; Mosser-Ruck, R.; Giffaut, E.; Michau, N. Sulphide mineral reactions in clay-rich rock induced by high hydrogen pressure. Application to disturbed or natural settings up to 250 °C and 30 bar. *Chem. Geol.* **2013**, *351*, 217–228. [CrossRef]

13. Truche, L.; Berger, G.; Destrigneville, C.; Pages, A.; Guillaume, D.; Giffaut, E.; Jacquot, E. Experimental reduction of aqueous sulphate by hydrogen under hydrothermal conditions: Implication for the nuclear waste storage. *Geochim. Cosmochim. Acta* **2009**, *73*, 4824–4835. [CrossRef]

14. Truche, L.; Berger, G.; Destrigneville, C.; Guillaume, D.; Giffaut, E. Kinetics of pyrite to pyrrhotite reduction by hydrogen in calcite buffered solutions between 90 and 180 °C: Implications for nuclear waste disposal. *Geochim. Cosmochim. Acta* **2010**, *74*, 2894–2914. [CrossRef]

15. Siantar, D.P.; Schreier, C.G.; Chou, C.-S.; Reinhard, M. Treatment of 1,2-dibromo-3-chloropropane and nitrate-contaminated water with zero-valent iron or hydrogen/palladium catalysts. *Water Res.* **1996**, *30*, 2315–2322. [CrossRef]

16. Smirnov, A.; Hausner, D.; Laffers, R.; Strongin, D.R.; Schoonen, M.A.A. Abiotic ammonium formation in the presence of Ni-Fe metals and alloys and its implication for the Hadean nitrogen cycle. *Geochem. Trans.* **2008**. [CrossRef] [PubMed]

17. Truche, L.; Berger, G.; Albrecht, A.; Domergue, L. Abiotic nitrate reduction induced by carbon steel and hydrogen: Implications for environmental processes in waste repositories. *Appl. Geochem.* **2013**, *28*, 155–163. [CrossRef]

18. Truche, L.; Berger, G.; Albrecht, A.; Domergue, L. Engineered materials as potential geocatalysts in deep geological nuclear waste repositories: A case study of the stainless steel catalytic effect on nitrate reduction by hydrogen. *Appl. Geochem.* **2013**, *35*, 279–288. [CrossRef]

19. Pintar, A.; Batista, J.; Levec, J.; Kajiuchi, T. Kinetics of the catalytic liquid-phase hydrogenation of aqueous nitrate solutions. *Appl. Catal. B Environ.* **1996**, *11*, 81–98. [CrossRef]

20. Pintar, A.; Setinc, M.; Levec, J. Hardness and Salt Effects on Catalytic Hydrogenation of Aqueous Nitrate Solutions. *J. Catal.* **1998**, *174*, 72–87. [CrossRef]

21. Bullister, J.L.; Guinasso, N.L., Jr.; Schink, D.R. Dissolved Hydrogen, Carbon Monoxide, and Methane at the CEPEX Site. *J. Geophys. Res.* **1982**, *87*, 2022–2034. [CrossRef]

22. Berta, M.; Dethlefsen, F.; Ebert, M.; Schäfer, D.; Dahmke, A. Geochemical Effects of Millimolar Hydrogen Concentrations in Groundwater: An Experimental Study in the Context of Subsurface Hydrogen Storage. *Environ. Sci. Technol.* **2018**, *52*, 4937–4949. [CrossRef] [PubMed]

23. Lagmöller, L.; Dahmke, A.; Ebert, M.; Metzgen, A.; Schäfer, D.; Dethlefsen, F. Geochemical Effects of Hydrogen Intrusions into Shallow Groundwater—An Incidence Scenario from Underground Gas Storage. Groundwater Quality 2019, Liège (B). 2019. Available online: https://www.uee.uliege.be/cms/c_4476800/fr/presentations-et-posters-gq2019 (accessed on 9 December 2019).

24. Légifrance. Arrêté du 11 Janvier 2007 Relatif aux Limites et Références de Qualité Des Eaux Brutes et Des Eaux Destinées à la Consommation Humaine Mentionnées aux Articles R. 1321-2, R. 1321-3, R. 1321-7 et R. 1321-38 du Code de la Santé Publique. Available online: https://www.legifrance.gouv.fr/affichTexte.do?cidTexte=JORFTEXT000000465574 (accessed on 9 December 2019).

25. Lacroix, E.; Lafortune, S.; de Donato, P.; Gombert, P.; Pokryszka, Z.; Liu, X.; Barres, O. Metrological assessment of on-site geochemical monitoring methods within an aquifer applied to the detection of H2 leakages from deep underground storages. *AGU American Geophysical Union-Fall Meeting 2020, AGU, Virtual Event.* 2020. Available online: https://search.proquest.com/openview/509d896f368e2b1f1c649ae4fc9dcb22/1?pq-origsite=gscholar&cbl=4882998 (accessed on 22 February 2021).

26. Ineris. E.Cenaris. Cloud Monitoring Solution for Observational Research and Monitoring Services Related to Underground Operations and Geostructures. Technical Sheet. 2018. Available online: https://cenaris.ineris.fr/SYTGEMweb/public/FichesProduit/FP-ficheA3_e-cenaris-def-an/FP-ficheA3_e-cenaris-def-an.html (accessed on 28 August 2020).

27. Labat, N.; Lescanne, M.; Hy-Billiot, J.; de Donato, P.; Cosson, M.; Luzzato, T.; Legay, P.; Mora, F.; Pichon, C.; Cordier, J.; et al. Carbon Capture and Storage: The Lacq Pilot (Project and Injection Period 2006–2013). Chap. 7: Environmental Monitoring and Modelling. 2015. Available online: https://www.globalccsinstitute.com/archive/hub/publications/194253/carbon-capture-storage-lacq-pilot.pdf (accessed on 13 December 2019).

28. Cai, S.; González-Vila, Á.; Zhang, X.; Guo, T.; Caucheteur, C. Palladium-coated plasmonic optical fiber gratings for hydrogen detection. *Opt. Lett.* **2019**, *44*, 4483–4486. [CrossRef] [PubMed]

29. Brouyère, S. Modelling the migration of contaminants through variably saturated dual-porosity, dual-permeability chalk. *J. Contam. Hydrol.* **2006**, *2*, 195–219. [CrossRef] [PubMed]

30. Gombert, P. Proposition de protocole de traçage appliqué au karst de la craie. *Eur. J. Water Qual.* **2007**, *38*, 61–78. [CrossRef]

Article

Thermodynamic Analysis of Compressed Air Energy Storage (CAES) Reservoirs in Abandoned Mines Using Different Sealing Layers

Laura Álvarez de Prado [1], Javier Menéndez [2,*], Antonio Bernardo-Sánchez [1], Mónica Galdo [3], Jorge Loredo [4] and Jesús Manuel Fernández-Oro [3]

[1] Department of Mining Technology, Topography and Structures, University of León, 24071 León, Spain; laura.alvarez@unileon.es (L.Á.d.P.); abers@unileon.es (A.B.-S.)
[2] Mining Department, SADIM Engineering, Avda. Galicia, 44, 33005 Oviedo, Spain
[3] Energy Department, University of Oviedo, 33271 Gijón, Spain; galdomonica@uniovi.es (M.G.); jesusfo@uniovi.es (J.M.F.-O.)
[4] Mining Exploitation Department, University of Oviedo, 33004 Oviedo, Spain; jloredo@uniovi.es
* Correspondence: javier.menendez@sadim.es

Abstract: Million cubic meters from abandoned mines worldwide could be used as subsurface reservoirs for large scale energy storage systems, such as adiabatic compressed air energy storage (A-CAES). In this paper, analytical and three-dimensional CFD numerical models have been conducted to analyze the thermodynamic performance of the A-CAES reservoirs in abandoned mines during air charging and discharging processes. Unlike other research works, in which the heat transfer coefficient is considered constant during the operation time, in the present investigation a correlation based on both unsteady Reynolds and Rayleigh numbers is employed for the heat transfer coefficient in this type of application. A tunnel with a 35 cm thick concrete lining, 200 m^3 of useful volume and typical operating pressures from 5 to 8 MPa were considered. Fiber-reinforced plastic (FRP) and steel were employed as sealing layers in the simulations around the fluid. Finally, the model also considers a 2.5 m thick sandstone rock mass around the concrete lining. The results obtained show significant heat flux between the pressurized air and the sealing layer and between the sealing layer and concrete lining. However, no temperature fluctuation was observed in the rock mass. The air temperature fluctuations are reduced when steel sealing layer is employed. The thermal energy balance through the sealing layer for 30 cycles, considering air mass flow rates of 0.22 kg s^{-1} (charge) and -0.45 kg s^{-1} (discharge), reached 1056 and 907 kWh for FRP and steel, respectively. In general, good agreements between analytical and numerical simulations were obtained.

Keywords: abandoned mines; underground reservoirs; energy storage; renewable energy; CAES; analytical modelling; numerical modelling; sealing layer

Citation: Prado, L.Á.d.; Menéndez, J.; Bernardo-Sánchez, A.; Galdo, M.; Loredo, J.; Fernández-Oro, J.M. Thermodynamic Analysis of Compressed Air Energy Storage (CAES) Reservoirs in Abandoned Mines Using Different Sealing Layers. *Appl. Sci.* **2021**, *11*, 2573. https://doi.org/10.3390/app11062573

Academic Editor: Luisa F. Cabeza

Received: 16 February 2021
Accepted: 11 March 2021
Published: 13 March 2021

Publisher's Note: MDPI stays neutral with regard to jurisdictional claims in published maps and institutional affiliations.

1. Introduction

Large scale energy storage systems are required to facilitate the penetration of variable renewable energies in the electricity grids [1–4]. Underground space from abandoned mines can be used as underground reservoirs for underground pumped storage hydropower (UPSH) and compressed air energy storage (CAES) systems [5–11]. Pumped storage hydropower (PSH) is the most mature large-scale energy storage technology, and the round trip efficiency is typically in the range of 70–80% [12,13]. Diabatic CAES (D-CAES) is an alternative to PSH that requires lower capital costs and the round trip energy efficiency is around 40–50% [14]. D-CAES systems use natural gas to heat the compressed air in the decompression period. However, Adiabatic CAES (A-CAES) allows the storage of the thermal energy during the air compression period, avoiding the consumption of natural gas, therefore increasing the round trip energy efficiency up to 70–75% [15–17].

During the operation of A-CAES plants, the air is stored in the underground reservoir at high pressures during the charge period. Then, in the discharge period, the pressurized air is released, heated (stored heat), and expanded to generate electricity as the pressure within the reservoir is reduced. During the storage period, between charging and discharging, the air pressure and temperature vary depending on the storage time. The amount of stored energy depends on the reservoir volume and the thermodynamic conditions. The air temperature variations depend on the thermal conductivity of the sealing layer, concrete lining and rock mass. Therefore, the air temperature and pressure fluctuations are essential to design the reservoir volume, to select the appropriate compressor and turbines and to ensure safe operating conditions. Kushnir et al. [18] developed a study to determine the temperature and pressure variations within compressed air energy storage caverns and showed that the heat transfer reduces the temperature and pressure variations during the compression and decompression periods, leading to a higher storage capacity. Thermodynamic models were performed to determine temperature and pressure variations within adiabatic caverns of compressed air energy storage plants [19]. The results showed that the storage volume is highly dependent on the air maximum to minimum pressure ratio. Safaei and Aziz [20] carried out a thermodynamic analysis of three compressed air energy storage systems. They concluded that A-CAES with physical heat storage is the most efficient option with an exergy efficiency of 69.5%.

A pilot-scale demonstration of A-CAES was built in Switzerland [21,22]. The cavern was 120 m long and 4.9 m in diameter with a usable volume of 1942 m^3. A packed-bed of rocks for thermal energy storage was located inside the cavern. Charging temperatures of 550 °C and air pressures between 0 and 0.7 MPa were employed in the test and constant heat transfer coefficients of 5 and 10 W m^{-2} K^{-1} were considered in the simulations. The estimated round trip efficiency of the pilot plant reached 63–74%. Schmidt et al. [23] analyzed the effect of the cyclic loading on the geomechanical performance of CAES systems for lined and unlined tunnels in closed mines. The cycling loading operation was simulated for 10,000 cycles considering an operating pressure within the reservoir from 4.5 to 7.5 MPa. They observed moderate deformations and small thickness of plastic zones, while an increase of the initial volume of less than 0.5% was reached.

Jiang et al. [24] carried out experimental and numerical investigations of a lined rock cavern to analyze the thermodynamic process within the underground cavern. The volume of the cavern was 28.8 m^3 with 5 m length and 2.9 m in diameter. A 2 cm thick RFP was employed as sealing layer around the compressed air. The air pressure reached 6 MPa at constant mass flow rate (0.12 kg s^{-1}). Zhou et al. [25] developed a numerical simulation for the coupled thermo-mechanical performance of a lined rock cavern for CAES systems. An air pressure range from 4.5 to 6.9 MPa was considered for 100 cycles with a thickness of concrete lining of 0 and 0.1 m. They concluded that significant air temperature fluctuation was observed in sealing layer and concrete lining and no fluctuation occurred in the rock mass. An analytical study was proposed to estimate the geomechanical performance caused by air temperature and pressure variations in CAES systems [26]. Khaledy et al. [27] studied the thermo-mechanical responses of CAES systems in rock salt caverns. They concluded that the stability of the salt cavern is affected by the internal operating conditions with the air temperature increasing the volume convergence. Li et al. [28] carried out a failure study for gas storage by thermo-mechanical modelling in salt caverns. The results showed an affected area by the air and temperature operating conditions up to 10 m in the rock mass from the cavern walls. Round trip efficiencies between 70.5 and 75.1% were estimated by Barbour et al. [29] in A-CAES systems for continuous operation with packed bed thermal energy storage. The thermodynamic and geomechanical performance was studied by Rutqvist et al. [30] in lined rock caverns. They found that 97% of the thermal energy injected in the charge period could be recovered in the discharge period. Deng et al. [31] designed a new CAES systems with constant gas pressure and temperature in closed coal mines. A power output of 18 MW and a generating time of 1.76 h were obtained for a mine tunnel with 10,000 m^3 of volume at 500 m depth. The ADELE project was studied in Germany to install an A-CAES plant with a storage capacity of 360 MWh and output power of 90 MW [2].

In this paper, abandoned mines are proposed as underground reservoirs for large scale energy storage systems. A 200 m^3 tunnel in an abandoned coal mine was investigated as compressed air reservoir for A-CAES plants, where the ambient air is stored at high pressure. The thermodynamic response of A-CAES reservoirs was analyzed considering three solids around the pressurized air: a 20 mm thick sealing layer, a 35 cm thick concrete lining, and a 2.5 m thick sandstone rock mass. One-dimensional analytical and three-dimensional CFD numerical models were conducted considering fiber-reinforced plastic (FRP) and steel as sealing layers and a typical operational pressure from 5 to 8 MPa. In the analytical model, air mass flow rates of 0.22 and -0.45 kg s^{-1} have been considered for 30 cycles of compression and decompression, respectively. To reduce the computational time, the numerical simulation has been conducted for five cycles with air mass flow rates of 50 and -75 kg s^{-1}. The air temperature and pressure fluctuations were estimated in the simulations. The temperature and the heat flux were also analyzed on the contact surfaces of the solids considering both FRP and steel sealing layers and the heat convection from the air to the sealing layers was calculated. Finally, the energy balance through the sealing layer was obtained during air charging and discharging.

2. Materials and Methods

2.1. Problem Statement

Underground space in abandoned mines may be used as compressed air storage systems for CAES plants. The simplified schematic diagram of the CAES system is shown in Figure 1. The compressor and turbine facilities are installed above the ground, while the compressed air reservoir is underground. The ambient air is compressed during off-peak periods and stored at high pressure in the underground reservoir (charge period). Then, when the electricity is required, the compressed air is released, heated, and expanded during peak hours in conventional gas turbines, driving a generator for power production (discharge period). The charge and discharge processes are carried out with a typical operating pressure range from 5 to 8 MPa. To reduce air leakage, two different sealing layers, FRP, and steel, have been employed in the present study. The air temperature and pressure fluctuations are estimated for both FRP and steel sealing layers.

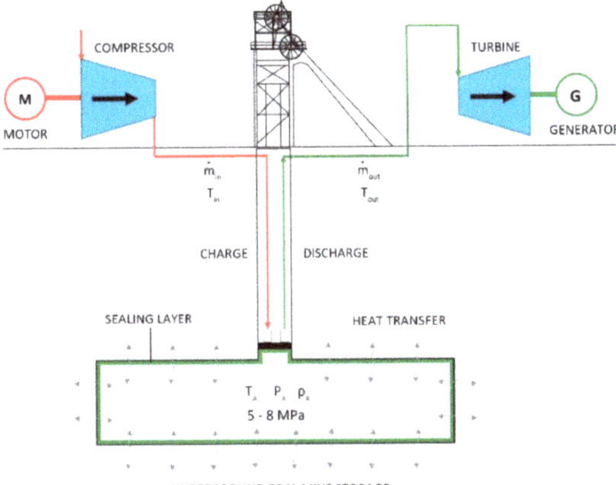

Figure 1. Schematic diagram of compressed air energy storage (CAES) system in abandoned underground mines. Compressor and turbine facilities installed on the surface and underground compressed air reservoir with an operating pressure range from 5 to 8 MPa.

2.2. Analytical Model

A one-dimensional analytical model has been developed in MATLAB to study the thermodynamic performance during the operation time of the CAES system in a closed mine. Figure 2 shows the scheme of the 50 m long model. A lined tunnel with a usable volume of 200 m^3 and a cross section of 8 m^2 has been considered. A 20 mm thick sealing layer, a 35 cm thick concrete lining and a 2.5 m thick sandstone rock mass have also been considered in the model around the fluid. The evolution of the temperature (T_a), density (ρ_a) and air pressure (Pa) over time has been analyzed considering different air mass flow rates ($\dot{m}$). The heat flux through the contact surfaces, i.e., air-sealing layer ($\dot{Q}_1$), sealing layer-concrete ($\dot{Q}_2$), and concrete-sandstone ($\dot{Q}_3$), has been estimated. Finally, the temperature on the sealing layer wall (T_1), concrete wall (T_2), and sandstone wall (T_3) have also been estimated during the operation time considering an external temperature (T_4) of 300 K.

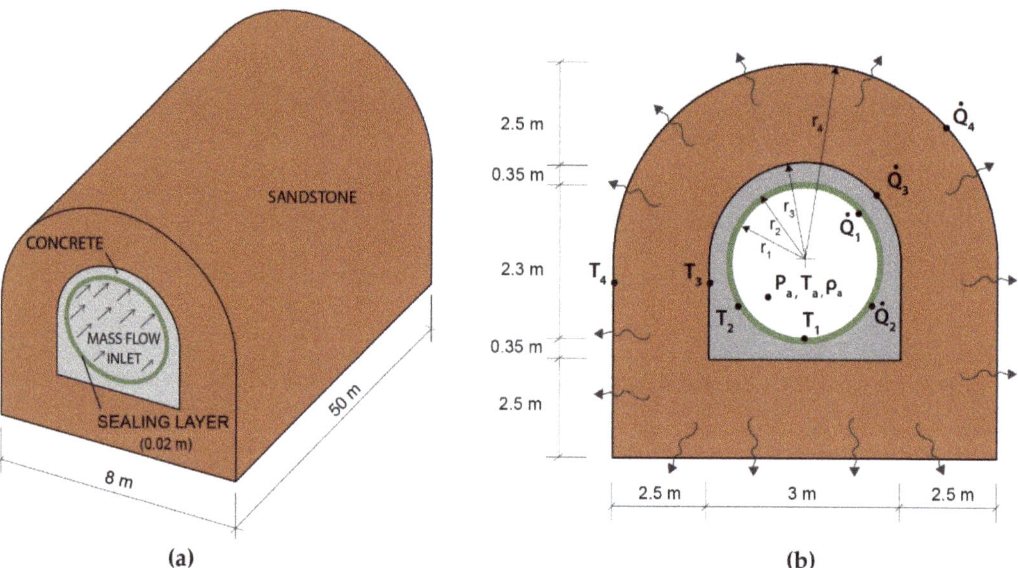

Figure 2. Scheme of the deep tunnel with a sealing layer, concrete lining and sandstone rock mass. (**a**) 3D model (not to scale); (**b**) Cross section.

The energy equation has been applied following Equation (1), where $\dot{Q}$ is the net heat transfer, $\dot{W}$ is the net work done, e is the specific energy, ρ is the density in kg m^{-3}, V is the reservoir volume, v is the air velocity and S is the cross section of the tunnel. In the air domain, the first term of Equation (2) represents the heat convection from the air to the sealing layer, which depends on the film coefficient of heat transfer (h), in Wm^{-2} K^{-1}, the sealing layer surface in contact with the fluid (A_1) and the temperature difference between the air and the concrete wall (T_a-T_1). The second term of Equation (2) corresponds to the energy variation within the reservoir where the air pressure increases, which depends on the air mass inside the reservoir (m), the specific heat at constant volume (C_v) and the air temperature (T_a). Finally, the third term represents the input and output of energy in the control volume due to the energy provided by the air jet in the form of heat (T_0) and kinetic energy. The air mass inside the reservoir at an instant of time, t, depends on the initial air mass (m_0) and the air mass flow rate ($\dot{m}$), and is obtained by applying Equation (3).

$$\dot{Q} - \dot{W} = \frac{\partial}{\partial t} \int \rho e \, dV + \oint \rho \hat{e} \, (\vec{v} dS) \tag{1}$$

$$-hA_1(T_a - T_1) = \frac{d}{dt}(mC_vT_a) - \dot{m}\left(C_vT_0 + RT_0 + \frac{v^2}{2}\right) \tag{2}$$

$$m = m_0 + \dot{m}\,t \tag{3}$$

After some algebra, the air temperature (T_a) is calculated by applying Equation (4). The air temperature and density increase as the air pressure increases in the reservoir, and therefore the wall temperature increases by heat convection.

$$\frac{dT_a}{dt} = \frac{\dot{m}}{m}\left[\frac{C_p}{C_v}T_0 + \frac{1}{2C_v}\left(\frac{\dot{m}}{m}L\right)^2 - T_a\right] - \frac{h2\pi r_1 L}{C_v m}(T_a - T_1) \tag{4}$$

The temperatures on the sealing layer wall (T_1), the concrete wall (T_2) and sandstone wall (T_3) are obtained applying the equations of non-stationary heat transfer in sealing layer, concrete and sandstone domains (solid regions) considering an external temperature (T_4) of 300 K and a temperature of the air mass flow inlet (T_0) of 310 K. The heat flux reaches the sealing layer by convection. A part of this heat is transmitted to the concrete by conduction while another part increases the temperature of the sealing layer. A similar mechanism takes place for the concrete lining. Equations (5)–(7) represent the heat transmission on the sealing layer, concrete and sandstone formation, respectively.

$$\frac{dT_1}{dt} = \frac{2U_h}{m_s C_{ps}}(T_a - T_1) - \frac{2U_s}{m_s C_{ps}}(T_1 - T_2) - \frac{dT_2}{dt} \tag{5}$$

$$\frac{dT_2}{dt} = \frac{2U_s}{m_c C_{pc}}(T_1 - T_2) - \frac{2U_c}{m_c C_{pc}}(T_2 - T_3) \tag{6}$$

$$\frac{dT_3}{dt} = \frac{2U_c}{m_{ss} C_{pss}}(T_2 - T_3) - \frac{2U_{ss}}{m_{ss} C_{pss}}(T_3 - T_4) \tag{7}$$

where U_h, U_s, U_c, and U_{ss} are the convection transmittance, sealing layer transmittance, concrete transmittance, and sandstone transmittance, respectively. These thermal coefficients are evaluated according to

$$U_h = h2\pi r_1 L \tag{8}$$

$$U_s = \frac{2\pi K_s L}{Ln\left(\frac{r_2}{r_1}\right)} \tag{9}$$

$$U_c = \frac{2\pi K_c L}{Ln\left(\frac{r_3}{r_2}\right)} \tag{10}$$

$$U_{ss} = \frac{2\pi K_{ss} L}{Ln\left(\frac{r_4}{r_3}\right)} \tag{11}$$

In these expressions, h is the film coefficient of heat transfer, in $\mathrm{Wm^{-2}\,K^{-1}}$; L is the tunnel length, in m; K_s, K_c and K_{ss} are the thermal conductivity of sealing layer, concrete and sandstone, in $\mathrm{Wm^{-1}\,K^{-1}}$; Cp_s, Cp_c, and Cp_{ss} are the specific heat at constant pressure for the sealing layer, reinforced concrete and sandstone, in $\mathrm{J\,kg^{-1}\,K^{-1}}$; and r_1, r_2, r_3, and r_4 are the (equivalent) radius of sealing layer, concrete, sandstone and external walls, in m (Figure 2b). Note that due to the geometrical characteristics of the tunnel, the heat transfer formulation in cylindrical coordinates has been used for Equations (8)–(11).

For the evaluation of the convection film coefficient, the correlation proposed by Woodfield et al. [32] and Heath et al. [33] for the measurement of heat transfer coefficients in high-pressure vessels during gas charging was considered. In particular, a mixed (natural and forced) convection is modelled as a combination of both unsteady Reynolds and Rayleigh numbers. For the present investigation, a slight correction has been introduced for the exponent of the Rayleigh number, resulting the Nusselt number in the following expression:

$$Nu = 0.56\,Re^{0.67} + 0.104\,Ra^{0.34} \tag{12}$$

where Re is defined as the Reynolds number of the incoming mass flow rate ($Re = 4\dot{m}/2\mu\pi r_1$) and Ra is computed from the instantaneous thermal properties of the air ($Ra = g\beta(T_a - T_1)C_p\rho^2 L^3/(\mu K)$), being g the gravity acceleration, β is the volumetric thermal expansion coefficient, Cp the specific heat coefficient at constant pressure, ρ the density, L the tunnel length, μ the dynamic viscosity and K the thermal conductivity. The film coefficient is thus computed as

$$h = Nu\frac{K}{L} \tag{13}$$

Finally, a backward Euler explicit discretization has been employed over Equations (3)–(7) to resolve the coupled system of equations. A typical time-step of 0.01 s was applied to provide an accurate temporal description of the different temperatures. The model was resolved iteratively until a maximum prescribe pressure was reached inside the reservoir.

In addition, according to Kushnir et al. [18] and Zhou et al. [25], the air has been considered as a real gas through a compressibility factor (Z) that it is computed using the Berthelot gas state equation. Finally, considering a density value uniformly distributed in the volume, the pressure value of the CAES may be obtained at any instant by applying Equations (14) and (15).

$$P_a = \left(\frac{R}{V}\right)ZmT_a \tag{14}$$

$$Z = 1 - \frac{9}{128}\left(\frac{P_a}{P_c}\right)\left(\frac{T_c}{T}\right)\left(\frac{6T_c^2}{T^2} - 1\right) \tag{15}$$

where T_c and P_c the air temperature and the air pressure at critical conditions, assumed to be 132.65 K and 3.76 MPa. The contact surfaces between air-sealing layer ($A_1 = 2\pi r_1 L$), sealing layer-concrete ($A_2 \sim 2\pi r_2 L$), concrete-sandstone ($A_3 \sim 2\pi r_3 L$), and sandstone-exterior ($A_4 \sim 2\pi r_4 L$), are 359, 365, 545, and 1610 m^2, respectively. Note that, in order to model a complete cycle of compression (storing energy) and expansion (releasing energy) of the CAES system, the energy balance for the air domain—Equation (4)—needs to be rewritten as a function of the discharge mass flow rate ($\dot{m}_{out} < 0$) and the air temperature within the reservoir, according to Equation (16):

$$\frac{dT_a}{dt} = \frac{\dot{m}_{out}}{m}\left[\left(\frac{C_p}{C_v} - 1\right)T_a + \frac{1}{2C_v}\left(\frac{\dot{m}_{out}}{m}L\right)^2\right] - \frac{h2\pi r_1 L}{C_v m}(T_a - T_1) \tag{16}$$

Air mass flow rates of 0.22 kg s^{-1} and -0.45 kg s^{-1} have been considered in the charge and discharge periods, respectively, for both RFP and steel sealing layers. The thermal properties and the volume of air, reinforced concrete, sandstone rock mass and sealing layers considered in the model are shown in Table 1 [24,25]. Note significant differences between the thermal conductivity of FRP and steel.

Table 1. Thermal properties and volume of air, concrete, sealing layers and sandstone rock mass.

Material	Specific Heat (J kg^{-1} K^{-1})	Thermal Conductivity (W m^{-1} K^{-1})	Volume (m^3)	Density KN m^{-3}
Air	1006	0.0242	200.53	0.0117
Reinforced concrete	1000	1.60	195.26	23
Sandstone	711	5.00	2605	25
Sealing layer (FRP)	384	0.40	7.24	8.82
Sealing layer (Steel)	500	45	7.24	76.5

2.3. CFD Numerical Model

2.3.1. Model Geometry, Mesh and Boundary Conditions

A 3D numerical model of a horseshoe-shaped tunnel located inside a closed mine was conducted to simulate the compression and decompression processes (air charging and discharging) that occur in the tunnel during the operation of the CAES plant. The computational domain is a tunnel with 8 m^2 of cross section and 50 m in length and includes both the fluid area and the solid areas around the fluid. The configuration of the simulated tunnel is illustrated in Figure 3. To improve the geomechanical performance, a 4 m^2 circular cross-section has been designed for the compressed air (blue area in Figure 3). The tunnel is reinforced with a 35 cm thick concrete lining and is finished with a dead-end. To avoid thermal leakage, a 20 mm thick sealing layer is considered between the air and concrete lining. Two types of sealing materials are considered: FRP and steel. The external part of the model corresponds to the 2.5 m thick sandstone rock mass existing in the mine. The total cross section of the model is 57.13 m^2 (8 × 8 m) and the model volume is 2587 m^3. In addition, the useful volume of air reaches 200.53 m^3. The entire geometry is meshed with 2,690,038 hexahedral and tetrahedral cells. Finer mesh was defined in the sealing layer, concrete and air zones, with higher grid density in those regions where the gradients of the flow characteristics are extremely important.

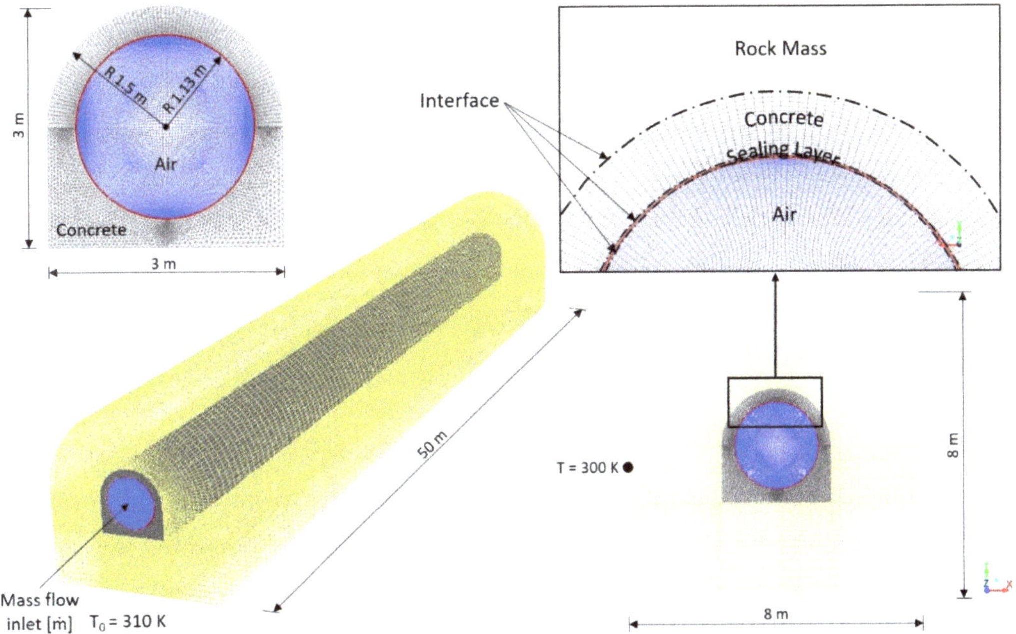

Figure 3. Computational domain. Mesh and boundary conditions used in CFD simulations.

Four different type of materials (air, sealing layer, concrete, and sandstone) are used to simulate the heat transfer process between the air inside the mine and the surrounding media. In addition, as in the analytical model, FRP and steel are considered as sealing layers. The air is defined as real-gas to allow the simulation of the compression process. To reproduce the concrete layer existing between the sealing layer and the sandstone, a solid material zone is defined. The sandstone and sealing layer zones are also defined as solid materials. In addition, a series of thermal properties are imposed on the sandstone, concrete and sealing zones to simulate the conduction heat transfer.

The numerical simulation of the compression/expansion process was carried out with the commercial software ANSYS Fluent® 16.2. This code was used to solve the full 3-D Unsteady Reynolds-Averaged Navier–Stokes (URANS) equations for compressible flow. The coupling between velocity and pressure was resolved by the SIMPLE algorithm. The spatial and temporal derivatives of the governing equations for the fluid flow were calculated by means of a second-order discretization, and the pressure interpolation PREssure STaggering Option (PRESTO) scheme has been used. Turbulent closure was established by the RNG k–ε model together with standard wall functions. In addition, the resolution of the energy equation for both fluid and solid volumes was activated. A constant mass flow rate was imposed as an inlet boundary condition for each model and an interface condition was established for solid/solid and solid/fluid surfaces. To reduce the computational time, air mass flow rates of 50 kg s^{-1} and -75 kg s^{-1} were considered in the numerical simulations. In Section 3.3, these mass flow rates are also employed in the analytical model to carry out the comparative analysis. In addition, a constant temperature of 300 K was considered in the sandstone surface. Moreover, a no-slip shear condition was imposed on the walls. The material properties used in the numerical model have been indicated in Table 1. A fixed time step of 5×10^{-2} s was selected for the simulations to ensure their stable convergence. The convergence criteria is based on the residual values of the solution solved in the numerical domain. A typical threshold of 10^{-5} was set for continuity, momentum and energy equations.

2.3.2. Mesh Sensitivity Study

A mesh sensitivity analysis has been carried out to optimize the numerical model results. Four different mesh sizes between 1.44×10^6 and 3.47×10^6 cells were used to simulate the air compression process from the ambient conditions to 8 MPa using FRP as sealing layer. The thermal energy balance through the sealing layer was analyzed for the first cycle of compression.

The results for the four scenarios are indicated in Table 2. Finally, a mesh model with 2.69×10^6 cells was selected to simulate the air charge and discharge periods in the underground reservoir. When a model with 3.47×10^6 cells is considered, an increase in the computational time of 26.72% is obtained regarding the selected mesh size.

Table 2. Mesh sensitivity analysis.

Mesh Cells $\times 10^6$	Thermal Energy (kWh)	Error (%)	Computational Time (h)
1.44	53.96	7.14	56
2.01	56.76	2.33	78
2.69	57.72	0.68	116
3.47	58.11	-	147

3. Results and Discussion

3.1. Analytical Model Results

The analytical model has been simulated for 30 cycles. The air is compressed from atmospheric conditions to 8 MPa in the first cycle. In the following cycles, the air pressure varies from 5 to 8 MPa at a stable mass flow rate considering an injected air temperature of 310 K and daily compression and decompression cycles. During an operation cycle, the air is compressed for 8 h at a mass flow rate of 0.22 kg s^{-1} and stored for 6 h. Then, the air is released for 4 h at a mass flow rate of -0.45 kg s^{-1} and stored for 6 h.

The variations in temperature for the air, sealing layer and concrete lining are shown in Figure 4 using FRP as sealing layer for 30 continuous cycles, which is equivalent to one month of operation. The results are also presented more in detail for the first cycle (Figure 4a) and the 30th cycle (Figure 4c). The air temperature increases up to 322 K during the first compression cycle. However, due to mixing with the injected air and heat exchange between the compressed air and the lining, the air temperature within the reservoir decreases to 307 K from the fifth cycle (five days). During the discharge period the air temperature is also stable, reaching minimum values of 294 K. The temperature of the FRP and concrete during the charge period reach 305 and 303 K, respectively, decreasing to 296 and 300 K during the discharge phase.

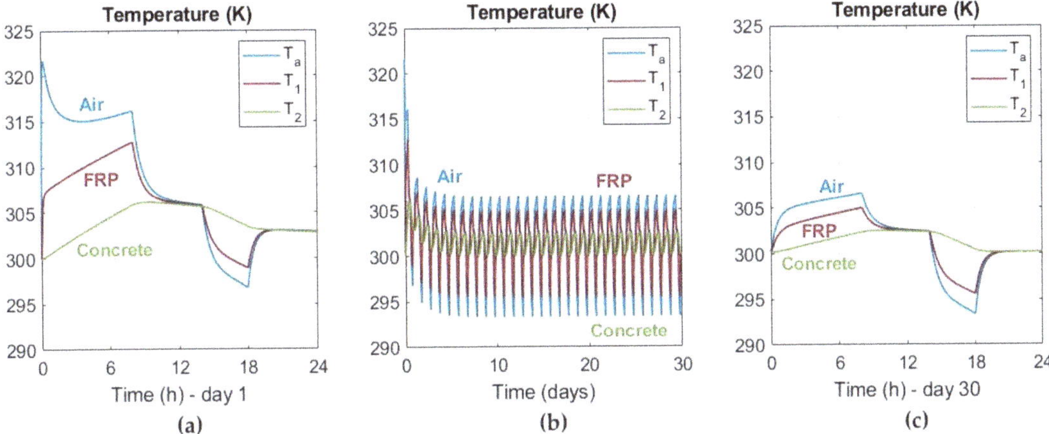

Figure 4. Temperature of air, fiber-reinforced plastic (FRP) and concrete lining for 30 charging and discharging cycles considering FRP as sealing layer. (**a**) First cycle; (**b**) 30 days of operation; (**c**) 30th cycle.

As shown in Figure 5, the air pressure varies from 5 to 8 MPa over the entire 30 days. Because of air temperature, the pressure decreases to 7.8 MPa during the storage period in the first cycle (Figure 5a). This reduction decreases to 7.9 MPa from the fifth cycle. The air pressure increases slightly in the storage period after the decompression stage.

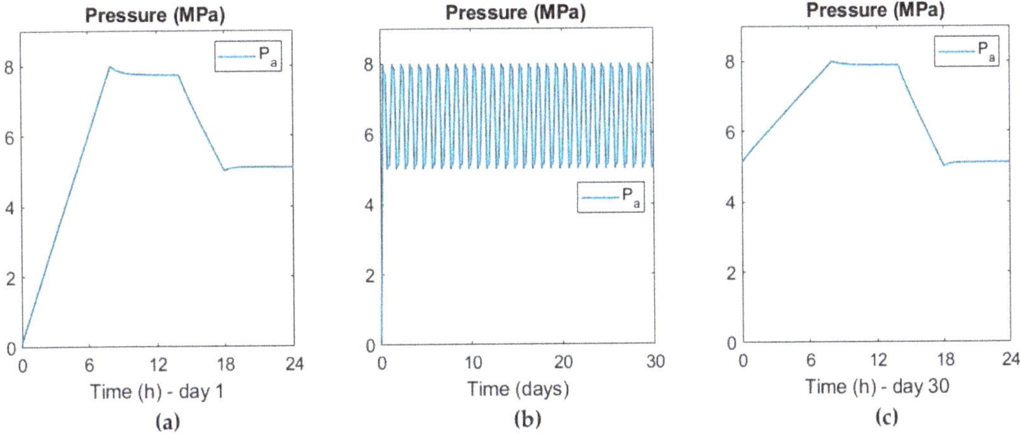

Figure 5. Air pressure for 30 charging and discharging cycles considering FRP as sealing layer. (**a**) First cycle; (**b**) 30 days of operation; (**c**) 30th cycle.

The surface heat flux by convection between the compressed air and the FRP and between the FRP and concrete is shown in Figure 6. Note the thickness of the sealing layer (20 mm). Due to the temperature effect, the surface heat flux increases in the first cycle to 150 and 140 W m^{-2} for FRP and concrete lining, respectively. The heat flux decreases to -90 W m^{-2} in the decompression stage. As observe in Figure 6c in more detail, from second cycle the heat flux through the FRP and concrete is stable, reaching maximum values of 50 and -100 W m^{-2} in the charge and discharge periods, respectively. The surface heat flux is decreasing rapidly during the storage periods. Due to the thickness of the sealing layer and the long air charging and discharging time, the surface heat flux in the sealing layer and the concrete lining is very similar.

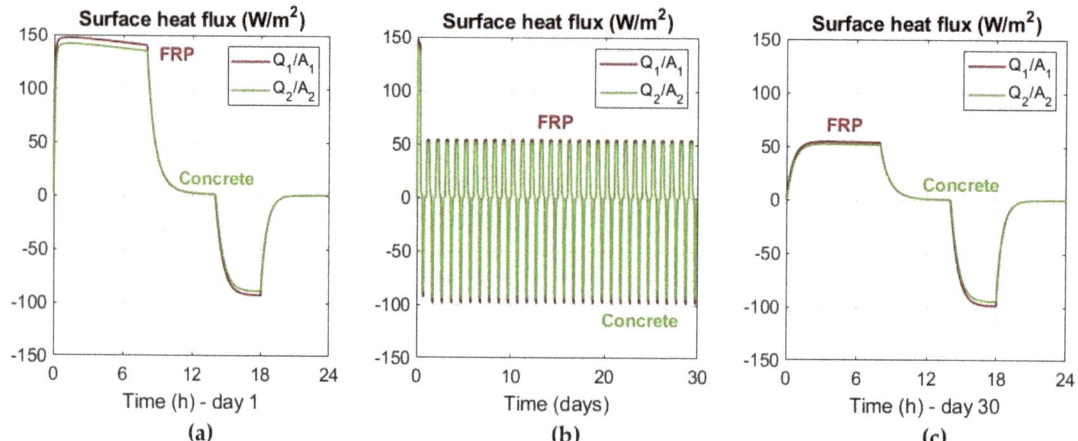

Figure 6. Surface heat flux of FRP and concrete lining for 30 charging and discharging cycles considering FRP as sealing layer. (**a**) First day; (**b**) 30 days of operation; (**c**) 30th cycle.

A comparative analysis for the air temperature, pressure and surface heat flux is shown in Figure 7 for the first cycle considering both FRP and steel as sealing layers. The results are also presented for cycle 30th in Figure 8. The air temperature fluctuation between the compression and decompression periods decreases when steel is employed as sealing layer. Because of the thermal conductivity, the temperature in the contact surface between the compressed air and sealing layer is lower when steel is employed as sealing layer (Figure 8a). The reduction in air pressure in the first cycle is greater during the storage period when FRP is used as sealing layer (Figure 7b). However, as shown in Figure 8b, this reduction is very similar from second cycle. Regarding the surface heat flux, the results are very similar for both sealing layers, reaching values 50 and -100 W m^{-2} in the compression and decompression stages (Figure 8c). The thermal energy balance in the 30th cycle (day 30) through the sealing layer reaches 25.27 and 20.48 kWh for FRP and steel, respectively. Although the system loses slightly more thermal energy in the compression stage when steel is used, the lining contributes with more energy to the compressed air in the decompression stage when steel is used as sealing layer (Figure 8c).

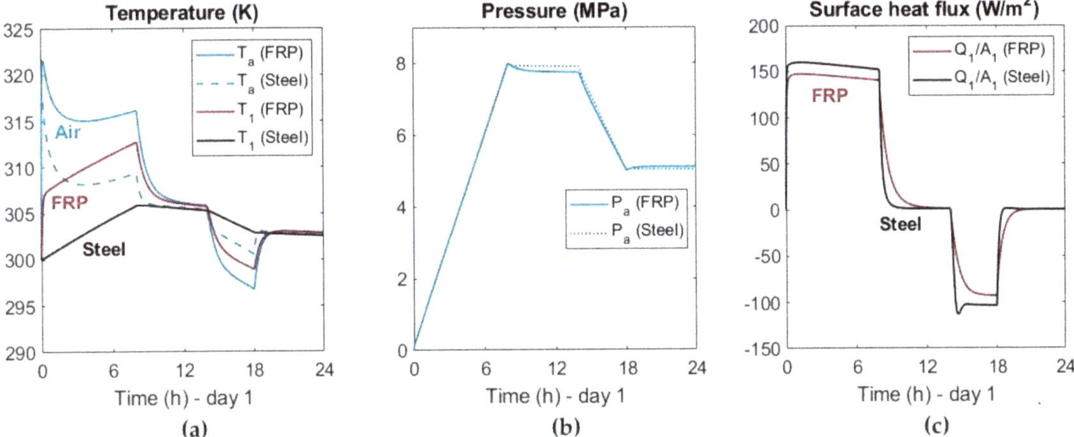

Figure 7. Comparative analysis between FRP and steel in the first cycle. (**a**) Temperature of air, FRP and steel; (**b**) Air pressure; (**c**) Surface heat flux of FRP and steel.

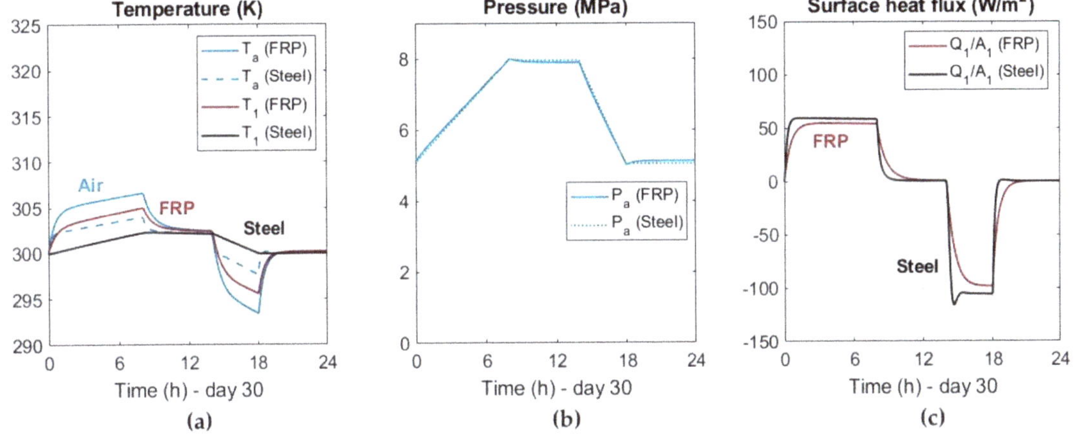

Figure 8. Comparative analysis between FRP and steel in the 30th cycle. (**a**) Temperature of air, FRP and steel; (**b**) Air pressure; (**c**) Surface heat flux of FRP and steel.

3.2. Numerical Model Results

To validate the results of the analytical model, a 3D CFD numerical model has been performed and the results are shown in Figure 9. The numerical model has been simulated for 5 cycles, considering an injected air temperature of 310 K and stable air mass flow rates of 50 kg s^{-1} and -75 kg s^{-1} in the charge and discharge periods, respectively. The model geometry and thermal properties of air and solids are the same as those used in the analytic model. As presented in Figure 9a, the maximum air temperature decreases from 410 to 340 K between the first and fifth cycle. The air temperature fluctuation between air charging and discharging reaches 45 K in the fifth cycle. Likewise, the FRP temperature decreases from 385 to 322 K. However, the concrete lining temperature is more constant throughout the process. The surface heat flux for FRP and concrete is shown in Figure 9c. Due to convective effects, the heat flux increases in FRP during the first cycle. Then, from third cycle is stabilized in a maximum value of 1000 W m^{-2} in the compression period and

-1000 W m^{-2} in the decompression period. The numerical model results using steel as sealing layer are shown for five cycles in Figure 10.

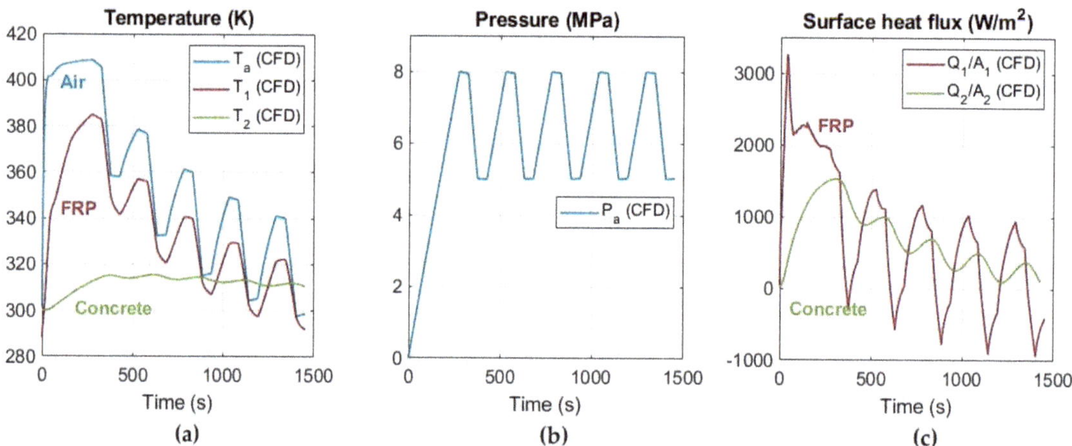

Figure 9. Numerical model results for 5 charging and discharging cycles considering FRP as sealing layer. (**a**) Temperature of air, FRP and concrete lining; (**b**) Air pressure; (**c**) Surface heat flux of FRP and concrete lining.

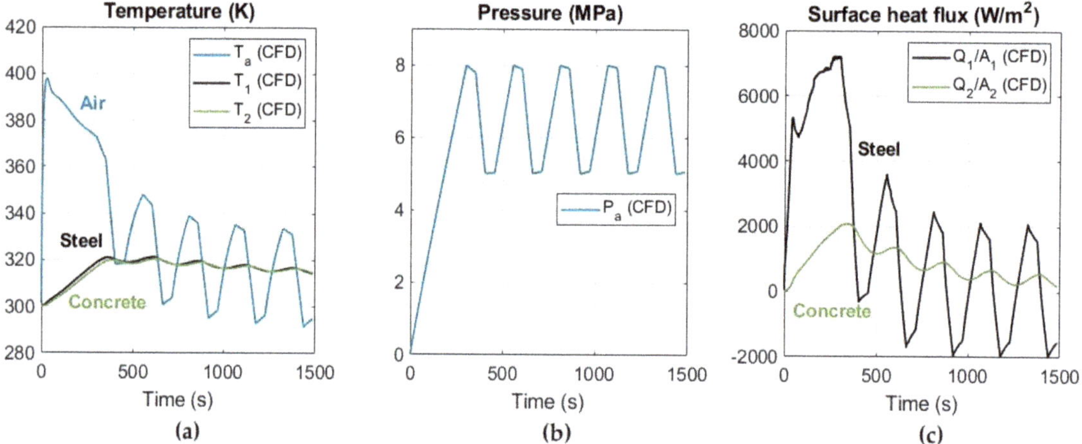

Figure 10. Numerical model results for 5 charging and discharging cycles considering steel as sealing layer. (**a**) Temperature of air, steel and concrete lining; (**b**) Air pressure; (**c**) Surface heat flux of steel and concrete lining.

As in the analytical model, the air temperature fluctuation between air compression and decompression periods is reduced down to 41 K when steel is employed as sealing layer (Figure 10a). In general, the air temperature is much lower than the previous scenario. The surface heat flux is presented in Figure 10c, varying in the steel surface between 2000 W m^{-2} and -2000 W m^{-2} for the charge and discharge periods, respectively. Due to the high thermal conductivity (45 W m^{-1} K^{-1}), the heat flux through the sealing layer is greater, and therefore the temperature in the steel surface is lower. The temperature in the steel coincides with the temperature on the concrete surface, reaching a maximum value of 320 K in the first cycle.

The distribution of the air temperature within de reservoir and the detail of the temperature in the sealing layer are shown in Figure 11 after the decompression period at 5 MPa in the first cycle, using FRP as sealing layer.

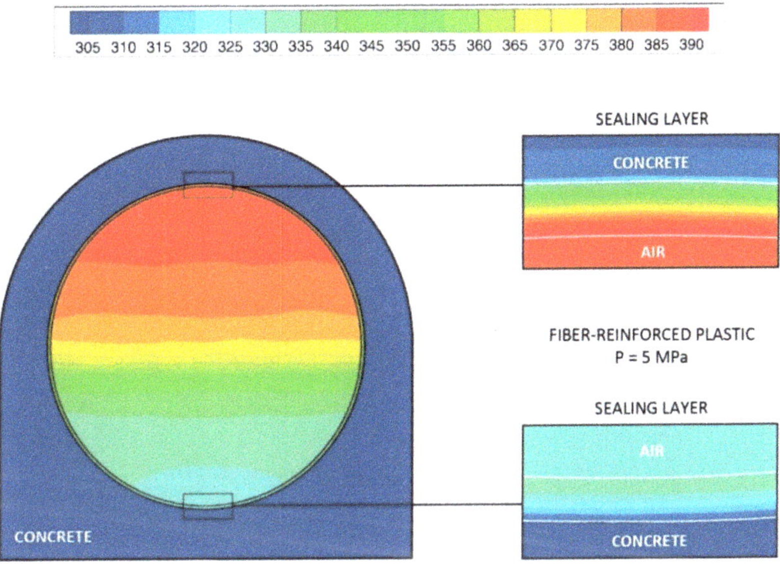

Figure 11. Numerical model results in the first cycle at 5 MPa (decompression). Distribution of air temperature within the reservoir and temperature detail in the FRP and concrete.

Due to the low thermal conductivity of FRP, the temperature reaches 390 K at the top of the reservoir. However, the temperature in the FRP decreases to 330 K at the bottom. A minimal increase in temperature is observed in the concrete lining. The distribution of the air temperature within de reservoir and the detail of the temperature in the sealing layer are shown in Figure 12 after the decompression period at 5 MPa in the first cycle, using steel as sealing layer. The temperature in the steel reaches 330 K at the top of reservoir. In this scenario, a temperature increase up to 320 K has been obtained for the concrete lining.

The distribution of the surface heat flux around the compressed air is shown in Figure 13 after the first compression cycle at 8 MPa for both FRP and steel sealing layers. The surface heat flux reaches a maximum value of 7500 W m^{-2} in the steel surface, while in the FRP surface the maximum value is 3000 W m^{-2}.

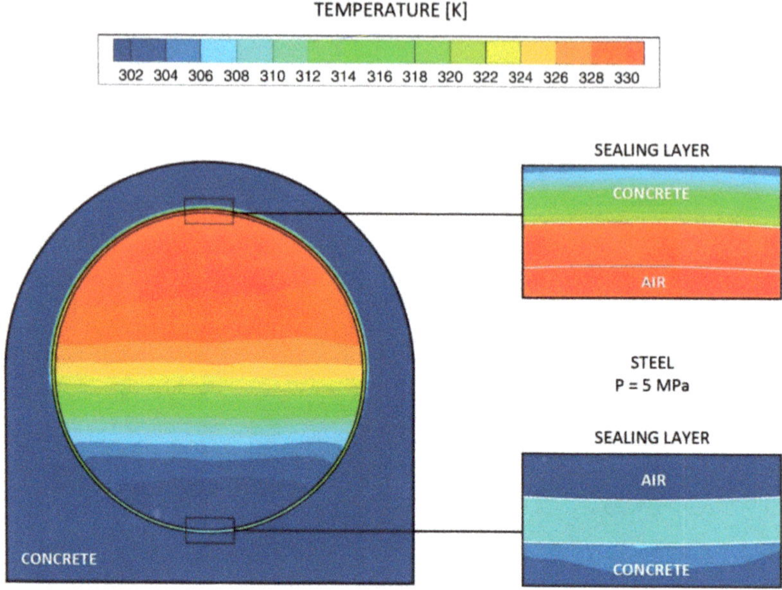

Figure 12. Numerical model results in the first cycle at 5 MPa (decompression). Distribution of air temperature within the reservoir and temperature detail in the steel and concrete.

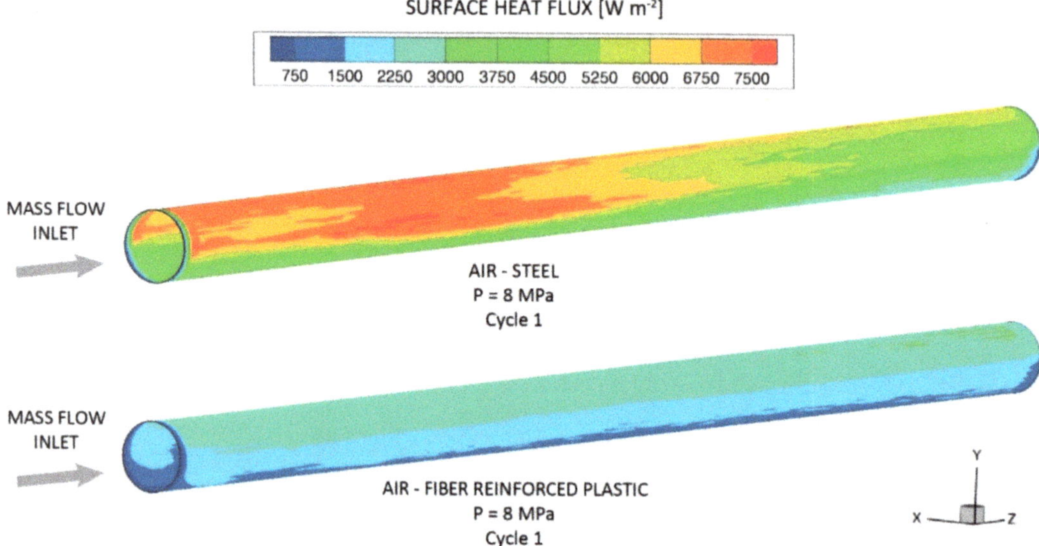

Figure 13. Numerical model results. Distribution of surface heat flux for FRP and steel in the first cycle at a pressure of 8 MPa in the contact surface air-sealing layer.

3.3. Comparative Analysis

A comparative analysis between the analytical and numerical models have been carried out. Air mass flow rates of 50 kg s^{-1} and -75 kg s^{-1} were also considered in the analytical model to carry out the comparative study. As indicated previously, although both analytical and numerical models use the same model geometry, the air mass flow rates of the numerical model are higher to reduce the computational time. Figure 14 shows a comparative analysis for five cycles between analytical and CFD results using FRP as sealing layer. A comparison between FRP and steel is shown afterwards in Figure 15 for one cycle. The obtained heat transfer coefficient is also indicated in Figure 15b. In general, good agreements have been obtained between both analytical and numerical simulations.

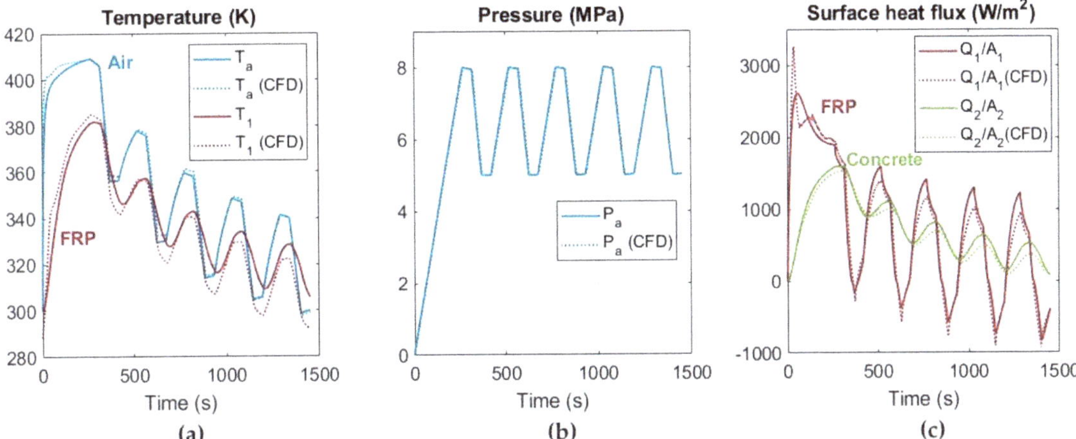

Figure 14. Comparative analysis between analytical and numerical models for 5 charging and discharging cycles considering FRP as sealing layer. (**a**) Temperature of air and FRP; (**b**) Air pressure; (**c**) Surface heat flux of FRP and concrete lining.

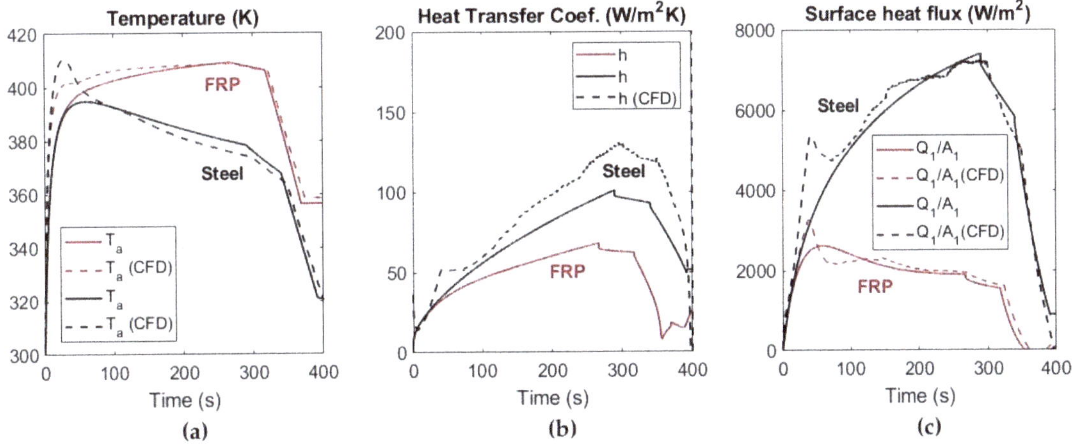

Figure 15. Comparative analysis between analytical and numerical models in the first cycle considering FRP and steel as sealing layers. (**a**) Air temperature; (**b**) Heat transfer coefficient; (**c**) Surface heat flux.

To justify the accuracy of the simulations, the results obtained in the analytical model have been compared with experimental and numerical models. Figure 16 shows the comparison of results obtained for air temperature (Figure 16a) and pressure (Figure 16b) during the first cycle with an experimental model developed by Jiang et al. [24]. Dot points, corresponding to the experimental values, are compared with the results obtained in solid blue lines in Figure 16. In addition, Figure 17 shows the comparison of results obtained for air and wall temperatures (Figure 17a) and air pressure (Figure 17b) during the first cycle with a numerical model conducted by Zhou et al. [25]. The analytical model (solid lines) has been employed to reproduce the numerical model considered by Zhou et al. (dashed lines), reporting good accuracy as shown in Figure 17.

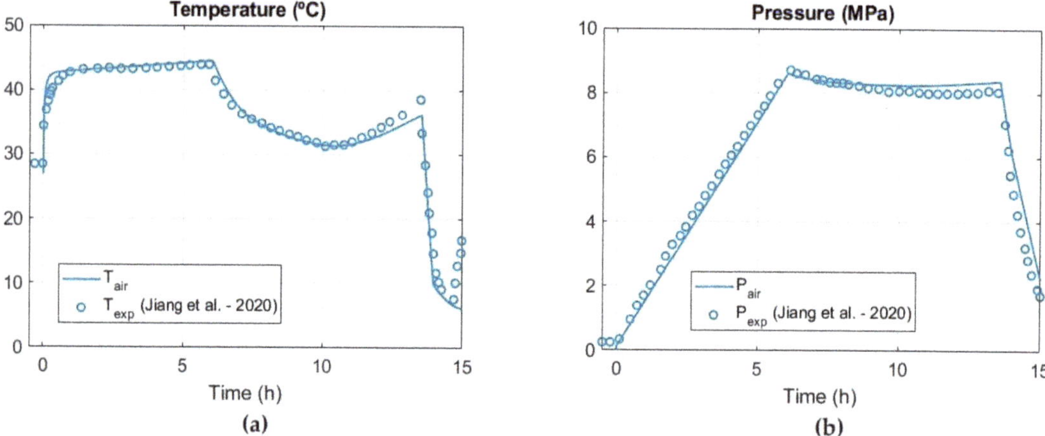

Figure 16. Comparison of results obtained during the first cycle with an experimental model developed by Jiang et al. [24]. (**a**) Air temperature; (**b**) Air pressure.

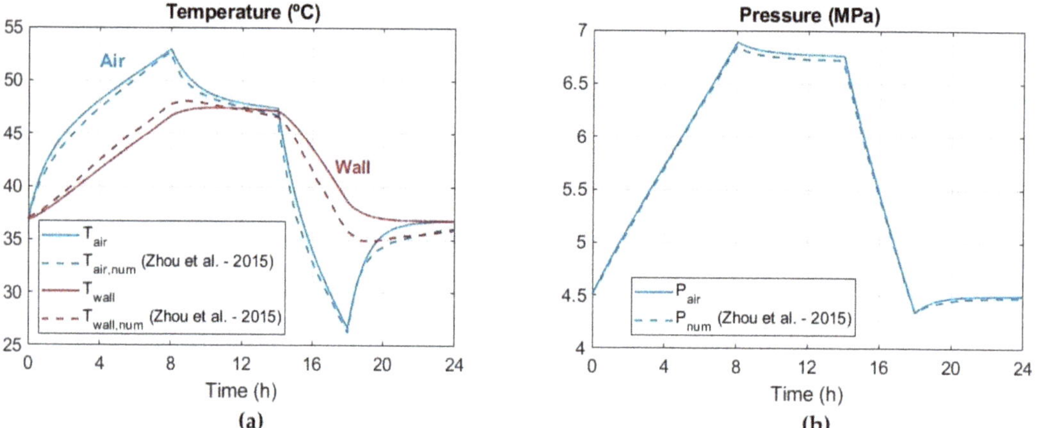

Figure 17. Comparison of results obtained during the first cycle with a numerical model conducted by Zhou et al. [25]. (**a**) Air and wall temperatures; (**b**) Air pressure.

4. Conclusions

Analytical and numerical models were established to investigate the thermodynamic performance of an underground reservoir located in abandoned mines for A-CAES plants. Analyzing air temperature and pressure fluctuations during the operation time in A-CAES plants is essential to design the underground reservoir volume and the turbomachinery. Typical operating pressures from 5 to 8 MPa were considered in the air charging and discharging periods. A 20 mm thick sealing layer, 35 cm thick concrete lining and 2.5 m thick sandstone rock mass have been considered around the compressed air. In addition, FRP and steel have been employed as sealing layer. Unlike other research works, in which the heat transfer coefficient is considered constant during the operation time, in the present investigation a correlation based on both unsteady Reynolds and Rayleigh numbers is employed for the heat transfer coefficient.

The results obtained show greater variations in air temperature between the air compression and decompression when FRP is used as sealing layer. Thus, significant temperature variations in the sealing layer and only a 15 cm thickness of the concrete lining is affected by a temperature rise during operation when steel is employed as sealing layer. The volume of concrete affected is reduced when FRP is used as sealing layer around the fluid. Regarding the sandstone rock mass, no temperature fluctuation was observed in the simulations. In addition, the heat flux increases and the air temperature within the reservoir is lower when steel is employed as sealing layer.

Finally, the thermal energy balance in the 30th cycle through the sealing layer considering air mass flow rates of 0.22 and -0.45 kg s^{-1} reached 25.27 and 20.48 kWh for FRP and steel, respectively. In general, good agreements were obtained between analytical and numerical simulations. Unlike a 1D analytical model, in the 3D CFD numerical model it is possible to analyze the distribution of the thermodynamic responses in the entire domain, assuming a significant advantage to design the reservoir.

Author Contributions: Conceptualization, L.Á.d.P., J.M., A.B.-S., and J.M.F.-O.; Methodology, L.Á.d.P., J.M., and J.M.F.-O.; Software, L.Á.d.P., J.M., M.G., and J.M.F.-O.; Investigation, L.Á.d.P., J.M., M.G., A.B.-S., and J.M.F.-O.; Validation, J.M., M.G., and J.F-O; Writing—original draft preparation, L.Á.d.P., J.M., and J.M.F.-O.; Writing—review and editing, J.M., M.G., A.B.-S., and J.M.F.-O.; Visualization, J.M.F.-O. and J.L.; Supervision, A.B.-S., J.L., and J.M.F.-O.; All authors have read and agreed to the published version of the manuscript.

Funding: This research received no external funding.

Institutional Review Board Statement: N/A.

Informed Consent Statement: N/A.

Conflicts of Interest: The authors declare no conflict of interest.

Nomenclature

A_1	Contact surface between air-sealing layer (m^2)
A_2	Contact surface between sealing layer-concrete (m^2)
A_3	Contact surface between concrete-sandstone (m^2)
A_4	Contact surface between sandstone-exterior (m^2)
C_{pc}	Specific heat at constant pressure for concrete (J kg^{-1} K^{-1})
C_{ps}	Specific heat at constant pressure for the sealing layer (J kg^{-1} K^{-1})
C_{pss}	Specific heat at constant pressure for sandstone layer (J kg^{-1} K^{-1})
C_v	Specific heat at constant volume (J kg^{-1} K^{-1})
e	Specific energy (J kg^{-1})
g	Gravity acceleration (m s^{-2})
h	Heat transfer coefficient (W m^{-2} K^{-1})
K_c	Thermal conductivity of sealing layer (W m^{-1} K^{-1})
K_s	Thermal conductivity of sealing layer (W m^{-1} K^{-1})
K_{ss}	Thermal conductivity of sandstone (W m^{-1} K^{-1})
L	Tunnel length (m)
$\dot{m}$	Air mass flow rate in the charge period (kg s^{-1})
$\dot{m}_{out}$	Air mass flow rate in the discharge period (kg s^{-1})
Nu	Nusselt number (-)
P_a	Air pressure (MPa)
P_c	Air pressure at critical conditions (MPa)
$\dot{Q}$	Surface heat transfer (W m^{-2})
r_1	Equivalent radius of sealing layer (m)
r_2	Equivalent radius of concrete lining (m)
r_3	Equivalent radius of sandstone (m)
r_4	Equivalent radius of external walls (m)

R_a Rayleigh number (-)
R_e Reynolds number (-)
S Tunnel cross section (m^2)
t Time (s)
T_1 Temperature on the sealing layer wall (K)
T_2 Temperature on the concrete lining wall (K)
T_3 Temperature on the sandstone rock mass wall (K)
T_4 External temperature (K)
T_a Air temperature (K)
T_c Air temperature at critical conditions (K)
U_c Concrete transmittance (W K^{-1})
U_h Convection transmittance (W m^{-2} K^{-1})
U_s Sealing layer transmittance (W K^{-1})
U_{ss} Sandstone transmittance (W K^{-1})
v Air velocity (m s^{-1})
$\dot{W}$ Net work (W)
V Tunnel volume (m^3)
Z Compressibility factor (-)
β Volumetric thermal expansion coefficient
μ Dynamic viscosity (kg m^{-1} s^{-1})
ρ_a Air density (kg m^{-3})

References

1. Gallo, A.; Simões-Moreira, J.; Costa, H.; Santos, M.; Dos Santos, E.M. Energy storage in the energy transition context: A technology review. *Renew. Sustain. Energy Rev.* **2016**, *65*, 800–822. [CrossRef]
2. Luo, X.; Wang, J.; Dooner, M.; Clarke, J. Overview of current development in electrical energy storage technologies and the application potential in power system operation. *Appl. Energy* **2015**, *137*, 511–536. [CrossRef]
3. Wang, J.; Lu, K.; Ma, L.; Wang, J.; Dooner, M.; Miao, S.; Li, J.; Wang, D. Overview of Compressed Air Energy Storage and Technology Development. *Energies* **2017**, *10*, 991. [CrossRef]
4. Sameer, H.; Johannes, L. A review of large-scale electrical energy storage. *Int. J. Energy Storage* **2015**, *39*, 1179–1195.
5. Pujades, E.; Willems, T.; Bodeux, S.; Orban, P.; Dassargues, A. Underground Pumped Storage Hydroelectricity Using Aban-doned Works (Deep Mines or Open Pits) and the Impact on Groundwater Flow. *Hydrogeol. J.* **2016**, *24*, 1531–1546. [CrossRef]
6. Menéndez, J.; Loredo, J.; Galdo, M.; Fernández-Oro, J.M. Energy storage in underground coal mines in NW Spain: Assessment of an underground lower water reservoir and preliminary energy balance. *Renew. Energy* **2019**, *134*, 1381–1391. [CrossRef]
7. Lutyński, M. An overview of potential benefits and limitations of Compressed Air Energy Storage in abandoned coal mines. *IOP Conf. Ser. Mater. Sci. Eng.* **2017**, *268*, 012006. [CrossRef]
8. Menéndez, J.; Fernández-Oro, J.M.; Loredo, J. Economic Feasibility of Underground Pumped Storage Hydropower Plants Providing Ancillary Services. *Appl. Sci.* **2020**, *10*, 3947. [CrossRef]
9. Menéndez, J.; Ordóñez, A.; Álvarez, R.; Loredo, J. Energy from closed mines: Underground energy storage and geothermal applications. *Renew. Sustain. Energy Rev.* **2019**, *108*, 498–512. [CrossRef]
10. Menéndez, J.; Schmidt, F.; Konietzky, H.; Fernández-Oro, J.M.; Galdo, M.; Loredo, J.; Díaz-Aguado, M.B. Stability analysis of the underground infrastructure for pumped storage hydropower plants in closed coal mines. *Tunn. Undergr. Space Technol.* **2019**, *94*, 103117. [CrossRef]
11. Menéndez, J.; Schmidt, F.; Konietzky, H.; Sánchez, A.B.; Loredo, J. Empirical Analysis and Geomechanical Modelling of an Underground Water Reservoir for Hydroelectric Power Plants. *Appl. Sci.* **2020**, *10*, 5853. [CrossRef]
12. Winde, F.; Kaiser, F.; Erasmus, E. Exploring the use of deep level gold mines in South Africa for underground pumped hy-droelectric energy storage schemes. *Renew. Sustain. Energy Rev.* **2016**, *78*, 668–682. [CrossRef]
13. Menéndez, J.; Fernández-Oro, J.M.; Galdo, M.; Loredo, J. Efficiency analysis of underground pumped storage hydropower plants. *J. Energy Storage* **2020**, *28*, 101234. [CrossRef]
14. Odukomaiya, A.; Abu-Heiba, A.; Gluesenkamp, K.R.; Abdelaziz, O.; Jackson, R.K.; Daniel, C.; Graham, S.; Momen, A.M. Thermal analysis of near-isothermal compressed gas energy storage system. *Appl. Energy* **2016**, *179*, 948–960. [CrossRef]
15. Tola, V.; Meloni, V.; Spadaccini, F.; Cau, G. Performance assessment of Adiabatic Compressed Air Energy Storage (A-CAES) power plants integrated with packed-bed thermocline storage systems. *Energy Convers. Manag.* **2017**, *151*, 343–356. [CrossRef]
16. Liu, J.L.; Wang, J.H. A comparative research of two adiabatic compressed air energy storage systems. *Energy Convers. Manag.* **2016**, *108*, 566–578. [CrossRef]
17. Yang, K.; Zhang, Y.; Li, X.; Xu, J. Theoretical evaluation on the impact of heat exchanger in Advanced Adiabatic Compressed Air Energy Storage system. *Energy Convers. Manag.* **2014**, *86*, 1031–1044. [CrossRef]

18. Kushnir, R.; Dayan, A.; Ullmann, A. Temperature and pressure variations within compressed air energy storage caverns. *Int. J. Heat Mass Transf.* **2012**, *55*, 5616–5630. [CrossRef]
19. Kushnir, R.; Ullmann, A.; Dayan, A. Thermodynamic Models for the Temperature and Pressure Variations Within Adiabatic Caverns of Compressed Air Energy Storage Plants. *J. Energy Resour. Technol.* **2012**, *134*, 021901. [CrossRef]
20. Safaei, H.; Aziz, M.J. Thermodynamic Analysis of Three Compressed Air Energy Storage Systems: Conventional, Adiabatic, and Hydrogen-Fueled. *Energies* **2017**, *10*, 1020. [CrossRef]
21. Geissbühler, L.; Becattini, V.; Zanganeh, G.; Zavattoni, S.; Barbato, M.; Haselbacher, A.; Steinfeld, A. Pilot-scale demonstration of advanced adiabatic compressed air energy storage, Part 1: Plant description and tests with sensible thermal-energy storage. *J. Energy Storage* **2018**, *17*, 129–139. [CrossRef]
22. Becattini, V.; Geissbühler, L.; Zanganeh, G.; Haselbacher, A.; Steinfeld, A. Pilot-scale demonstration of advanced adiabatic compressed air energy storage, Part 2: Tests with combined sensible/latent thermal-energy storage. *J. Energy Storage* **2018**, *17*, 140–152. [CrossRef]
23. Schmidt, F.; Menéndez, J.; Konietzky, H.; Pascual-Muñoz, P.; Castro, J.; Loredo, J.; Sánchez, A.B. Converting closed mines into giant batteries: Effects of cyclic loading on the geomechanical performance of underground compressed air energy storage systems. *J. Energy Storage* **2020**, *32*, 101882. [CrossRef]
24. Jiang, Z.; Li, P.; Tang, D.; Zhao, H.; Li, Y. Experimental and Numerical Investigations of Small-Scale Lined Rock Cavern at Shallow Depth for Compressed Air Energy Storage. *Rock Mech. Rock Eng.* **2020**, *53*, 2671–2683. [CrossRef]
25. Zhou, S.-W.; Xia, C.-C.; Zhao, H.-B.; Mei, S.-H.; Zhou, Y. Numerical simulation for the coupled thermo-mechanical performance of a lined rock cavern for underground compressed air energy storage. *J. Geophys. Eng.* **2017**, *14*, 1382–1398. [CrossRef]
26. Zhou, S.-W.; Xia, C.-C.; Du, S.-G.; Zhang, P.-Y.; Zhou, Y. An Analytical Solution for Mechanical Responses Induced by Temperature and Air Pressure in a Lined Rock Cavern for Underground Compressed Air Energy Storage. *Rock Mech. Rock Eng.* **2014**, *48*, 749–770. [CrossRef]
27. Khaledi, K.; Mahmoudi, E.; Datcheva, M.; Schanz, T. Analysis of compressed air storage caverns in rock salt considering thermo-mechanical cyclic loading. *Environ. Earth Sci.* **2016**, *75*, 1–17. [CrossRef]
28. Li, W.; Miao, X.; Yang, C. Failure analysis for gas storage salt cavern by thermo-mechanical modelling considering rock salt creep. *J. Energy Storage* **2020**, *32*, 102004. [CrossRef]
29. Barbour, E.; Mignard, D.; Ding, Y.; Li, Y. Adiabatic Compressed Air Energy Storage with packed bed thermal energy storage. *Appl. Energy* **2015**, *155*, 804–815. [CrossRef]
30. Rutqvist, J.; Kim, H.-M.; Ryu, D.-W.; Synn, J.-H.; Song, W.-K. Modeling of coupled thermodynamic and geomechanical performance of underground compressed air energy storage in lined rock caverns. *Int. J. Rock Mech. Min. Sci.* **2012**, *52*, 71–81. [CrossRef]
31. Deng, K.; Zhang, K.; Xue, X.; Zhou, H. Design of a New Compressed Air Energy Storage System with Constant Gas Pressure and Temperature for Application in Coal Mine Roadways. *Energies* **2019**, *12*, 4188. [CrossRef]
32. Woodfield, P.L.; Monde, M.; Mitsutake, Y. Measurement of Averaged Heat Transfer Coefficients in High-Pressure Vessel during Charging with Hydrogen, Nitrogen or Argon Gas. *J. Therm. Sci. Technol.* **2007**, *2*, 180–191. [CrossRef]
33. Heath, M.; Woodfield, P.L.; Hall, W.; Monde, M. An experimental investigation of convection heat transfer during filling of a composite-fibre pressure vessel at low Reynolds number. *Exp. Therm. Fluid Sci.* **2014**, *54*, 151–157. [CrossRef]

*applied
sciences*

Article

Is the Large-Scale Development of Wind-PV with Hydro-Pumped Storage Economically Feasible in Greece?

Anna Dianellou [1], Theofanis Christakopoulos [1], George Caralis [1,*], Vassiliki Kotroni [2], Konstantinos Lagouvardos [2] and Arthouros Zervos [1]

[1] School of Mechanical Engineering, Fluids Section, National Technical University of Athens (NTUA), Zografou, 15771 Athens, Greece; annadnl@hotmail.gr (A.D.); fanischrist@hotmail.com (T.C.); zervos@fluid.mech.ntua.gr (A.Z.)
[2] National Observatory of Athens, Institute for Environmental Research, Penteli, 15236 Athens, Greece; kotroni@meteo.noa.gr (V.K.); lagouvar@noa.gr (K.L.)
* Correspondence: gcaralis@mail.ntua.gr

Abstract: The achievement of the long-term national energy targets in Greece for large-scale integration of wind and solar energy may be facilitated by the development of hydro-pumped storage projects. In light of the above, technical aspects related with the operation of the Greek power system and its ability to absorb renewable energy are analyzed in connection with the role of hydro-pumped storage and relative economic aspects. The aim of this work is to assess the potential contribution of hydro-pumped storage projects and estimate the capacity magnitude order to support large-scale wind and photovoltaic (PV) integration in Greece. For this purpose, scenarios for the Greek power system with focus on Wind and PV development, in conjunction with hydro-pumped storage capacity, are developed, and results for current situation and reference years 2030 and 2050 are presented. For the simulation, among others, high resolution mesoscale wind data for a typical year in the whole Greek territory are used for the steady state simulation of the Greek power system, in order to better estimate the power that could be generated from installed wind turbines, taking into consideration technical characteristics of a typical commercial wind turbine. Results indicate the need of gradual development of hydro-pumped storage in parallel with the large-scale integration of wind and PV capacity into the Greek power system. In addition, the feasibility of the examined scenarios is supported from the low cost of wind and PV generation. In the case of Greece, thanks to the complex morphology and hydraulic conditions of the country, hydro-pumped storage composes an efficient and low-cost storage solution.

Keywords: wind energy; photovoltaics; wind curtailment; mesoscale atmospheric model; hydro-pumped storage

Citation: Dianellou, A.; Christakopoulos, T.; Caralis, G.; Kotroni, V.; Lagouvardos, K.; Zervos, A. Is the Large-Scale Development of Wind-PV with Hydro-Pumped Storage Economically Feasible in Greece?. *Appl. Sci.* **2021**, *11*, 2368. https://doi.org/10.3390/app11052368

Academic Editor: Jorge Loredo and Javier Menéndez

Received: 17 December 2020
Accepted: 2 March 2021
Published: 7 March 2021

Publisher's Note: MDPI stays neutral with regard to jurisdictional claims in published maps and institutional affiliations.

1. Introduction

1.1. Background

The consequences of climate change due to energy production from traditional energy sources, the growing energy demand and the EU decarbonization targets lead to a need for replacement of conventional resources with renewable ones.

However, in the power sector, large-scale integration of intermittent output renewable sources is a great challenge, especially in non-interconnected or saturated grids. For this reason, the growth of large-scale wind and solar integration is a prerequisite for the simultaneous development of energy storage infrastructure.

Storage infrastructure may be a decisive factor for the size of wind and solar energy integration in the power system, as energy storage can act as a balancing component of the system. The energy storage technology that has met the biggest development, is applicable at large scale and has a considerable rate of efficiency is hydro-pumped storage.

Hydro-pumped storage attracts the attention of the scientific community. In parallel, the role of storage solutions is drawing towards large-scale non-dispatchable renewable energy penetration [1]. The policy framework for large-scale electricity storage to use wind energy surplus has been comparatively analyzed in France and Germany for 2020 and 2030 [2]. Additionally, hydro-pumped storage has been widely examined as a solutionto reduce wind energy curtailment in the cases of Ireland [3], China [4,5], Greece [6–8] and in some North European Smart Islands [9]. A special research interest on the combination of hydro-pumped storage with wind energy was shown for autonomous islands, like Azores [10], Gran Canaria [11], El Hierro [12], Crete [13,14] and other Greek islands [7,13,15].

Concerning this paper's case study, the Greek energy supply system is characterized by the oddity that it is not a cohesive system. In particular, it consists of the mainland's electric grid and small non-interconnected power systems on the islands. The interconnections of Greece with the neighboring countries have a relatively small capacity, and the power supply of the system is mainly based on the production of lignite and natural gas power plants, with lignite power plants being gradually phased out and natural gas power gaining a larger production share.

Moreover, the potential of energy production from renewable sources is considered to be high due to the geographical location and the weather conditions that prevail across the country. The instability and the deficiency that could be caused to the Greek power system with the higher integration of renewable sources, inhibit their integration, as the system depends mainly on inland production, and security of supply cannot be compromised. For that reason, hydro-pumped storage or other energy storage systems should be integrated into the power system, in order to facilitate the share increase of power produced from renewable sources and in parallel diversify the energy mix of the country, introducing storage capacity.

A literature review of barriers related to large-scale market integration of Variable Renewable Energy Sources in European electricity markets design has been included in the discussion of the storage facilities' importance underlying the barriers of their high capital cost and the unsound business case due to the lack of scarcity price [16]. However, the role of storage is analyzed in large-scale wind and PV integration in Germany showing that integration level up to 50% could be achieved if flexible back-up power plants are used [1]. Different storage technologies (batteries, pumped hydro storage, adiabatic compressed air energy storage, thermal energy storage, and power-to-gas technology) and their role have been investigated in the transition path towards a 100% renewable energy power sector by 2050 in Europe [17]. The synergy between storage and balancing is analyzed in a fully or highly renewable pan-European power system, based on 8 years solar—wind demand data, with focus on the forms of hydro-pumped storage and hydrogen [18].

The sitting of hydro-pumped storage is a critical parameter from the social, environmental and energy points of view. A review of the existing types of pumped-hydro storage plants, highlights the advantages and disadvantages of each configuration and proposes innovative arrangements in order to increase the possibility to find suitable locations for building large-scale reservoirs for long-term energy and water storage [19]. A global analysis identifies 616,000 sites for closed-loop off-river hydro-pumped storage, based on high resolution global digital elevation models [20]. Off-river pumped hydro energy storage together is also proposed as an effectual solution to support 100% renewable energy in East Asia [21]. Finally, underwater pumped storage has been proposed as an alternative emerging technology with significant potential [22].

1.2. Recent Situation and Trends

In 2019, 60 GW of wind power capacity was added globally, with the cumulative capacity reaching 650 GW. Costs have fallen rapidly for both onshore and offshore wind power, increasing dramatically the amount of bid prices in auctions around the world, especially in the past year. This development can be explained considering the constant technology innovation, the reduced financing costs and the ever-growing competition in the industry, with electric utilities and large oil and gas companies continuing to invest further in that direction. Additionally, incentives such as robust regulatory reforms, wind energy's cost-competitiveness and its potential environmental benefits show that there is no expected deceleration in the wind power sector progress. Finally, the world's first commercial floating wind park has been commissioned in Scotland, while the sizes of turbines continued to increase, with some manufacturers producing turbines of up to 10 MW.

Meanwhile, development in solar PV sector is even faster, with almost 100 GW of new capacity added and increased the global capacity to 580 GW at the end of 2019. This growth can be explained mainly due to the raised awareness of the technology's potential to alleviate pollution, reduce carbon dioxide emissions and provide energy access to developing countries. Subsequently, intense competition, increasing efficiencies and reduction in energy costs have led to record-low auction prices.

Hydro-pumped storage advancements over the past few years continue to upgrade an already proven and reliable technology that represented more than 95% of all energy storage solutions globally, in terms of cumulative capacity. These include improved efficiencies with modern reversible pump-turbines, adjustable-speed pumped turbines, advanced equipment controls such as static frequency converters and generator insulation systems, as well as innovative underground construction methods and design capabilities, leading to faster response times of load follow for intermittent renewables more efficiently and cost effectively. Globally, there are approximately 270 pumped storage plants either operating or under construction, representing a combined generating capacity of over 127,000 MW. Of these total installations, 36 units consist of adjustable speed machines, 17 of which are currently in operation, totaling 3569 MW and 19 of which are under construction, totaling 4558 MW [23].

2. The Greek Power System

2.1. Current Power System Situation in Greece

The annual electricity demand in the Interconnected System of Greece, with reference year 2019, was 52.17 TWh. The total nominal capacity of conventional plants was 8806 MW (December 2019), comprising 3904 MW of lignite power plants and 4902 MW of natural gas [24]. Conventional power plants possess a significant share of electricity production which reaches approximately 51% on the annual electricity production (2019). Wind installations represent a capacity of 3283 MW, while the capacity of photovoltaics including roof-top PVs reaches 2639 MW [25]. Both technologies contributed in 2019 with 14% of generated power [24]. Moreover, the nominal capacity of large and small hydroelectric power plants was 3411 MW, biomass units accounted for 87 MW, while Combined Heat and Power (CHP) 105.47 MW [25]. During 2019, the interconnections balance, i.e., net imports, contributed to the 19% of electricity consumption in Greece. Overall, lignite units are gradually being phased out, while the share of electricity produced by natural gas and lignite power plants combined has been reduced compared to past years. Renewable sources have a smaller share in the energy mix, increased compared to the past, and will further increase in the future [25].

2.2. Prospects for the Future

The National Energy and Climate Planning (NECP) was issued in 2019 with the view to present the long-term energy targets set for the country's energy sector, in accordance with the recent European Directives. Among the targets set is the reduction of greenhouse gases up to 56% (with reference levels of year 2005), the gradual decommissioning of lignite power plants by 2028 and the increase of the share of Renewable Energy Sources (RES) in gross electricity consumption up to 60% [26].

The installed capacity of hydropower plants is not expected to increase significantly, since hydropotential has been already exploited to a great extent and environmental constraints prevent its further development in Greece. However, there is a potential for development of reverse hydro-pumped storage, since reverse pumped storage units could be constructed within the facilities of existing hydropower plants without raising major environmental issues. The installation of storage facilities is essential, in view of the excess energy due to the higher penetration of intermittent renewable energy sources and may provide more flexibility to the system. The interconnection of mainland's power system with the small autonomous ones on the islands is essential, in order to construct a unified system, exploit further the abundant Renewable Energy Sources (RES) potential, withdraw local oil stations and secure power supply on the islands. In parallel, it is expected that the power supply in Greece will be based more on renewable energy sources, while the role of lignite will be reduced and eventually eliminated by 2028. Overall, the main target set towards 2050 is the gradual decarbonization of the power generation sector and its transition to a more sustainable future through the higher integration of renewable sources [26].

2.3. Wind and PV Development

Renewable energy sources shall substantially contribute towards the achievement of decarbonization targets set for the energy sector. One of the main objectives of energy planning is the increase of renewable energy sources' share in the energy mix, in order to substitute part of the conventional units' production. In the renewable sources capacity development scenarios, reference is mainly made to large penetration of wind and photovoltaic systems. In the corresponding studies, wind installations' capacity is projected to reach, according to Ministry of Environment and Energy in 2030, approximately 7 GW, while by 2050 11–18 GW (off-shore wind installation included) [27]. According to a study conducted by European Commission, by 2030, wind installations' capacity may reach 6 GW and by 2050 7.8 GW [28]. In addition, World Wide Fund for Nature (WWF) estimates that the wind installations will account for 5 GW, in 2030, and 6.7 GW, in 2050 [29]. Moreover, the increase of PV capacity is projected, according to the Energy Roadmap by the Ministry of Energy (the National Energy and Climate Plan and the Long-term strategy towards 2050), for 2030 at 7.7 GW, while for 2050 it is estimated at 8–12 GW, depending on the scenario considered [26,27]. According to a study conducted by the European Commission for Greece, in 2030, it is estimated that 5.6 GW of PV will have been constructed, while by 2050 9 GW [28]. Furthermore, according to WWF's projections, PV capacity will reach 4.8 GW by 2030 and 7.1 GW by 2050 [29]. Table 1 presents forecasts on the cumulative capacity of PV and Wind in Greece by 2030 and 2050 [26–29].

Table 1. Forecasts for the cumulative capacity of PV and Wind capacity in Greece by 2030 and 2050 [26–29].

	Ministry of Environment and Energy		WWF		European Commission	
	PV (GW)	Wind (GW)	PV (GW)	Wind (GW)	PV (GW)	Wind (GW)
2030	7.7	7	4.8	5.0	5.6	6
2050	8–12	11–18	7.1	6.7	8.9	7.8

There are significant differences between the figures predicted for wind and PV capacity by the competent bodies. In order to achieve a comparative assessment of the various development scenarios, the ratio of the forecasted wind and photovoltaic capacity by the annual mean load of the country is presented in Figure 1. In 2030 the wind and PV capacity lies between 150–205% of the annual mean load demand, while in 2050 between 200–255%. The ratio between Wind and PV in terms of installed capacity is almost 1 by 1 in most of the forecasts. More PV than Wind capacity is expected by 2050 according to the Energy Roadmap of the Ministry of Energy.

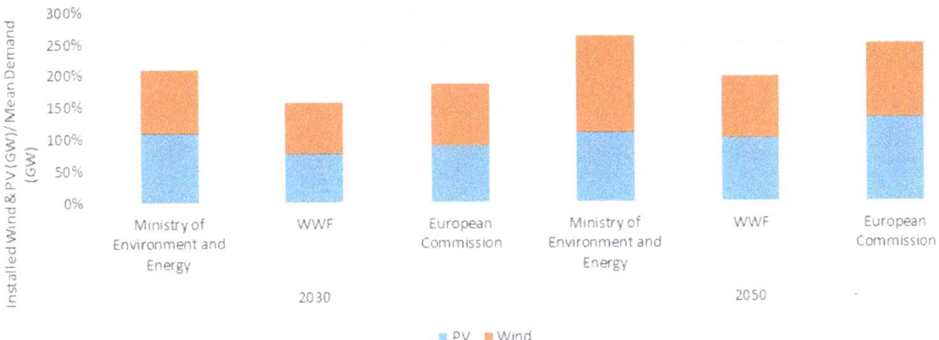

Figure 1. Normalized PV and Wind capacity (by the annual mean load) [26–29].

3. Methodological Approach

3.1. A Simplified Simulation of the Greek Power System

In this connection, a simplified simulation of the Greek power system is proposed to calculate the RES electricity that can be absorbed directly, the energy surplus and the ability of Hydro-pumped storage to exploit it.

Methods for solving units commitment and optimization algorithms for economic dispatch in power systems with high wind penetration have been extensively analysed and proposed in previous works [30–33]. In most of these cases, wind generation is obtained by the use of forecasting tools and unit scheduling is estimated in order to meet demand with a high probability over the scheduling horizon. It is out of the aims of the current approach to optimize the load dispatch and units commitment. The proposed methodology is based on the analysis of the steady-state operation of the Greek power system and takes into account the specific characteristics of demand, the technical features of conventional and hydropower plants and the technical constraints for the smooth and safe operation of the system. Hourly time series are used instead of a probabilistic approach [34], due to the fact that the sequence of events is essential for the simulation of hydro-pumped storage [35].

The large-scale Wind and PV integration is challenging for the power system operator due to their variable power output, the difficulty to predict wind power output accurately and the limited capacity of Greek power system with neighboring countries. However, flexible conventional and hydro units are able to cover any sudden or expected deficit, which may occur to the system. Additionally, the landscape of Greece constitutes an ideal topography for the construction of hydro-pumped storage units which could contribute significantly to the stability of the system and increase renewable peak demand supply.

For the simulation of the Greek power system a steady state analysis is performed based on hourly timeseries data. Alternative scenarios for reference years 2019, 2030 and 2050 are comparably presented in terms of energy flows and Levelized Cost of Energy (LCOE).

For the simplified simulation all the corresponding data of the power units (load factors, technical constraints, maintenance periods, stochasticity) are taken into consideration. The main target of this simulation is the accomplishment of energy balance, giving priority to renewable energy sources. All the relative assumptions are conservative in order to ensure the safe operation of the power system, with respect to all the technical constraints. The contribution of interconnections is not taken into consideration due to their limited capacity, and the Greek power system is considered as a remote power system. This assumption gives results on the safe side, while in reality, the management of the system could be easier with interconnections.

The innovation of the analysis is based on the use of simultaneous mesoscale typical wind year data for the representation of the wind resource in the whole Greek territory. Ninety points are used to provide high geographical resolution and cover current and future possible sites for wind farms. At each point, hourly wind power output is calculated for one year on the basis of hourly wind speed and wind power installed capacity. Then the aggregated hourly wind power in the whole Greek territory is used as an input for the simulation of the Greek power system. Typical wind turbines power curves are used to calculate wind power output, given the wind installed capacity and the time series of wind speed at each point under consideration. Then the aggregated wind power output $\left(P_{W_{prod}} \right)$ for every hour is resulted.

The main principles of the methodology, units' commitment and load dispatch are presented in this section in four steps. In the first step PV power is absorbed in priority; in the second one, conventional units commitment is defined; in the third one, wind power absorption is calculated, and in the last one, Hydro and Hydro-pumped storage output is resulted.

3.1.1. Step 1. PV Power Absorbed in Priority

In the step 1 PV power is absorbed in priority. Wind and PV will gradually represent the mainstream of electricity in the Greek power system; then the management rules and operational principles of the power system should be reformed in comparison with the recent practices [35]. One of the prime assumptions of the current approach is that PV energy is absorbed in priority. PV production is considered as a predictable source of energy, and it is allowed to be absorbed in priority without any constraints, unless it surpasses the load demand. In a large area under consideration, like Greece, geographical distribution of PV has a positive effect on the smoothing of PV power output fluctuations and on the predictability of PV aggregated power output. For every hour i of the year (i = 1 to 8760), given the initial actual demand P_D and subtracting the PV power output P_{PV}, the residual load $P_{D\text{-}PV}$ is calculated (Equation (1)) to be used as the basis for the commitment of conventional units:

$$P_{D\text{-}PV(i)} = P_{D(i)} - P_{PV(i)} \tag{1}$$

3.1.2. Step 2. Conventional Units Commitment

Lignite power plants are used today as a base load units. They are not flexible to switch on and switch off, and their operation should be scheduled in medium term. The number of lignite power plants to be committed is estimated with respect to the expected demand of next days, taking into consideration their non-flexible features and their high technical minimums of the order of 50%. The criterion is that the technical minimums of lignite units to be committed should not exceed the minimum expected residual load of the following seven days. Initially, there are set to operate at their technical minimums or more in case of power deficit. Given this simulation approach, the operation of lignite power units in the Greek power system will be gradually reduced as far as the PV capacity increases. It should be noted that lignite power plants are considered only for reference year 2019, since by 2028 according to national targets the lignite units shall have been decommissioned.

Natural gas units are more flexible than the lignite units, and as a result, they are dispatched in order to support the balance between supply and demand in the short term. The number of natural gas units operating is defined in order to be able to cover the mean residual load of the next twenty-four hours. Initially, they are set to operate at their technical minimums (30% of the nominal output) or more in case of power deficit. The flexibility of natural gas units allows the efficient and rapid response to the fluctuations of power supply.

The combined heat and power (CHP) plants and the biomass units are dispatched hereafter. The total nominal power of these units is relatively small, and their contribution is limited. The produced energy of both is defined using an average capacity factor of 70%.

3.1.3. Step 3. Wind Power Absorbed and Curtailed

After the definition of conventional units commitment, the absorbed P_{Wabs} and curtailed P_{Wcurt} wind energy are calculated. The wind power allowed to be absorbed by the system is limited by technical restrictions, in order to ensure the stability of the system. A percentage of instantaneous penetration δ of wind power to the grid is applied in the calculations [35,36]. This is a dynamic limit which ensures the safe operation of the system in case that all the wind power is lost, and it is related with the ability of other units (i.e., conventional, hydro and storage) to supply load demand [36]. On the other hand, wind power absorption should respect the technical minimums of the committed conventional power plants [35]. Technical minimums of lignite, natural gas, CHP and biomass are taken into consideration. These two technical restrictions are related with power system and define the ability of the power system to absorb wind power. Finally, the actual wind power absorbed $\left(P_{W_{abs}}\right)$ cannot exceed the available wind power $\left(P_{W_{prod}}\right)$, as it has been calculated using the mesoscale data. Then, the actual absorbed wind energy is resulted (Equation (2)):

$$P_{W_{abs}(i)} = \min \left\{ \begin{array}{c} \delta \cdot P_{D\text{-}PV(i)} \\ P_{D\text{-}PV(i)} - P_{TM(i)} \\ P_{W_{prod}(i)} \end{array} \right\} \tag{2}$$

Meanwhile, the wind energy curtailment is calculated (Equation (3)):

$$P_{W_{curt}(i)} = P_{W_{prod}(i)} - P_{W_{abs}(i)} \tag{3}$$

3.1.4. Step 4. Hydropower Output and Hydro-Pumped Storage

Hydropower plants supply the peaks of the load curve. The power produced annually from hydropower plants is highly dependent on the hydrological year. In a good year in terms of hydrological conditions, the annual production could reach 5 TWh. A special methodological approach is used for hydropower [35]. Before the integration of PV in the Greek power system, hydropower was used for peak demand supply. At that time peak demand occurred especially in summer period. Today, hydropower has a more complex role. The integration of PV has shifted the need for peak supply from summer to winter. Hydropower will continue to provide peak power supply whenever peak load demand occurs. Additionally, hydropower will balance the variability of rest RES power generation. Renewable power surplus and wind curtailment may occur in low demand periods or in windy ones. During low demand hours hydropower plants are switched off. During peak demand periods, if there is wind power surplus, hydropower plants may reduce their operation saving water for peak demand periods of low wind. So, wind power plants could save water in the hydro plants' reservoirs and hydro generation will not constrain wind power absorption. Good and bad hydrologic years occur in Greece, with an average annual hydro energy output in Greece of 5 TWh. This energy amount is split into the hours of peak demand using a simplified iterative approach, which finally defines the threshold of power demand for hydropower generation. The operation of hydroelectric power plants is defined by the "peak-shaving" method; it is defined by a minimum power demand threshold (P_{T_H}), when this load demand limit is surpassed hydroelectric power is produced. This limit is determined after an iterative procedure (goal seek function in MS Excel is used) for the whole year, so that the integral of hydro annual production will be equal with 5 TWh (typical hydraulic year). In this step, the remaining load demand after subtracting PV, lignite, natural gas, CHP, biomass and wind power absorbed is used $(P_{D\text{-}PV\text{-}CONV\text{-}W})$. Then, hydropower output is defined for every hour of the year (Equations (4) and (5)) with respect to the nominal power of hydro plants in Greece. This approach is very close to the reality, since existing hydro-plants are operated today in the way of seasonal storage and peak supply. Additionally, the PV's summer peak supply match well with the dry period of Greece from May to September.

$$P_{H(i)} = \begin{cases} 0, \ if \ P_{D\text{-}PV\text{-}CONV\text{-}W(i)} < P_{T\_H} \\ Min\left\{P_{H\_nom}, \ P_{D\text{-}PV\text{-}CONV\text{-}W(i)} - P_{T\_H}\right\}, \ if \ P_{D\text{-}PV\text{-}CONV\text{-}W(i)} > P_{T\_H} \end{cases} \tag{4}$$

$$\sum_{i=1}^{8760} P_{H(i)} = 5 \ \text{TWh} \tag{5}$$

By this approach, the power surplus (wind curtailment and PV surplus) $P_{surplus}$ is minimized in most of the conservative scenarios. Obviously, for large-scale wind and PV integration, due to the technical constraints, power surplus occurs and could be transformed through hydro-pumped storage to useful peak demand supply. The aforementioned excess energy is stored in hydro-pumped storage units, by pumping water to the upper reservoir when there is a surplus of energy, while it is recovered through the hydro-turbines operation when other renewable sources are not available.

Due to the distribution of the curtailed power, it is not economically feasible to exploit 100% of this energy. Such a scenario would require enormous installed capacity of pumps and volume of the upper reservoirs which would be used only few hours per year. A sufficient degree of annual exploitation of excess system's power can be considered to be 70% [6]. On this basis, the required installed capacity of pumps P_{P_nom} is defined in order to achieve the target of 70% exploitation of the annual energy surplus. The hydro turbines capacity is considered to be equal with the pumps' nominal output. Therefore, the nominal power of the pumps is calculated (Equations (6) and (7)):

$$P_{Pump(i)} = Min\left\{P_{P\_nom}, \ P_{surplus(i)}\right\} \tag{6}$$

$$\sum_{i=1}^{8760} P_{Pump(i)} = 70\% \cdot \sum_{i=1}^{8760} P_{surplus(i)} \tag{7}$$

where P_{pump} is the power used for pumping, P_{P_nom} is the nominal power of pumps and $P_{surplus}$ is the available power surplus (wind curtailment and PV surplus). The volume of the upper reservoir is defined to ensure a 24-h operation of the turbines at their nominal capacity.

3.2. Wind Data—Mesoscale Modeling

Simultaneous information on wind statistics over every potential area for wind farm development is required for this analysis. Even if a large number of wind measurements are available, it is practically difficult to represent simultaneous data series and cover every potential area of interest. Installation of a mast network for this purpose could lead to rather prohibitive technical and economic restrictions. Additionally, existing wind monitoring networks are relatively large and can provide large spatial coverage but not necessarily high resolution [37]. On the other hand, use of wind potential maps is not a solution since they only provide an estimation of the spatial distribution of the mean wind speed without any information on its temporal variation. Application of a Numerical Weather Prediction (NWP) model can effectively provide the information required.

In this connection, high resolution analytical wind data timeseries for typical wind year are used. These data have been produced by the systematic application of a numerical weather prediction model. Analytical presentation and description of the approach was given in [38]. In Figure 2, the relative high resolution wind atlas of Greece is presented in terms of power density and parameter c of Weibull distribution, which are the most common ways to present Aeolian maps. The wind atlas of Greece was based on a typical wind year and 12 months of weather model simulations for grid boxes 2 × 2 km^2 in size. The numerical weather prediction model used is "MM5" which is run operationally at the

National Observatory of Athens since 2002 [39] and has been verified [40,41] for its forecast skill over the area of interest.

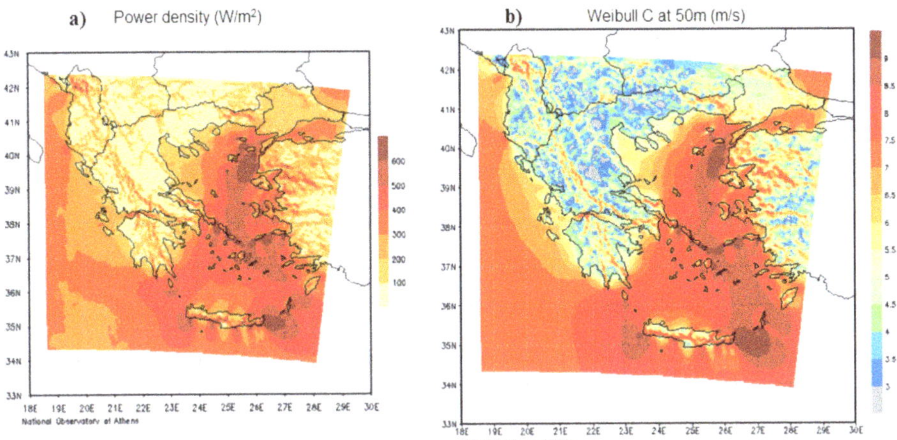

Figure 2. High resolution wind atlas for a typical wind year [38]: (**a**) power density (W/m²), (**b**) parameter c (m/s) of Weibull distribution.

The selection of 90 representative points in the Greek territory is based on the location of wind farms as it is expressed by investors' interest and it is depicted in the Regulatory Authority for Energy (RAE) geographical information system in the Greek territory (Figure 3). For future scenarios, spatial dispersion of wind farms and existing plans for interconnections of Greek islands with the power system of mainland Greece are considered. Data for the selected 90 points are presented in Table 2. The duration curve of wind power output is dependent on the spatial distribution of wind farms. Historical data of single wind turbine power production typically show extended time periods with zero or rated production. However, as the spatial dispersion is increased and more wind farms are introduced, the time periods with cumulative zero or rated production are reduced [34]. The aggregated hourly wind power output is calculated on the basis of the mesoscale wind data, and the selected points in the whole Greek territory for the current wind energy development (2019) and the scenarios of installed capacity in each point under consideration for 2030 and 2050 (Figure 4). In Figure 2, high resolution wind atlas are presented for the typical wind year [38]. The indexes of Power density (W/m²) and Parameter c of Weibull distribution (m/s) are presented. Both are widely used in Aeolian maps for representation of the wind potential [42,43]

Figure 3 shows the overview of wind farms applications in the Greek territory and selection of 90 representative points in the whole Greek territory.

In Table 2 the details of the 90 representative points are presented (name, location, k and c of the Weibull distribution and wind power density) [38].

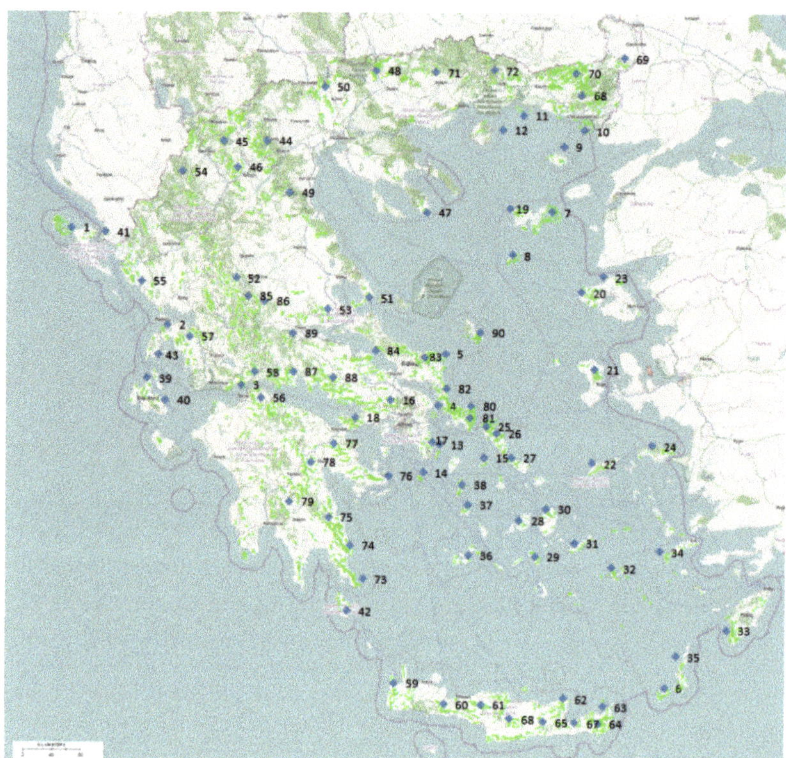

Figure 3. Overview of wind farms applications in the Greek territory and selection of representative points in the whole Greek territory.

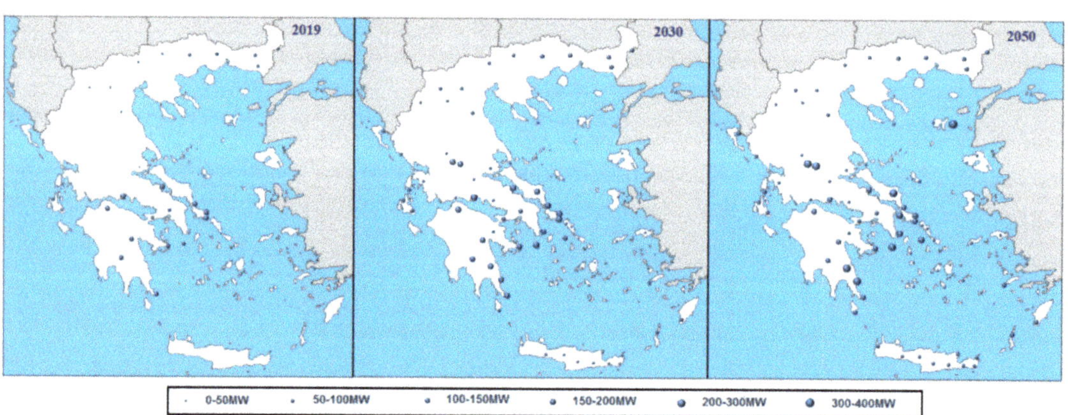

Figure 4. Location of wind farms by 2019, 2030 and 2050.

Table 2. Mesoscale typical wind year data of the representative points [38].

Point	Site	Lat (°)	Long (°)	k (-)	c (m/s)	Wind Power Density (W/m^2)	Point	Site	Lat (°)	Long (°)	k (Weibull Coefficient)	c (m/s)	Wind Power Density (W/m^2)
1	Othonoi	39.786	19.422	1.59	5.99	232	46	Kozani	40.368	21.545	1.58	4.78	116
2	Lefkada	38.867	20.652	1.71	5.37	146	47	Chalkidiki	39.947	23.972	1.73	5.35	143
3	Krioneri	38.292	21.602	1.21	5.78	339	48	Serres	41.296	23.317	1.49	4.63	116
4	Petalioi	38.103	24.133	1.8	7.41	369	49	Veroia	40.127	22.214	1.59	4.47	92
5	Kimi	38.599	24.221	1.75	6.68	278	50	Kilkis	41.134	22.663	1.39	4.53	127
6	Karpathos	35.417	27.039	1.91	8.92	592	51	Trikeri	39.126	23.237	1.77	5.33	137
7	Lemnos	39.951	25.573	1.69	7.62	431	52	Trikala	39.312	21.525	1.51	6.16	273
8	Ai-Stratis	39.536	25.071	1.8	7.25	344	53	Anavra	39.017	22.705	1.65	5.06	128
9	Samothraki	40.575	25.714	1.5	5.96	251	54	Ioannina	40.327	20.837	1.72	5.8	186
10	Alexandroupoli	40.726	25.977	1.63	6.4	272	55	Perdika	39.278	20.328	1.52	4.21	83
11	Fanari	40.867	25.205	1.52	5.07	150	56	Kalavrita	38.751	20.933	1.5	5.18	164
12	Thasos	40.733	24.946	1.55	5.09	145	57	Mitikas	38.417	21.773	1.75	5.2	129
13	Makronisos	37.732	24.150	1.66	7.22	379	58	Naupaktos	38.174	21.856	1.55	5.81	221
14	Ai-Giorgis	37.465	23.942	1.68	6.77	307	59	Kissamos	35.455	23.587	1.63	6.55	290
15	Gyaros	37.609	24.723	1.61	7.4	422	60	Sfakia	35.258	24.225	1.43	7.78	554
16	Erithres	38.151	23.525	1.56	6.23	268	61	Psiloreitis	35.250	24.697	1.59	7.39	429
17	Lavrio	37.755	24.058	1.7	7.25	372	62	Elounta	35.312	25.740	1.7	8.2	526
18	Geraneia	37.985	23.067	1.43	5.87	261	63	Toplou	35.239	26.250	1.74	7.34	374
19	Lemnos (Mirina)	39.983	25.043	1.73	5.95	199	64	Ziros	35.078	26.188	1.59	8.6	638
20	Lesvos (Eresos)	39.183	25.947	1.86	6.52	239	65	Malia	35.090	25.492	1.54	8.24	594
21	Chios	38.447	26.126	1.85	7.45	363	66	Irakleio (south)	35.117	25.068	1.59	6.52	299
22	Ikaria	37.559	26.096	1.37	8.26	662	67	Mochlos	35.080	25.896	1.49	8.73	699
23	Lesvos (Mantamados)	39.331	26.228	1.76	6.57	263	68	Alexandroupoli	41.063	25.942	1.78	6.74	280

Table 2. *Cont.*

Point	Site	Lat (°)	Long (°)	k (-)	c (m/s)	Wind Power Density (W/m^2)	Point	Site	Lat (°)	Long (°)	k (Weibull Coefficient)	c (m/s)	Wind Power Density (W/m^2)
24	Samos	37.724	26.865	1.66	7.37	402	69	Orestiada	41.418	26.481	1.45	4.39	104
25	Andros (Kalivbarni)	37.903	24.750	1.77	7.23	349	70	Dokos	41.273	25.867	1.91	6.36	214
26	Andros (Arnas)	37.840	24.885	1.56	7.86	518	71	Drama	41.280	24.083	1.49	4.12	80
27	Tinos	37.607	25.073	1.64	7.42	417	72	Xanthi	41.298	24.828	1.63	5.52	173
28	Paros	37.009	25.178	1.61	7.07	372	73	Velanidi	36.448	23.176	1.66	7.4	408
29	Ios	36.664	25.375	1.75	7.74	433	74	Monemvasia	36.768	23.013	1.52	6.9	374
30	Naxos	37.117	25.514	1.81	7.54	386	75	Kosmas	37.038	22.730	1.49	6.37	308
31	Amorgos	36.797	25.873	1.82	8.32	510	76	Argolida	37.431	23.497	1.61	6.48	286
32	Astipalaia	36.559	26.354	1.81	7.54	385	77	Sofiko	37.737	22.794	1.44	5.02	161
33	Rodos	35.976	27.840	1.72	6.68	285	78	Tripoli	37.562	22.504	1.44	5.81	252
34	Kos	36.715	26.972	1.87	7.79	408	79	Kalamata	37.184	22.229	1.64	4.95	121
35	Karapthos (Mesochori)	35.718	27.190	1.84	8.89	610	80	Kavo Doro	38.099	24.556	1.62	7.31	405
36	Milos	36.678	24.525	1.84	7.17	324	81	Potami	37.989	24.546	1.71	7.64	430
37	Sifnos	37.158	24.520	1.58	7.49	446	82	Stira	38.264	24.240	1.62	6.64	305
38	Serifos	37.347	24.444	1.73	7.34	376	83	Kimi	38.556	23.957	1.7	6.92	322
39	Kefalonia (Atheras)	38.352	20.402	1.63	5.22	145	84	Larimna	38.622	23.322	1.63	4.98	125
40	Kefalonia (Ainos)	38.146	20.633	1.55	5.21	158	85	Karpenisi	39.141	21.686	1.66	6.35	258
41	Kerkira	39.747	19.856	1.37	4.34	114	86	Fourna	39.098	21.899	1.75	4.77	98
42	Kithira	36.144	22.974	1.72	7.81	455	87	Lidoriki	38.421	22.269	1.39	5.88	277
43	Lefkada	38.572	20.549	1.74	5.96	198	88	Thisvi	38.366	22.785	1.68	5.6	173
44	Naousa	40.618	21.922	1.69	4.91	112	89	Mpralos	38.791	22.264	1.72	5.63	169
45	Florina	40.614	21.360	1.69	5.48	159	90	Skyros	38.798	24.666	1.88	7.13	311

3.3. Load Demand Data

Actual time series of load demand data for the interconnected power system have been used (Power Public Corporation S.A. data, https://www.dei.gr/en, accessed on 15 January 2021). Corresponding adjustments to the demand time series were realized in order to formulate the corresponding timeseries for the years 2030 and 2050. The base year for the load demand time-series is 2006. This is the last year before the start-up of PV development in Greece. The forecasts for annual electricity demand and peak power demand are based on relevant studies that have been conducted for the power system of Greece. The comparative study of the researches carried out for Greece's future demand concludes that, in 2030, a modest estimation is considered to be 57.2 TWh with a peak demand of 10.5 GW, while for 2050, 74 TWh with a peak demand of 13 GW. The studies that were taken into consideration were conducted by the Ministry of Environment and Energy [27], the European Commission [28] and WWF [29]. A comparative representation of the electricity demand forecasts is presented in Figure 5.

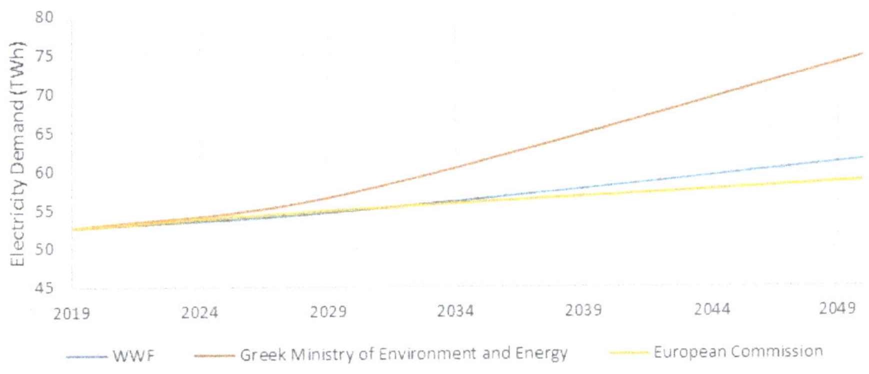

Figure 5. Electricity demand development scenarios [27–29].

3.4. PV Data

In terms of PV production, actual time series provided by CRES (Centre of Renewable Energy Sources) were used with the appropriate adjustments. The adjustment is based on PVGIS estimated annual output [44]. Table 3 present the information used from PVGIS and representative duration curves of PV power output for four cases.

Table 3. Annual PV production [44].

	Peloponnese	Central Greece	Thessaly-Epirus	Macedonia
Annual PV production (kWh/kWp)	1510	1550	1300	1360
C.F.	17%	18%	15%	15%

4. Application—Results

4.1. Scenarios

By 2030, the electricity demand in the country is expected to reach 57.15 TWh, with a peak of 10 GW and by 2050, 74 TWh with a peak of 11.3 GW [27].

Lignite power plants are expected to be decommissioned by 2028, in view of the achievement of decarbonization targets. The nominal output of natural gas units for reference years 2030 and 2050 is expected to reach 6.97 GW [27] and 7.1 GW [29].

In this connection, the reference scenario by 2030 for wind and PV capacity refers to a cumulative capacity of up to 16 GW (8 GW wind, 8 GW PV). By 2050, it is assumed that the renewable energy sources' capacity, i.e., wind and PV, will account for 18 GW aggregated, with wind installations of 9 GW and photovoltaic of 9 GW. The normalized wind and PV capacities (by the average annual demand) are 1.83 by 2030 and 2.48 by 2050, close to the relative normalized capacities in the corresponding studies discussed and presented in Figure 1.

The percentage of instantaneous wind penetration (δ) is considered to be 50% for 2030 and 60% for 2050. The increase of the instantaneous penetration (δ) in 2050 is based on the fact that the management of the grid renewable energy will have matured, and weather load forecast models will be widely used operationally. However, both figures are considered as very conservative approaches, which will keep the results on the safe side.

According to IPTO adequacy study for 2020–2030, as of December 2019, 700 MW of hydroelectric plants have been licensed, including 590 MW of hydro-pumped storage facilities [45].

The nominal output of power plants using as feedstock biomass is considered to reach 300 MW by 2030 and 600 MW by 2050 [27].

4.2. Energy Mix by 2030 and 2050

By 2030 and 2050, considering the developments and assumptions mentioned above, the final energy mix could be formed as depicted in Figure 6.

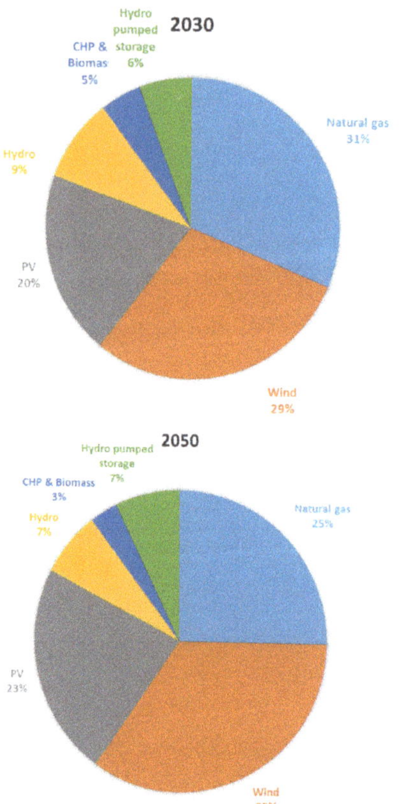

Figure 6. Energy share by technology for reference years 2030 and 2050.

The estimated nominal output of hydro-pumped storage units reaches 1500 MW for the reference year 2030. The capacity factor of pumped-storage units, expressed by the rated power of the turbines, reaches 19%. This value of capacity factor is below the lower acceptable limit which renders the investment economically viable (a benchmark of the capacity factor could be 25% [6]). In this study, this value is considered acceptable, since, according to current data for Greece, large hydro units' capacity factor varies to similar levels (15–17%); therefore, this value could also be considered for reverse hydro-pumped storage projects. The capacity factor of hydropower units may be lower than benchmark, because they are used as a safety net, for cases of emergency, so they are over-dimensioned in order to be able to support the system when a deficit occurs.

The energy contribution of wind and PV installations reaches 49%, and the energy contribution of conventional units is lowered to 31%. The electricity sector is characterized by a higher renewable share, and electricity production is partially decarbonized, compared to the current situation. Hydro-pumped storage units contribute with 6% to the annual electricity demand, by exploiting 70% of curtailed energy.

In 2050, the installed capacity of wind and photovoltaic installations is considered to be 24 GW in total, which results in a reduction of the conventional power plants energy contribution to 25%. Renewable energy sources possess the highest share of production, accounting for approximately 60%. The energy surplus increases due to the higher integration of renewables, and the hydro-pumped storage capacity is estimated at 2700 MW. The share of hydro-pumped storage projects reaches 7% of the total energy produced. The capacity factor of hydro-pumped storage units is estimated at 23%. This higher capacity factor could be attributed to the higher integration of RES, which results in the increase of curtailed power. Consequently, the higher integration of RES contributes to the enhancement of the economically viable operation of hydro-pumped storage units.

4.3. Economic Assessment

To answer the question whether large-scale integration of Wind—PV with the parallel development of hydro-pumped storage is feasible, the levelized cost of energy is used. The levelized cost of energy is calculated as follows:

$$\text{LCOE} = \frac{\text{CAPEX·CRF} + \text{OPEX}}{E}, \quad \text{CRF} = \frac{\text{DR}}{1 - (1 + \text{DR})^{-N}} \tag{8}$$

where, CAPEX is the capital expenditure, CRF the annuity factor, E the annual energy produced, OPEX the operation and maintenance (O&M) expenditures, DR the discount rate and N the lifetime of the investment.

The levelized cost of energy is a valuable cost indicator because of its relative simplicity which allows the comparison of different technologies. At the same time, it is a useful tool for benchmarking the cost of different technology units, considering the lifetime of the units.

The levelized cost of energy has been calculated for the current power supply system and for the examined scenarios for 2030 and 2050, based on the simulation results and the data collected [26–29,46–65]. Studies conducted and demonstrating projections for costs data, such as CAPEX and OPEX, have been considered for the calculation of the LCOE. The final values used constitute an average value of the costs identified in other publications. A discount rate DR equal to 7% is considered in all cases to secure that estimations of LCOE are modest and relative risks are considered. After a long period of economic uncertainty in Greece due to the economic crisis, in the last 3 years, Greek government bonds had constantly relatively low rates, less than 3%. Tables 4–6 show the details about LCOE components, data, assumptions and results for the years 2019, 2030 and 2050, respectively. All the prices presented in the three tables are not harmonized. In some cases, different sources are available and mean values are used. The extracted value of LCOE evaluates not only the relative CAPEX and OPEX for each type of units but also their utilization, as it is introduced by the energy output in the denominator. Then, the results of LCOE are useful for the specific cases studies in Greece, under the assumptions of the current methodological approach and conclusions cannot be universalized. Additionally, the various parameters of cost in case of hydro and hydro-pumped storage are site dependent, and then, results are not useful for other case studies.

Table 4. Data used for LCOE calculation for 2019.

	2019						
Technology	Lignite	Natural Gas	PV	Wind	Hydro	CHP	Biomass
Nominal power P_{nom} (MW)	3904 [27]	4900 [27]	2639 [27]	3283 [27]	3411 [27]	105 [27]	86.89 [27]
Produced Energy (GWh)	10,418	16,228	3249	6565	4052	186	362
CAPEX (€/kW)	1850 [46–48]	700 [46,47]	1100 [48–50]	1300 [47,49,50]	1800 [49,53]	1100 [53]	2650 [53]
Fixed O&M (€/KW)	35 [54]	21 [49]	22 [49]	52 [49,55]	18 [49,56]	40 [57]	79.5 [58]
Variable O&M (€/MWh)	2 [48]	2 [49]	-	-	2.05 [54]	2	4 [58]
Cost (€/$t_{lignite}$ | €/m^3 Natural gas)	17.7 [59]	0.3	-	-	-	0.3	130 [58,60]
Lower Heating Value (GJ/ $t_{lignite}$ | GJ/ m^3 Natural gas)	5.3 [59]	0.03	-	-	-	0.03	19 [58,61]
Efficiency (%)	45% [46]	60% [46]	-	-	-	35% [62]	30% [63]
Heat rate (GJ/MWh)	8.0	6.0	-	-	-	10.3	12.0
Fuel unit cost (€/MWh)	27	60	-	-	-	103	82
CO_2 emissions (t/Mwh)	1.5	0.5 [49]	-	-	-	-	-
CO_2 cost (€/t)	24 [26]	5	-	-	-	-	-
Carbon Cost (€/MWh)	36	2.5	-	-	-	-	-
Life time (years)	30	30	25	25	50	30	30
CRF	0.081	0.081	0.086	0.086	0.072	0.081	0.081
LCOE (€/MWh)	133.7	97.3	94.5	81.8	127.0	161.2	156.3
Weighted average LCOE (€/MWh)	107.59						

Table 5. Data used for LCOE calculation for 2030.

	2030						
Technology	Natural Gas	PV	Wind	PHS	Hydro	CHP	Biomass
Nominal power P_{nom} (MW)	6970	8000	8000	1500	3411	125	300
Produced Energy (GWh)	17,086	11,293	18,624	2545	5203	705	1692
CAPEX (€/kW)	700	700 [64,65]	1000 [49]	1200 [51]	1800	1100	2650
Fixed O&M (€/KW)	21	14	40	18	18	40	79.5
Variable O&M (€/MWh)	2	-	-	-	2.05	2	3.2
Cost (€/$t_{lignite}$ | €/m^3 Natural gas)	0.3	-	-	-	-	0.3	130
Lower Heating Value (GJ/ $t_{lignite}$ | GJ/ m^3 Natural gas)	0.03	-	-	-	-	0.03	19
Efficiency (%)	60%	-	-	-	-	35%	30%
Heat rate (GJ/MWh)	6.0	-	-	-	-	10.3	12.0
Fuel unit cost (€/MWh)	60	-	-	-	-	103	82
CO_2 emissions (t/Mwh)	0.5	-	-	-	-	0.5	-
CO_2 cost (€/t)	30 [27]	-	-	-	-	0	-
Carbon Cost (€/MWh)	15	-	-	-	-	0	-
Life time (years)	30	25	25	50	50	30	30
CRF	0.081	0.086	0.086	0.072	0.072	0.081	0.081
LCOE (€/MWh)	108.5	52.5	54	61.9	99.4	127	137.2
Weighted average LCOE (€/MWh)	77.86						

Table 6. Data used for LCOE calculation for 2050.

Technology	2050						
	Natural Gas	PV	Wind	PHS	Hydro	CHP	Biomass
Nominal power P_{nom} (MW)	7100	12,000	12,000	2700	3565	125	600
Produced Energy (GWh)	18638	16,924	25,620	5334	5203	402	1929
CAPEX (€/kW)	700	500 [64,65]	800 [49]	1200	1800	1100	2650
Fixed O&M (€/KW)	21	10	32	18	18	40	79.5
Variable O&M (€/MWh)	2	-	-	-	2.05	2	3.2
Cost (€/$t_{lignite}$ | €/m^3 $_{Natural\ gas}$)	0.3	-	-	-	-	0.3	130
Lower Heating Value (GJ/ $t_{lignite}$ | GJ/ m^3 $_{Natural\ gas}$)	0.03	-	-	-	-	0.03	19
Efficiency (%)	60%	-	-	-	-	35%	35%
Heat rate (GJ/MWh)	6.0	-	-	-	-	10.3	10.3
Fuel unit cost (€/MWh)	60	-	-	-	-	103	70
CO_2 emissions (t/Mwh)	0.5	-	-	-	-	0.5	-
CO_2 cost (€/t)	88 [27]	-	-	-	-	0	-
Carbon Cost (€/MWh)	44	-	-	-	-	0	-
Life time (years)	30	25	25	50	50	30	30
CRF	0.081	0.086	0.086	0.072	0.072	0.081	0.081
LCOE (€/MWh)	**135.5**	**37.5**	**47.1**	**53.1**	**102.6**	**138.9**	**164.6**
Weighted average LCOE (€/MWh)	75.06						

In Figure 7, the size of the bubbles corresponds to the amount of energy generated by each type of technology, whereas on the horizontal axis the LCOE of each unit type is presented, and on the vertical axis, the share of the cost of the different types of technologies in the total cost of the system is depicted.

In the current power supply system, the participation of lignite and natural gas units is crucial in order to meet electricity demand. The LCOE of the system is estimated to 107.59 €/MWh, and the largest share of the cost is due to the conventional units. Although CHP and biomass units are characterized by high levelized costs due to their limited nominal capacity, they do not participate significantly to the final composition of the cost of the system.

The weighted average LCOE of the power system in Greece by 2030, according to the calculations, is decreased to 79 €/MWh. This reduction could be attributed to the decommissioning of lignite units, which present a higher LCOE compared to RES (wind and PV installations) in 2019. Natural gas units gradually replace lignite ones by 2030. Carbon dioxide allowances are expected to rise to a higher level by 2030; therefore, the operational expenditures of conventional units are increased, resulting in a higher LCOE for natural gas units. The expected decrease of the CAPEX required for RES installations contribute to the significant decrease of the weighted average LCOE of the system.

By 2050, a further reduction of the system's cost is expected; the levelized cost is approximately 75 €/MWh, reduced by 30% compared to current levels. Increasing the participation of renewable energy sources in the energy production with the simultaneous significant reduction of their CAPEX leads to a decrease in the levelized cost of the system. Hydro-pumped storage is among the technologies presenting the lowest LCOE, following wind and PV. The CO_2 prices are significantly higher compared to 2030 resulting in the increase of the levelized cost of energy of natural gas power plants.

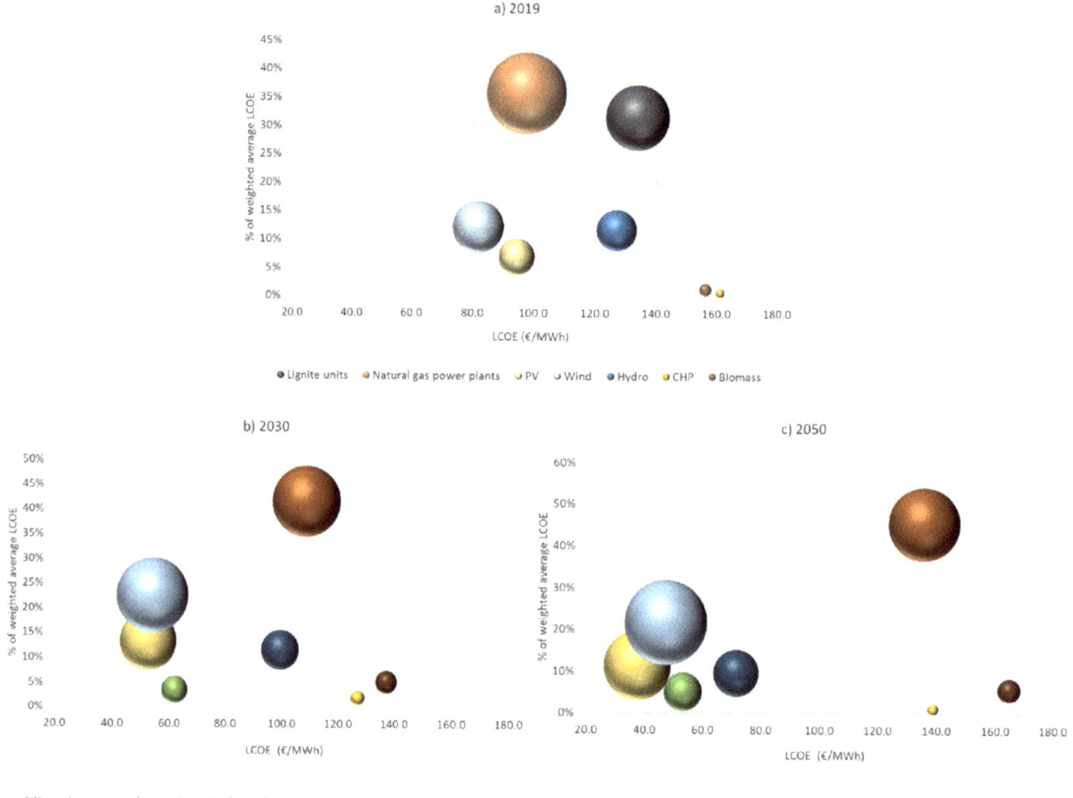

Figure 7. Levelized Cost of Energy and contribution to the weighted average cost of the system by technology (the size of the bubbles corresponds to power generation by each type of technology).

5. Discussion and Conclusions

High renewable penetration could lead to a significant reduction of the system's cost, while hydro-pumped storage systems may contribute to the peak demand supply.

The size of the reverse hydro units should be proportional to the development of Wind and PV installed capacity. The over-dimensioning of reverse hydroelectric projects may lead to installed hydroelectric pumps and turbines which will be used only for a few hours per year without ensuring their economic feasibility, while under-dimensioning will lead to insufficient exploitation of the potential. Furthermore, the higher integration of renewable sources (wind and PV) across the country (spatial dispersion) results in significant energy surplus and curtailment during the year, therefore energy that could be stored in hydro-pumped storage systems is increased. In this case, hydro-pumped storage units can contribute significantly to the energy balance, as analyzed in the scenarios.

The economic assessment of the future power system of Greece, based on the scenarios examined in this connection, demonstrates that the increase of RES contribution may lead to lower costs for the system. Greater penetration of photovoltaic and wind installations results in a reduction of the levelized cost of energy of the system, since capital expenditures required for wind and PV systems are expected to decrease significantly over the years. The cost of thermal units in reference years 2030 and 2050 is expected to be higher due to the increase in the cost of carbon dioxide emission allowances. The reduction of the energy production share of the latter in combination with the higher integration of RES prevents the increase of the weighted average LCOE of the power system.

Author Contributions: Conceptualization, G.C. and A.Z.; methodology, G.C. and A.D.; software and formal analysis, A.D.; investigation, A.D.; data curation, V.K. and K.L.; writing—original draft preparation, A.D., T.C.; writing—review and editing, A.D. and G.C.; supervision, G.C.; All authors have read and agreed to the published version of the manuscript.

Funding: This research received no external funding.

Institutional Review Board Statement: Not applicable.

Informed Consent Statement: Not applicable.

Data Availability Statement: Not applicable.

Conflicts of Interest: The authors declare no conflict of interest.

References

1. Weitemeyer, S.; Kleinhans, D.; Vogt, T.M.; Agert, C. Integration of Renewable Energy Sources in future power systems: The role of storage. *Renew. Energy* **2015**, *75*, 14–20. [CrossRef]
2. Loisel, R.; Mercier, A.; Gatzen, C.; Elms, N.; Petric, H. Valuation framework for large scale electricity storage in a case with wind curtailment. *Energy Policy* **2010**, *38*, 7323–7337. [CrossRef]
3. Tuohy, A.; O'Malley, M. Pumped storage in systems with very high wind penetration. *Energy Policy* **2011**, *39*, 1965–1974. [CrossRef]
4. Zhang, N.; Lu, X.; McElroy, M.B.; Nielsen, C.P.; Chen, X.; Deng, Y.; Kang, C. Reducing curtailment of wind electricity in China by employing electric boilers for heat and pumped hydro for energy storage. *Appl. Energy* **2016**, *184*, 987–994. [CrossRef]
5. Kong, Y.; Kong, Z.; Liu, Z.; Wei, C.; Zhang, J.; An, G. Pumped storage power stations in China: The past, the present, and the future. *Renew. Sustain. Energy Rev.* **2017**, *71*, 720–731. [CrossRef]
6. Anagnostopoulos, J.S.; Papantonis, D.E. Study of pumped storage schemes to support high RES penetration in the electric power system of Greece. *Energy* **2012**, *45*, 416–423. [CrossRef]
7. Caralis, G.; Rados, K.; Zervos, A. On the market of wind with hydro-pumped storage systems in autonomous Greek islands. *Renew. Sustain. Energy Rev.* **2010**, *14*, 2221–2226. [CrossRef]
8. Tigas, K.; Giannakidis, G.; Mantzaris, J.; Lalas, D.; Sakellaridis, N.G.; Nakos, C.; Vougiouklakis, Y.; Theofilidi, M.; Pyrgioti, E.; Alexandridis, A.T. Wide scale penetration of renewable electricity in the Greek energy system in view of the European decarbonization targets for 2050. *Renew. Sustain. Energy Rev.* **2015**, *42*, 158–169. [CrossRef]
9. Cannistraro, G.; Cannistraro, M.; Trovato, G. Island "Smart Energy" for Eco-Sustainable Energy—A Case Study "Favignana Island". *Int. J. Heat Technol.* **2017**, *35*. [CrossRef]
10. Lopes, A.S.; Castro, R.; De Jesus, J.F. Contributions to the preliminary assessment of a Water Pumped Storage System in Terceira Island (Azores). *J. Energy Storage* **2016**, *6*, 59–69. [CrossRef]
11. Padrón, S.; Medina, J.F.; Rodriguez, A. Analysis of a pumped storage system to increase the penetration level of renewable energy in isolated power systems. Gran Canaria: A case study. *Energy* **2011**, *36*, 6753–6762. [CrossRef]
12. Bueno, C.; Carta, J. Technical–economic analysis of wind-powered pumped hydrostorage systems. Part II: Model application to the island of El Hierro. *Sol. Energy* **2005**, *78*, 396–405. [CrossRef]
13. Caralis, G.; Zervos, A. Analysis of the combined use of wind and pumped storage systems in autonomous Greek islands. *IET Renew. Power Gener.* **2007**, *1*, 49–60. [CrossRef]
14. Katsaprakakis, D.A.; Christakis, D.G.; Zervos, A.; Papantonis, D.; Voutsinas, S. Pumped storage systems introduction in isolated power production systems. *Renew. Energy* **2008**, *33*, 467–490. [CrossRef]
15. Krajačić, G.; Duić, N.; Tsikalakis, A.; Zoulias, M.; Caralis, G.; Panteri, E.; da Graça Carvalho, M. Feed-in tariffs for promotion of energy storage technologies. *Energy Policy* **2011**, *39*, 1410–1425. [CrossRef]
16. Hu, J.; Harmsen, R.; Crijns-Graus, W.; Worrell, E.; Broek, M.V.D. Identifying barriers to large-scale integration of variable renewable electricity into the electricity market: A literature review of market design. *Renew. Sustain. Energy Rev.* **2018**, *81*, 2181–2195. [CrossRef]
17. Child, M.; Bogdanov, D.; Breyer, C. The role of storage technologies for the transition to a 100% renewable energy system in Europe. *Energy Procedia* **2018**, *155*, 44–60. [CrossRef]
18. Rasmussen, M.G.; Andresen, G.B.; Greiner, M. Storage and balancing synergies in a fully or highly renewable pan-European power system. *Energy Policy* **2012**, *51*, 642–651. [CrossRef]
19. Hunt, J.D.; Zakeri, B.; Lopes, R.; Barbosa, P.S.F.; Nascimento, A.; De Castro, N.J.; Brandão, R.; Schneider, P.S.; Wada, Y. Existing and new arrangements of pumped-hydro storage plants. *Renew. Sustain. Energy Rev.* **2020**, *129*, 109914. [CrossRef]
20. Stocks, M.; Stocks, R.; Lu, B.; Cheng, C.; Blakers, A. Global Atlas of Closed-Loop Pumped Hydro Energy Storage. *Joule* **2021**, *5*, 270–284. [CrossRef]

21. Cheng, C.; Blakers, A.; Stocks, M.; Lu, B. Pumped hydro energy storage and 100 % renewable electricity for East Asia. *Glob. Energy Interconnect.* **2019**, *2*, 386–392. [CrossRef]

22. Dubbers, D. Comparison of underwater with conventional pumped hydro-energy storage systems. *J. Energy Storage* **2021**, *35*, 102283. [CrossRef]

23. Challenges and Opportunities for New Pumped Storage Development—A White Paper Developed by NHA's Pumped Storage Development Council. Available online: http://www.hydro.org/wp-content/uploads/2012/07/NHA_PumpedStorage_071212 b1.pdf (accessed on 15 September 2017).

24. Greek Operator of Electricity Market. *DAS Monthly Report*; Greek Operator of Electricity Market: Athens, Greece, 2019. (In Greek)

25. DAPEEP. *RES and CHP Report*; DAPEEP: Athens, Greece, 2019.

26. Ministry of Environment and Energy. *National Energy and Climate Plan*; Ministry of Environment and Energy: Athens, Greece, 2019.

27. Greek Ministry of Energy and Environment. *National Energy Planning, 2050 Energy Roadmap*; Greek Ministry of Energy and Environment: Athens, Greece, 2019.

28. Capros, P.; de Vita, A.; Tasios, N.; Siskos, P.; Kannavou, M.; Petropoulos, A.; Evangelopoulou, S.; Zampara, M.; Papadopoulos, D.; Nakos, L.; et al. *EU Reference Scenario 2016, Energy, transport and GHG emissions Trends to 2050*; European Commission: Luxembourg, 2016.

29. Moirasgentis, S.; Sarafidis, G.; Georgopoulou, E. *Long Term Plan for the Greek Energy System*; WWF: Athens, Greece, 2017. (In Greek)

30. Tuohy, A.; Meibom, P.; Denny, E.; O'Malley, M. Unit Commitment for Systems with Significant Wind Penetration. *IEEE Trans. Power Syst.* **2009**, *24*, 592–601. [CrossRef]

31. Ruiz, P.A.; Philbrick, C.R.; Sauer, P.W. Modeling approaches for computational cost reduction in stochastic unit commitment formulations. *IEEE Trans. Power Syst.* **2010**, *25*, 588–589. [CrossRef]

32. Lujano-Rojas, J.; Osório, G.; Catalão, J. New probabilistic method for solving economic dispatch and unit commitment problems incorporating uncertainty due to renewable energy integration. *Int. J. Electr. Power Energy Syst.* **2016**, *78*, 61–71. [CrossRef]

33. Osório, G.; Lujano-Rojas, J.; Matias, J.; Catalão, J.P.S. A probabilistic approach to solve the economic dispatch problem with intermittent renewable energy sources. *Energy* **2015**, *82*, 949–959. [CrossRef]

34. Caralis, G.; Perivolaris, Y.; Rados, K.; Zervos, A. On the Effect of Spatial Dispersion of Wind Power Plants on the Wind Energy Capacity Credit in Greece. *Environ. Res. Lett.* **2008**, *3*, 015003. [CrossRef]

35. Caralis, G.; Papantonis, D.; Zervos, A. The role of pumped storage systems towards the large scale wind integration in the Greek power supply system. *Renew. Sustain. Energy Rev.* **2012**, *16*, 2558–2565. [CrossRef]

36. Papathanassiou, S.A.; Boulaxis, N.G. Power limitations and energy yield evaluation for wind farms operating in island systems. *Renew. Energy* **2006**, *31*, 457–479. [CrossRef]

37. Archer, C.L.; Jacobson, M.Z. Spatial and temporal distributions of U.S. winds and wind power at 80 m derived from measurements. *J. Geophys. Res. Space Phys.* **2003**, *108*, 9. [CrossRef]

38. Kotroni, V.; Lagouvardos, K.; Lykoudis, S. High-resolution model-based wind atlas for Greece. *Renew. Sustain. Energy Rev.* **2014**, *30*, 479–489. [CrossRef]

39. Kotroni, V.; Lagouvardos, K. Evaluation of MM5 High-Resolution Real-Time Forecasts over the Urban Area of Athens, Greece. *J. Appl. Meteorol.* **2004**, *43*, 1666–1678. [CrossRef]

40. Kotroni, V.; Lagouvardos, K. Precipitation forecast skill of different convective parameterization and micro-physical schemes: Application for the cold season over Greece. *Geoph. Res. Let.* **2001**, *108*, 1977–1980. [CrossRef]

41. Akylas, E.; Kotroni, V.; Lagouvardos, K. Sensitivity of high-resolution operational weather forecasts to the choice of the planetary boundary layer scheme. *Atmos. Res.* **2007**, *84*, 49–57. [CrossRef]

42. Troen, I.; Lundtang Petersen, E. European Wind Atlas. Risø National Laboratory. 1989. Available online: Orbit.dtu.dk/files/1121 35732/European_Wind_Atlas.pdf (accessed on 15 February 2021).

43. Caralis, G.; Gao, Z.; Yang, P.; Huang, M.; Zervos, A.; Rados, K. Development of Aeolian map of China using mesoscale atmospheric modelling. *Renew. Energy* **2015**, *74*, 60–69. [CrossRef]

44. Photovoltaic Geographical Information System (PVGIS). Geographical Assessment of Solar Resource and Performance of Photovoltaic Technology. Available online: http://re.jrc.ec.europa.eu/pvgis/ (accessed on 15 September 2017).

45. Independent Power Transmission Operator. *Power Adequacy Study for Years 2020–2030*; Independent Power Transmission Operator: Athens, Greece, 2019.

46. Papaefthymiou, G.; Grave, K.; Dragoon, K. *Flexibility Options in Electricity Systems*; Ecofys: Berlin, Germany, 2014.

47. Larsson, S. *Reviewing Electricity Generation*; Uppsala Universitet: Uppsala, Sweden, 2012.

48. Schröder, A.; Kunz, F.; Meiss, J.; Mendelevitch, R.; Hirschhausen, C.V. *Current and Prospective Costs of Electricity Generation until 2050*; DIW: Berlin, Germany, 2013.

49. Joint Research Centre. *Energy Technology Reference Indicator Projections for 2010–2050*; European Union: Luxembourg, 2014.

50. International Energy Agency; Nuclear Energy Agency. *Projected Costs of Generating Electricity*; IEA: Paris, France; NEA: Paris, France, 2015.

51. Zach, K.; Auer, H.; Lettner, G. *Report summarizing the current Status, Role and Costs of Energy Storage Technologies*; WIP Renewable Energies: Munich, Germany, 2012.

52. ECOFYS. *Subsidies and Costs of EU Energy, Annex 4–5*; European Commission: Brussels, Belgium, 2014.

53. Ministry of Environment & Energy. *Description of Functional Support Scheme in the Renewable Energy Sources and Cogeneration Power and Heat High Performance*; Ministry of Environment & Energy: Athens, Greece, 2016.
54. Official Government Gazette of Hellenic Republic. *Determination of the Minimum Bid Price of Auctioned Energy for the Year 2017*; n. B' 2278/04.07.2017; Government of Greece: Athens, Greece, 2017; p. 23014. (In Greek)
55. International Renewable Energy Agency (IRENA). *Renewable Energy Technologies: Cost Analysis Series, Wind Power*; IRENA: Bonn, Germany, 2012.
56. Lako, P. *Hydropower*; IEA ETSAP: Paris, France, 2010.
57. IEA ETSAP. *Combined Heat and Power, Technology Brief E04*; IEA ETSAP: Paris, France, 2010.
58. International Renewable Energy Agency. *Renewable Energy Technologies: Cost Analysis Series, Biomass for Power Generation*; IRENA: Bonn, Germany, 2012.
59. Booz & Company. *Understanding Lignite Generation Costs in Europe*; Public Power Corporation S.A.: Athens, Greece, 2012.
60. Eleutheriadis, I. *Biomass Potential and Solid Biofuels*; Center for Renewable Energy Sources and Saving: Athens, Greece, January 2013.
61. EUBIA, European Biomass Industry Association. 2017. Available online: http://www.eubia.org/cms/wiki-biomass/biomass-characteristics/ (accessed on 15 June 2017).
62. Official Government Gazette of Hellenic Republic. *Measures to Support and Develop the Greek Economy in the Framework of Application of Law 4046/2012 and Other Provisions*; n. 4254/07.04.2014; Government of Greece: Athens, Greece, 2014. (In Greek)
63. IEA. *IEA Energy Technology Essentials*; IEA: Paris, France, 2007.
64. Chiantore, P.V.; Gordon, I.; Hoffmann, W.; Perezagua, E.; Philipps, S.; Roman, E.; Sandre, E.; Sink, W.; Simonot, E.; Martínez, A. *Future Renewable Energy Costs: Solar Photovoltaics*; KIC InnoEnergy: Eindhoven, The Netherlands, 2015.
65. Mayer, J.N. *Current and Future Cost of Photovoltaics*; Agora Energiewende: Berlin, Germany, 2015.

applied sciences

Article

The Impact of Hydrogeological Features on the Performance of Underground Pumped-Storage Hydropower (UPSH)

Estanislao Pujades [1,*], Angelique Poulain [2], Philippe Orban [3], Pascal Goderniaux [4] and Alain Dassargues [3] 

1   Institute of Environmental Assessment and Water Research (IDAEA), Severo Ochoa Excellence Center of the Spanish Council for Scientific Research (CSIC), Jordi Girona 18-26, 08034 Barcelona, Spain
2   Avignon University, UMR EMMAH, University of Avignon, 84000 Avignon, France; poulain.angelique@outlook.fr
3   Hydrogeology & Environmental Geology, Urban and Environmental Engineering research unit, University of Liège (Belgium), 4000 Liège, Belgium; P.Orban@uliege.be (P.O.); Alain.Dassargues@uliege.be (A.D.)
4   Geology and Applied Geology, Polytech Mons, University of Mons, 7000 Mons, Belgium; Pascal.Goderniaux@umons.ac.be
*   Correspondence: estanislao.pujades-garnes@geologist.com

Featured Application: This work evaluates the influence of groundwater exchanges occurring in the context of underground pumped storage hydropower using abandoned mines on the efficiency and on the environment. The findings are useful to define (1) design criteria of future underground pumped storage hydropower plants and (2) screening methodologies to choose the best places to construct them.

check for
updates

Citation: Pujades, E.; Poulain, A.; Orban, P.; Goderniaux, P.; Dassargues, A. The Impact of Hydrogeological Features on the Performance of Underground Pumped-Storage Hydropower (UPSH). *Appl. Sci.* **2021**, *11*, 1760. https://doi.org/10.3390/app11041760

Academic Editors: Jorge Loredo and Javier Menéndez

Received: 21 January 2021
Accepted: 13 February 2021
Published: 17 February 2021

Publisher's Note: MDPI stays neutral with regard to jurisdictional claims in published maps and institutional affiliations.

Abstract: Underground pumped storage hydropower (UPSH) is an attractive opportunity to manage the production of electricity from renewable energy sources in flat regions, which will contribute to the expansion of their use and, thus, to mitigating the emissions of greenhouse gasses (GHGs) in the atmosphere. A logical option to construct future UPSH plants consists of taking advantage of existing underground cavities excavated with mining purposes. However, mines are not waterproofed, and there will be an underground water exchange between the surrounding geological medium and the UPSH plants, which can impact their efficiency and the quality of nearby water bodies. Underground water exchanges depend on hydrogeological features, such as the hydrogeological properties and the groundwater characteristics and behavior. In this paper, we numerically investigated how the hydraulic conductivity (K) of the surrounding underground medium and the elevation of the piezometric head determined the underground water exchanges and their associated consequences. The results indicated that the efficiency and environmental impacts on surface water bodies became worse in transmissive geological media with a high elevation of the piezometric head. However, the expected environmental impacts on the underground medium increased as the piezometric head became deeper. This assessment complements previous ones developed in the same field and contributes to the definition of (1) screening strategies for selecting the best places to construct future UPSH plants and (2) design criteria to improve their efficiency and minimize their impacts.

Keywords: energy storage; renewable energy; hydropower; mine; groundwater; numerical modelling; environmental impacts; efficiency

1. Introduction

Renewable energies, such as solar or wind, may not be sufficiently efficient since they are intermittent and random, and consequently, their production of electricity is not adapted to the demand [1–4]. For this reason, they must be combined with energy storage systems (ESSs) [5] that allow for balancing the production and the demand [6]. ESSs are useful to store the surplus of electricity during periods of low demand and to generate electricity when the demand increases. Pumped storage hydropower (PSH) is the most worldwide used EES [7] because it allows for the storage and production of large amounts of electricity [8]. For example, about 95% of the utility-scale energy storage in the United States is PSH [9], and up to 99% in the European Union [10].

PSH plants consist of two reservoirs placed at different elevations (upper and lower reservoirs). The excess of electricity during low demand periods is stored in the form of potential energy by pumping water from the lower to the upper reservoir. Later, during high demand periods, electricity is produced by discharging the water through turbines from the upper to the lower reservoir [11]. Despite its extensive use, PSH has limitations [12–14], the most important being that a specific topography is required as both reservoirs must be located at different elevations [15]. Consequently, PSH plants can only be installed in relatively steep areas [16].

Underground pumped storage hydropower (UPSH) [17] is an opportunity to increase the capacity of managing the electrical production in areas where a conventional PSH is not possible. In addition, UPSH avoids some of the adverse environmental impacts related to PSH, and to hydropower in general, such as modifying the flow discharge in a river or changing the seasonal flow regime [18,19]. UPSH uses an underground cavity as the lower reservoir (underground reservoir) and constructs the upper reservoir at the surface [20] or, alternatively, at a shallow depth.

While the underground reservoir can be specifically excavated [21], the most inexpensive (i.e., efficient) option can be to take advantage of abandoned underground mines [22,23]. In addition, there are numerous mines that could be potentially used for UPSH. For example, in France, there are up to 4710 active mines and 101,616 abandoned mines [24], and, in Belgium, there are 964 active mines and more than 5000 abandoned mines [25]. Clearly, not all of these mines are suitable for constructing an UPSH plant; however, the objectives of electricity production and storage could be reached by using only a small portion of them.

For example, in France, these objectives could be reached by using 0.1% of the total available mines [26], and, in Belgium, it would be possible to obtain up to e 4896 MWh considering only mines with suitable characteristics for UPSH [27]. However, since mines are generally not waterproofed, it is expected that water exchanges will occur between the underground reservoir of UPSH plants and the surrounding groundwater systems [28]. This fact may entail negative consequences in terms of the environmental impacts [29–31] and for the efficiency (η) of UPSH [32]. We refer to η as the ratio between the energy used for pumping water from the underground reservoir and the energy generated when water is discharged from the upper reservoir under ideal conditions. Thus, energy losses due to conversion issues are not considered.

Recently, researchers observed that water exchanges may progressively fill the underground reservoir, reducing η. Occasionally, a volume of pumped water could actually not be fully discharged into the underground reservoir because the latter has been partially filled by underground water exchanges [33]. In addition, this volume of water must then be discharged into surface water systems, which could alter their quality because mine water is often not of an appropriate quality. If the released water is to impact the quality of the water bodies, it must be treated before its release to fulfill the current regulations concerning the water quality, such as the Water Framework Directive [34].

This decision should be taken based on the chemical composition of the water pumped from the mine and the expected reactions when mixing with surface water. If a treatment is needed, the additional investment required negatively affects the overall efficiency of the UPSH. Therefore, water exchanges with the surrounding geological medium are of paramount importance and must be investigated. Theoretically, these water exchanges depend on the local hydrogeological characteristics. Therefore, these features should play an essential role in the performance of UPSH influencing η and the potential environmental impacts. However, no studies were found that were focused on analyzing how hydrogeological properties influence water exchanges and their associated consequences. This information, however, appears to be crucial to define screening methodologies and to determine the best locations, in terms of η and the environmental impacts, where future UPSH plants could be constructed.

Thus, the objective of this paper was to determine the role of hydrogeological features (i.e., hydraulic conductivity and piezometric head) on the groundwater exchanges occurring in the context of UPSH and how they influence the efficiency of UPSH plants and their associated environmental impacts. This objective was reached by comparing the numerical results of different simulated scenarios based on an abandoned mine in Belgium that potentially could be used for constructing an UPSH plant.

The objective of this investigation was not to ascertain the system behavior at a specific site. The final goal was to provide a set of criteria to be considered during the design of future UPSH plants in these types of mining exploitation, to increase their efficiency and decrease the potential environmental impacts. Therefore, although the investigation was based on a real abandoned mine, the numerical models were purposely simplified to allow for determining the role of the different variables in the system behavior and extrapolating the main findings.

The main novelty of this work is that we investigated how the η of UPSH plants, and their associated environmental impacts vary depending on the hydraulic conductivity (K) of the surrounding medium and on the relative elevation of the piezometric head. This information, which has not yet been considered, will be crucial for designing future UPSH plants by taking advantage of abandoned mines.

2. Materials and Methods

2.1. Problem Statement

The groundwater model was based on the characteristics of an abandoned mine located in Martelange in south-east Belgium (Figure 1). This abandoned mine could be used for the construction of a UPSH plant.

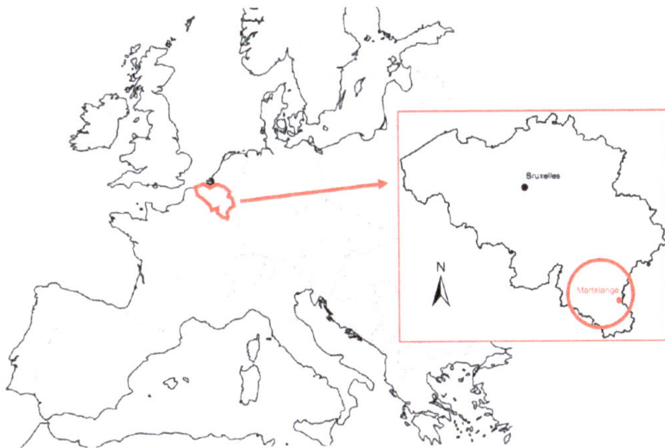

Figure 1. General view of Europe with Belgium highlighted with a red line (on the left) and a detailed view of Belgium (on the right) indicating the location of the considered mine (Martelange).

The mine of Martelange was developed to extract metamorphic slates from lower Devonian formations in the Ardenne anticlinorium. Specifically, from the "Formation de La Roche". The formation of these fractured slates started at the Lower Devonian, when the transgressive seas of the lower Devonian were at their maximum and clays and silts were deposited. Afterward, the clays and silt deposits were affected by different stages of deformation and metamorphism to become a dark fractured slate containing a thin bed of quartzites. The main slate cleavage (schistosity) was induced orthogonally to the main stress conditions during metamorphism phases but was not actually parallel to the bedding plane.

The exploitable layers had a dip between 55° and 66° [35]. Concerning the hydrogeological characteristics of the site, reference data was derived from previous works since, unfortunately, we did not have the opportunity to carry out hydraulic tests. According to previous works, these slates have a low global K ($\approx 10^{-7}$ m/s [36]), and groundwater flows through preferential flow channels in multiple fractures. Thus, the hydrogeological behavior of the formation depends strongly on the aperture, density, and connectivity of the fractures.

The specific storage coefficient was 10^{-4} m^{-1}, the saturated water content was 0.05, and the residual water content was 0.01. These parameters are typical of slate mines [22,37]. When the mining activities ceased, the piezometric head recovered, flooding the mine because its natural position is near the top of the mined cavities. The terms "hydraulic head" and "piezometric head" are used from this point forward to refer the water head inside the underground reservoir and the groundwater head, respectively.

The underground cavity roughly consists of nine adjacent and vertical chambers (CH) that are connected through galleries. The volume of the chambers varies as they have different heights. Their width and length are, approximately, 15 and 45 m, respectively, while their heights vary from 70 to 110 m. The top of all chambers is located at the same depth (40 m below the surface), whilst their base depth decreases progressively from CH1 to CH9, with a decrement of about 5 m (Figure 2). Thus, the bases of chambers CH1, CH2, CH3, CH4, CH5, CH6, CH7, CH8, and CH9 are 150, 145, 140, 135, 130, 125, 120, 115, and 110 m deep, respectively.

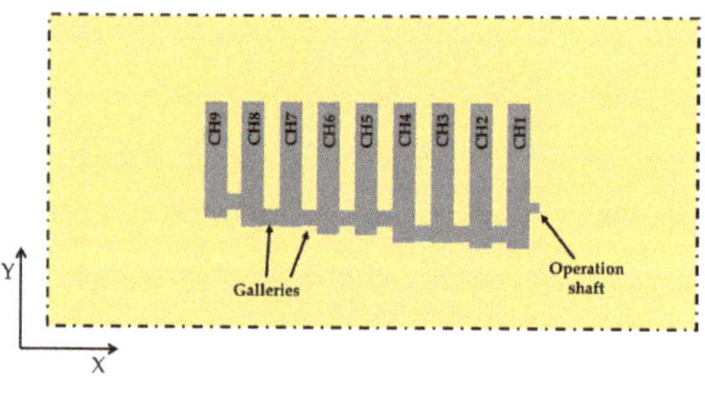

(a)

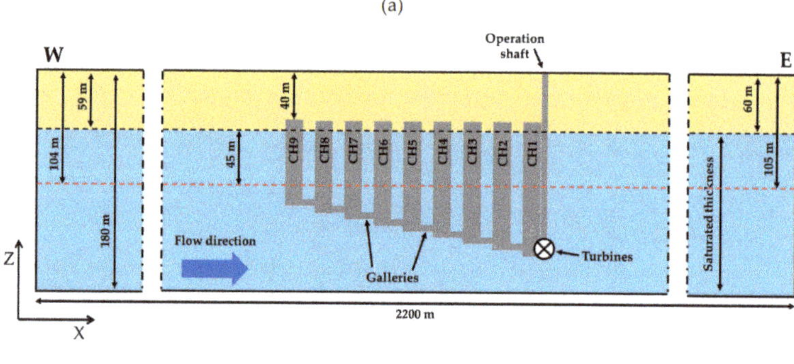

(b)

Figure 2. Schematic plan view (**a**) and cross section (**b**) of the mine in Martelange that is considered in this study. The black and red dashed lines in (**b**) indicate the natural position of the piezometric head considered in the two simulated scenarios. The black line indicates the scenarios called TOP, whilst the red line indicates the scenarios denoted as MIDDLE. The pictures taken inside the chambers can be checked at http://tchorski.morkitu.org/2/martelange-02.htm.

A vertical 170-m-deep extraction shaft connects the base of CH1 with the surface [38]. Approximately, we calculated that a volume of 400,000 m^3 could be potentially used for UPSH. This value was obtained by considering that (1) the top of the chambers is not exceeded by the hydraulic head, and (2) 10% of the maximum available volume is not used (i.e., pumped) to avoid total emptying of the reservoir (the underground reservoir is not totally emptied to avoid the pumps and turbines being out of the water). Consequently, this mine has a high water capacity.

If a surface reservoir was constructed strategically 500 m away in the northwest direction [38], it could be possible to reach a mean effective hydraulic head difference of 215 m between the underground and the upper reservoirs. Thus, a large amount of electricity could be stored and produced. Assuming an average pumping–discharge rate of 6 m^3/s, the available power may reach up to 10^4 MW (this value may vary depending on the considered efficiency for the pumps and turbines). Figure 2 shows a simplified plain view (2a) and cross section (2b) of the modeled mine.

2.2. Description of the Numerical Model

2.2.1. Code

SUFT3D [39,40] is the finite element numerical code we used to develop the groundwater numerical model. This code solves the groundwater flow equation (Equation (1)) based on a mixed formulation of Richard's equation proposed by Celia et al. [41] using the control volume finite element (CVFE):

$$\frac{\partial \theta}{\partial t} = \nabla \cdot \underline{\underline{K}}(\theta)\nabla h + \nabla \cdot \underline{\underline{K}}(\theta)\nabla z + q, \tag{1}$$

where t is the time [T], θ is the water content [-], z is the elevation [L], h is the pressure head [L], q is a source/sink term [T^{-1}], and $\underline{\underline{K}}$ is the hydraulic conductivity tensor [LT^{-1}] defined as

$$\underline{\underline{K}} = K_r \underline{\underline{K}}_S, \tag{2}$$

where $\underline{\underline{K}}_S$ is the saturated permeability tensor [LT^{-1}], and K_r is the relative hydraulic conductivity [-] that varies from a value of 1 for full saturation to a value of 0 when the water phase is considered immobilized [42]. In the partially saturated zone, the value of K_r evolves according with the following equations [43]:

$$\theta = \theta_r \frac{(\theta_s - \theta_r)}{(h_b - h_a)}(h - h_a), \tag{3}$$

$$K_r(\theta) = \frac{\theta - \theta_r}{\theta_s - \theta_r} \tag{4}$$

where θ_r is the residual water content [-], θ_s is the saturated water content [-], h_a is the pressure head at which the water content is just lower than the saturated one [L], and h_b is the pressure head at which the water content is the same as the residual one [L]. The K_r varies linearly between the unsaturated and saturated zones as can be observed in Equations (3) and (4). This adopted linearity for defining the transition between saturated and unsaturated zones does not alter the results of the model, because this work is focused on processes that occurred only in the saturated portion of the soil, while this contributed to mitigate the convergence errors that are common when non-linear expressions are used.

The main reason we choose SUFT3D is because it has certain capabilities specifically designed for modelling underground mines, improving the realism and the results of the groundwater numerical model. Specifically, underground cavities were simulated as linear reservoirs using the hybrid finite element mixing cell (HFEMC) method [39,40] implemented in the SUFT3D code [44–46]. This method combines physically-based and spatially distributed models as well as black-box models. The domain can be divided into different subdomains depending on their characteristics with specific behaviors assigned depending on their nature.

Groundwater processes through unmined areas, which were modeled with finite elements, were computed according to the flow equation in variably saturated porous media (Equation (1). Single mixing cells were used to discretize the underground cavities (i.e., chambers) that are modelled as linear reservoirs, which is similar to the box model approaches that consider a mean hydraulic head for the whole cell. The groundwater exchange between the domains modelled as linear reservoirs and those modelled as porous medium varies linearly [47] and is governed by the following internal dynamic Fourier boundary condition (BC) [40]:

$$Q_i = \alpha' A \left(h_{aq} - h_{ur} \right), \tag{5}$$

where h_{aq} is the piezometric head in the aquifer [L], h_{ur} is the hydraulic head in the underground reservoir [L], Q_i is the exchanged flow [L^{3}T^{-1}], A is the exchange area [L^2], and α' is the exchange coefficient [T^{-1}]. It is important to highlight that the water velocity inside the mixing cell is not considered; however, the influence of this particularity on the results was minimized by using different linear reservoirs for modelling the different chambers.

SUFT3D was also used because it allows adopting virtual connections that are essential for the purpose of this investigation. Virtual connections, that are also named "by-pass", allow the establishment of hydraulic connections between non-adjacent subdomains that are modeled as linear reservoirs. Virtual connections are defined by a first-order transfer equation (Equation (5) that can be switched off or on according to the hydraulic head difference between the two connected subdomains:

$$Q_{vr} = \alpha_{vr} \left(h_{SDj} - h_{SDi} \right), \tag{6}$$

where h_{SDj} and h_{SDi} are the hydraulic heads inside each of the connected linear reservoirs, Q_{vr} is the flow between reservoirs, and α_{vr} is the exchange coefficient of the virtual connection [L^{2}T^{-1}]. Virtual connections allow constraining the maximum and minimum hydraulic heads into the underground reservoir. They are crucial for the development of the model, as it is not possible to anticipate the water exchanges and, therefore, when the underground reservoir will be full or empty.

Consequently, it is not possible to predict if a pumping or discharge can be carried out. Two types of virtual connections are implemented. The first one extracts immediately and automatically the discharged water when the underground reservoir is full. The virtual connection is switched off for most of the time, and it is only switched-on when the underground reservoir is completely full. At this moment, a value of $10^6\ m^2/d$ is adopted for α_{vr} to force the surplus of discharged water to be extracted. The second virtual connection is used to avoid the hydraulic head being lower than the bottom of each chamber. In this case, the connection is switched on most of the time ($\alpha_{vr} = 10^6\ m^2/d$ to allow the hydraulic connection between chambers). Only when the hydraulic head is at a lower elevation than the bottom of a chamber is the virtual connection deactivated, thus, disconnecting individually each chamber from the rest of the underground reservoir if the hydraulic head is too low.

2.2.2. Characteristics of the Model

The numerical model is a squared domain with a thickness of 180 m and a side of 2200 m (Figures 2 and 3). The simulated mine is placed just in the center of the model. Thus, the distance between the underground reservoir and the outer boundaries is of 1000 m. This distance allows to minimize the effects of the outer BCs on the results. The underground reservoir is represented by the nine underground chambers as described above (CH1 to CH9) that are hydraulically connected by galleries. The operation shaft is modeled using a rectangular prism adjacent to CH1 that connects the mine to the surface (Figures 2 and 3).

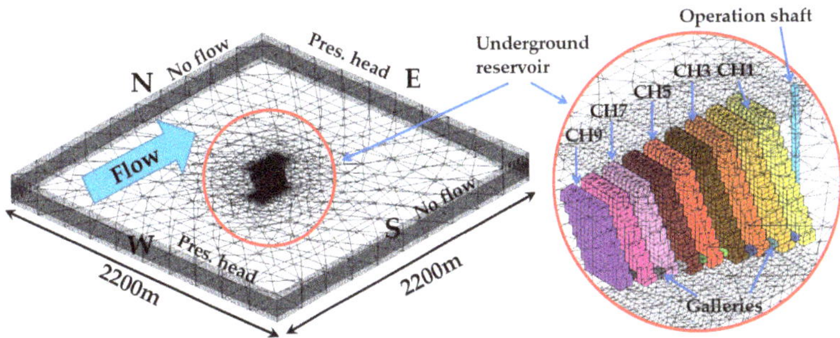

Figure 3. A detailed view of the modelled mine is shown on the right while a general view of the numerical model is shown on the left side of the figure. It is possible to observe how the bottom of the chambers increases progressively from west to east.

The chambers and the operation shaft are modelled as independent domains that behave as linear reservoirs. Hydrogeologically, the groundwater flow behavior in the Martelange site is governed by fractures. However, it is known that a fracture-based groundwater flow is particularly difficult to implement in classical groundwater numerical models. In these types of cases, the underground medium is modeled using an equivalent porous medium (EPM) approach [42]. EPM approaches have been proven to be suitable to estimate the global groundwater behavior by numerous authors, including [48,49], particularly in sites with the presence of a high density of fractures as in the study site.

Concerning the hydraulic parameters, the K was modified in four different scenarios to observe its influence on the system behavior (note that the given EPM approach is considered, and K refers to the equivalent hydraulic conductivity). The values adopted for K in the different scenarios are specified in Section 2.2.3. The other hydraulic parameters were the same in all scenarios: the saturated and residual water contents were 0.05 and 0.01, respectively, while the specific storage coefficient was 10^{-4} m^{-1}. The chosen parameters are typical of slate mines [22,37].

The spatial and temporal discretizations were as follows: The domain was divided vertically in 29 layers and horizontally discretized using 3D prismatic elements. The maximum horizontal size of the elements was 150 m at the outer boundaries, and the mesh was refined around the mine where the horizontal size was about 5 m (Figure 3). The model was composed of 64,844 nodes and 38,680 elements. The total simulation length was one year and was divided into constant time steps of 15 minutes. Larger time steps would induce convergence problems and errors.

Three different types of BCs were implemented. First, Dirichlet BCs were used to prescribe the piezometric head on the W and E outer boundaries of the model. Two different hypotheses (scenarios TOP and MIDDLE) were considered to assess the influence of the elevation of the piezometric head on the system performance. In the TOP scenarios, the piezometric head was fixed at an elevation (i.e., with respect to the bottom of the model) of 121 and 120 m on the upgradient (W) and downgradient (E) sides, respectively.

Consequently, the mine was totally full of water in the natural conditions. In the MIDDLE scenarios, groundwater head was prescribed at 76 and 75 m on the upgradient (W) and downgradient (E) sides, respectively. In this case, only half of the volume of the mine was full of water under natural conditions. As a result of the prescribed heads, the hydraulic gradient was 4.6×10^{-4} for both hypotheses, and the groundwater flowed from W to E (Figure 2). The BCs implemented at the outer boundaries did not change through the simulations. Secondly, no-flow BCs were assigned to the top and the bottom of the modeled domain and to the N and S boundaries.

Finally, pumping and discharge operations were simulated by use of a Neuman BC prescribing discharged or pumped water from the underground reservoir through the operation shaft. The value of the pumping and discharge rates were 5.94 m^3/s, which is

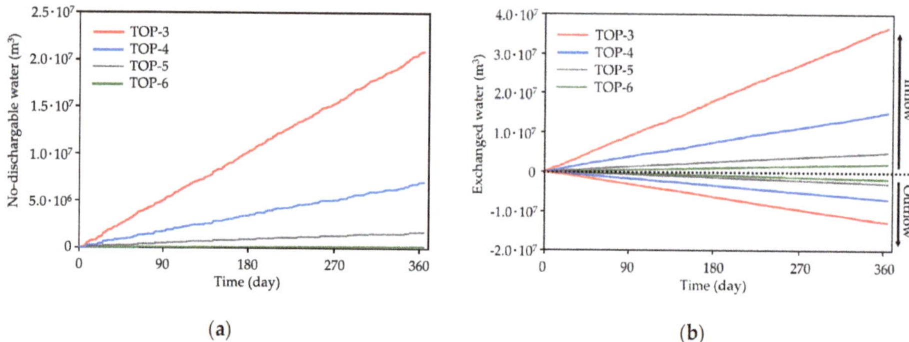

Figure 5. Volume of water that cannot be discharged into the underground reservoir (a) and accumulated difference of water exchanges between the mine and the surrounding groundwater system (b) for the TOP scenarios.

Therefore, the total volume of water that inflows into the underground reservoir was larger than that flowing out (Figure 5b). Consequently, the underground reservoir was filled partially, and a portion of the pumped water could not be discharged. As expected, the water exchanges and, therefore, the volume of non-dischargeable water increased with K (Figure 5b). The accumulated volume of non-dischargeable water over a year varied between 2.1×10^7 and 2.2×10^5 m^3 for the scenarios TOP-6 and TOP-3.

Considering that non-dischargeable water should be discharged into surface water systems, the environmental impacts would be considered as increasing as the surrounding medium becomes more permeable. Environmental impacts on the underground medium would be also higher for scenario TOP-3, since water exchanges between the mine and the surrounding groundwater system are also higher, i.e., changes in the hydraulic head elevation or in the water chemistry are easily transmitted to the groundwater in the surrounding medium.

3.1.2. MIDDLE Scenarios (Influence of k)

Figure 6 shows the evolution of the non-dischargeable volume of water (a) and the accumulated volume of water exchanges in both directions (i.e., the inflows and outflows) (b) for the MIDDLE scenarios. In this case, the influence of K was lower than in TOP scenarios and the non-dischargeable volume of water did not evolve proportionally with k. This behavior was related to the elevation of the hydraulic head with respect to the piezometric head in the surrounding medium. Given that the piezometric head was located at a half depth, the hydraulic head in the mine was sometimes higher and sometimes lower, depending on the operation schedule of the plant.

Therefore, in contrast to the TOP scenarios, similar volumes of water were exchanged in both directions (toward the underground medium and toward the underground reservoir) (Figure 6b). Thus, the inflows and outflows were more equilibrated than in the TOP scenarios, and less water was accumulated at the surface reservoir. In scenario MID-3, non-dischargeable water was not accumulated because the surrounding medium was so permeable that pumping and discharge did not modify the hydraulic head greatly because pumped water is quickly replaced by water from the surrounding medium or discharged water is transferred quickly to it. In addition, when the hydraulic head is modified by consecutive pumping or discharge periods, it returns quickly to the elevation of the piezometric head after the cessation of these periods.

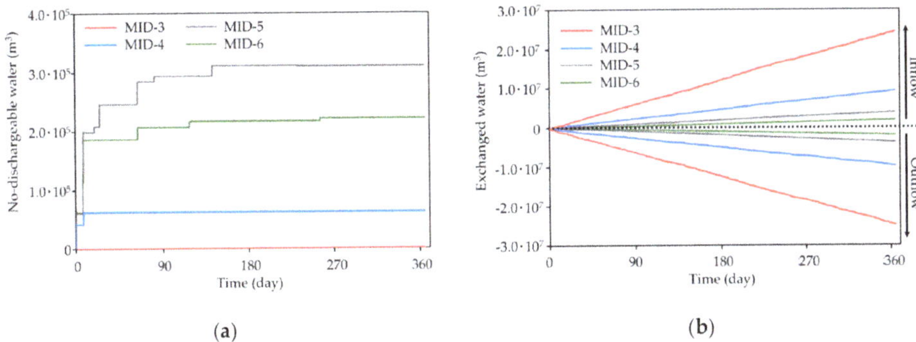

Figure 6. Volume of water that cannot be discharged into the underground reservoir (**a**) and the accumulated difference of water exchanges between the mine and the surrounding groundwater system (**b**) for the MIDDLE scenarios.

As a result, the hydraulic head never reaches the top of the underground reservoir and water can be always discharged. This behavior is also reflected in Figure 6b where it is possible to observe the high volume of water exchanged in both directions. In scenario MID-4, the volume of non-dischargeable water increased slightly since the water exchanges were constrained by the K of the surrounding groundwater system, and the top of the chambers was reached occasionally as a consequence of consecutive discharge periods. This is the same reason for which a volume of non-dischargeable water is accumulated in scenario MID-5; however, in this case, the final volume was larger than in scenario MID-4 because the K was lower, and thus, the water exchanges were more constrained.

However, the trend changed for scenario MID-6 since the non-discharged volume of water decreased with respect to the scenario MID-5. Contrary to the observed trend in scenarios MID-3, MID-4, and MID-5, the non-discharged volume of water decreased when K was reduced more than 10^{-5} m/s (i.e., scenario MID-6). In scenario MID-6, the water exchanges were very low (Figure 6b) due to the value of k, and the system response was more similar to that of an isolated underground reservoir. Given that the operation scenarios were designed considering an isolated underground reservoir, the head evolved as expected and the top of the underground reservoir was not exceeded during most of the simulated time. If the K was reduced lower than 10^{-6} m/s, the non-discharged volume of water would decrease even more.

Concerning the environmental impacts into surface water bodies, the largest impacts were expected for scenario MID-5; however, they were much smaller than those in the TOP scenarios. Nevertheless, the largest environmental impacts in the underground medium were expected for scenario MID-3. Despite the fact that all the pumped water can be discharged into the underground reservoir, the results showed that water exchanges were higher than in the other scenarios, which indicates that the interaction between the UPSH plant and the surrounding materials, and therefore the impact on the groundwater, was high.

3.1.3. Influence of the Piezometric Head Elevation

The comparison between Figures 5 and 6 shows that the volumes of non-dischargeable water were higher in the TOP scenarios. This means that expected environmental impacts on surface water bodies increased with the higher elevation of the piezometric head. Concerning the impacts on the surrounding groundwater head distribution, they would be similar with different elevations of the piezometric head. The magnitude of the produced oscillations should depend only on the value of K (magnitudes are proportional to k) but not on the initial depth of the piezometric head.

However, in the TOP scenarios, the piezometric head will oscillate systematically below the elevation of the natural piezometric head, whilst, in the MIDDLE scenarios, the piezometric head will oscillate around the natural position of the piezometric head. With regard to the impacts on the groundwater quality, they were expected to be larger when the piezometric head was lower since the magnitude of the outflow increased, which can be deduced by comparing Figures 5b and 6b, and therefore, the spreading of mine water in the surrounding medium also increased. If the piezometric head is at the top, the hydraulic head is mostly located below it, and therefore, groundwater from the aquifer tends to flow toward the mine minimizing the spreading of mine water in the surrounding medium.

3.2. Variations in η

In this section, we analyze the influence of the water exchanges on the global η of UPSH plants. Considering that the pumped volume of water was the same in all scenarios, the variation of η over a year was computed by comparing the total discharged water obtained from the simulations by the total discharged water under ideal conditions (i.e., without the existence of groundwater exchanges). Table 1 shows the variation in the η for all the simulated scenarios with respect to the ideal reference scenario. Overall, the η was less affected by groundwater exchanges in the MIDDLE compared with in the TOP scenarios. In the TOP scenarios, the η was clearly related to the value of k, decreasing up to 37.3% in scenario TOP-3 with respect to the ideal reference scenario.

Table 1. Variations in the efficiency ($\Delta\eta$) with respect to an ideal reference scenario.

TOP Scenario	$\Delta\eta$ (%)	MIDDLE Scenario	$\Delta\eta$ (%)
TOP-3	−37.3	MID-3	0
TOP-4	−12.6	MID-4	−0.1
TOP-5	−3.1	MID-5	−0.6
TOP-6	−0.4	MID-6	−0.4

The results show that the variation in the η was acceptable for values of K smaller than 10^{-5} m/s; however, if K was increased above this value, the η decreased dramatically. The differences in η were much smaller in the MIDDLE scenarios, and all of them were acceptable. The η decreases in some scenarios were a consequence of exceptional large periods of discharge during which the underground reservoir was totally filled. In terms of η, a low piezometric head in the surrounding medium would be more favorable. However, additional problems could arise if the piezometric head was too low, as sometimes there would not be enough water to pump in the underground reservoir, and a portion of the excess of electricity could not be stored, decreasing the η of the plant.

3.3. Previous Works and Future Investigations

Previous studies have investigated the role of the hydrogeological parameters on the potential impacts induced by UPSH in the underground medium [16,30]. The present investigation goes further and shows how the K controls the potential environmental impacts on surface water bodies and on the global η of the plant. While the results are useful to understand the system behavior and to define screening methodologies, further investigation should be done regarding not only the influence of other hydrogeological parameters, such as the effective porosity, but also considering heterogeneities in the underground medium. In addition, the importance of the shape and volume of the available underground cavity should be assessed.

The results showed that, under certain circumstances, the discharge of large volumes of the pumped groundwater into surface water bodies would be needed. In addition, the effects on the piezometric head of pumping and discharge may alter the interactions between the groundwater and nearby surface water bodies. Consequently, it would be advisable to investigate how UPSH may influence the e-flows of nearby rivers [51] by considering UPSH plants during the environmental flow assessment processes.

This paper also investigated, for the first time, the relevance of the elevation of the piezometric head in terms of η and the environmental impacts. The results showed that the relative elevation of the piezometric head was crucial for the operation of a UPSH plant, and for this reason, it should certainly deserve more investigation in the future.

Finally, we assessed the influence of the water exchanges on the global η of UPSH plants. To date, only one study had investigated the relation between groundwater exchanges and η [32]. However, this previous investigation only considered the influence of groundwater exchanges on the η of pumps and turbines. While the findings of this previous study were relevant, the reported variations of η were much lower than those obtained in this paper.

The current investigation went further and considered the variations in the global η of UPSH by comparing the energy consumed for pumping water from the underground reservoir and that obtained by discharging water from the upper reservoir. Considering that the η of the pumps and turbines depends on the head difference between reservoirs, it is advisable to investigate the effect on their η of different positions for the piezometric head as in the present paper.

4. Conclusions

This is the first paper to assess the role of hydrogeological features (i.e., the K and piezometric head) on the groundwater exchanges that occur in the context of UPSH using abandoned mines and, therefore, on the η of UPSH plants and their associated environmental impacts. Despite the relevant η issues, those related to potential environmental impacts are paramount importance, as UPSH plants must fulfill the current regulations concerning water bodies. For example, in the European context, the Water Framework Directive [34] states that countries must preserve the "good state" of water bodies.

Considering that, under some scenarios, the water quality may be deteriorated under the influence of UPSH (e.g., if abandoned coal mines are used as underground reservoirs [29]) additional water treatments may be required before releasing the water surplus into surface water bodies. Overall, the final goal of this work was to improve the knowledge about the interactions between an UPSH and the groundwater to complement previous studies developed in this field and, in this way, to contribute to establish criteria for (1) the selection of abandoned mines that are most suitable for UPSH and (2) designing future UPSH plants. The main conclusions of this paper are the following.

- The K of the surrounding medium drove the groundwater exchanges. Consequently, K played an important role concerning the η of UPSH and its associated environmental impacts. Thus, K should be considered in the selection process of abandoned mines when constructing future UPSH plants.
- The influence exerted by K depended on the elevation of the piezometric head with respect to the mine. If the piezometric head was located at a high elevation, high values of K were harmful for the η and the environmental impacts. When the natural piezometric head was located at the half elevation of the mine, the K did not affect the η of UPSH nor the environmental impacts over surface water bodies; however, the impact could become important in terms of the groundwater quality, increasing with high values of K.
- The elevation of the piezometric head was relevant and must be considered when designing an UPSH plant. The results showed that, for the same values of K, the η was higher, and the environmental impacts over surface water bodies were lower if the piezometric head was located at a low elevation. However, the potential impacts on the groundwater increased, since the outflows from the underground reservoir to the surrounding geological medium increased with low piezometric heads. Consequently, an agreement between the η, the environmental impacts into surface water bodies, and those generated in nearby aquifer systems will be needed in order to choose potential sites to implement UPSH.

The pumping–discharge frequencies must be adapted according to the K and the position of the piezometric head in order to increase the efficiency of UPSH plants. If the hydrogeological parameters are not considered, large volumes of water could be not discharged into the underground reservoir under certain circumstances (i.e., large values K and high piezo metric heads). This would decrease the η and increase the environmental impacts on surface water bodies.

The model used in this investigation was purposely simplified to obtain general results applicable to other sites with similar features, which was the primary objective of this work. However, the consideration of a specific mine for constructing a UPSH plant requires site specific, detailed, and realistic numerical models. These models must consider all the characteristics of the site, such as the heterogeneity, the seasonal variations or the presence of fractures and faults. In addition, these models should be in 3D and should simulate the system behavior over a large period of years.

Author Contributions: Conceptualization, E.P.; methodology, E.P.; software, P.O.; investigation, E.P., A.P., P.O., P.G. and A.D.; resources, P.O., A.D.; data curation, E.P.; writing—original draft preparation, E.P., A.P., P.O., P.G. and A.D.; writing—review and editing, E.P., A.P., P.O., P.G. and A.D.; visualization, E.P.; supervision, A.D.; project administration, A.D, E.P.; funding acquisition, E.P. and A.D. All authors have read and agreed to the published version of the manuscript.

Funding: IDAEA-CSIC is a Centre of Excellence Severo Ochoa (Spanish Ministry of Science and Innovation, Project CEX2018-000794-S). This research was funded by the Public Service of Wallonia–Department of Energy and Sustainable Building through the Smartwater project. E.P. was also funded by the Barcelona City Council through the Award for Scientific Research into Urban Challenges in the City of Barcelona 2020 (20S08708).

Institutional Review Board Statement: Not applicable.

Informed Consent Statement: Not applicable.

Data Availability Statement: All analyzed data in this study has been included in the manuscript.

Conflicts of Interest: The authors declare no conflict of interest. The funders had no role in the design of the study; in the collection, analyses, or interpretation of data; in the writing of the manuscript; or in the decision to publish the results.

References

1. Hu, Y.; Bie, Z.; Ding, T.; Lin, Y. An NSGA-II Based Multi-Objective Optimization for Combined Gas and Electricity Network Expansion Planning. *Appl. Energy* **2016**, *167*, 280–293. [CrossRef]
2. Mileva, A.; Johnston, J.; Nelson, J.H.; Kammen, D.M. Power System Balancing for Deep Decarbonization of the Electricity Sector. *Appl. Energy* **2016**, *162*, 1001–1009. [CrossRef]
3. Okazaki, T.; Shirai, Y.; Nakamura, T. Concept Study of Wind Power Utilizing Direct Thermal Energy Conversion and Thermal Energy Storage. *Renew. Energy* **2015**, *83*, 332–338. [CrossRef]
4. Menéndez, J.; Fernández-Oro, J.M.; Loredo, J. Economic Feasibility of Underground Pumped Storage Hydropower Plants Providing Ancillary Services. *Appl. Sci.* **2020**, *10*, 3947. [CrossRef]
5. Gebretsadik, Y.; Fant, C.; Strzepek, K.; Arndt, C. Optimized Reservoir Operation Model of Regional Wind and Hydro Power Integration Case Study: Zambezi Basin and South Africa. *Appl. Energy* **2016**, *161*, 574–582. [CrossRef]
6. Delfanti, M.; Falabretti, D.; Merlo, M. Energy Storage for PV Power Plant Dispatching. *Renew. Energy* **2015**, *80*, 61–72. [CrossRef]
7. Zhang, N.; Lu, X.; McElroy, M.B.; Nielsen, C.P.; Chen, X.; Deng, Y.; Kang, C. Reducing Curtailment of Wind Electricity in China by Employing Electric Boilers for Heat and Pumped Hydro for Energy Storage. *Appl. Energy* **2016**, *184*, 987–994. [CrossRef]
8. Steffen, B. Prospects for Pumped-Hydro Storage in Germany. *Energy Policy* **2012**, *45*, 420–429. [CrossRef]
9. Pumped-Storage Hydropower. Available online: https://www.energy.gov/eere/water/pumped-storage-hydropower (accessed on 8 October 2020).
10. Menéndez, J.; Fernández-Oro, J.M.; Galdo, M.; Loredo, J. Efficiency Analysis of Underground Pumped Storage Hydropower Plants. *J. Energy Storage* **2020**, *28*, 101234. [CrossRef]
11. Hadjipaschalis, I.; Poullikkas, A.; Efthimiou, V. Overview of Current and Future Energy Storage Technologies for Electric Power Applications. *Renew. Sustain. Energy Rev.* **2009**, *13*, 1513–1522. [CrossRef]
12. Wong, I.H. An Underground Pumped Storage Scheme in the Bukit Timah Granite of Singapore. *Tunn. Undergr. Space Technol.* **1996**, *11*, 485–489. [CrossRef]
13. Menéndez, J.; Fernández-Oro, J.M.; Galdo, M.; Loredo, J. Transient Simulation of Underground Pumped Storage Hydropower Plants Operating in Pumping Mode. *Energies* **2020**, *13*, 1781. [CrossRef]
14. Menéndez, J.; Schmidt, F.; Konietzky, H.; Fernández-Oro, J.M.; Galdo, M.; Loredo, J.; Díaz-Aguado, M.B. Stability Analysis of the Underground Infrastructure for Pumped Storage Hydropower Plants in Closed Coal Mines. *Energies* **2019**, *94*, 103117. [CrossRef]

15. Mueller, S.C.; Sandner, P.G.; Welpe, I.M. Monitoring Innovation in Electrochemical Energy Storage Technologies: A Patent-Based Approach. *Appl. Energy* **2015**, *137*, 537–544. [CrossRef]
16. Pujades, E.; Willems, T.; Bodeux, S.; Orban, P.; Dassargues, A. Underground Pumped Storage Hydroelectricity Using Abandoned Works (Deep Mines or Open Pits) and the Impact on Groundwater Flow. *Hydrogeol. J.* **2016**, *24*, 1531–1546. [CrossRef]
17. Uddin, N.; Asce, M. Preliminary Design of an Underground Reservoir for Pumped Storage. *Geotech. Geol. Eng.* **2003**, *21*, 331–355. [CrossRef]
18. Kuriqi, A.; Pinheiro, A.N.; Sordo-Ward, A.; Garrote, L. Water-Energy-Ecosystem Nexus: Balancing Competing Interests at a Run-of-River Hydropower Plant Coupling a Hydrologic–Ecohydraulic Approach. *Energy Convers. Manage.* **2020**, *223*, 113267. [CrossRef]
19. Suwal, N.; Huang, X.; Kuriqi, A.; Chen, Y.; Pandey, K.P.; Bhattarai, K.P. Optimisation of Cascade Reservoir Operation Considering Environmental Flows for Different Environmental Management Classes. *Renew. Energy* **2020**, *158*, 453–464. [CrossRef]
20. Menéndez, J.; Schmidt, F.; Konietzky, H.; Bernardo Sánchez, A.; Loredo, J. Empirical Analysis and Geomechanical Modelling of an Underground Water Reservoir for Hydroelectric Power Plants. *Appl. Sci.* **2020**, *10*, 5853. [CrossRef]
21. Madlener, R.; Specht, J.M. An Exploratory Economic Analysis of Underground Pumped-Storage Hydro Power Plants in Abandoned Coal Mines. *SSRN J.* **2013**. [CrossRef]
22. Bodeux, S.; Pujades, E.; Orban, P.; Brouyère, S.; Dassargues, A. Interactions between Groundwater and the Cavity of an Old Slate Mine Used as Lower Reservoir of an UPSH (Underground Pumped Storage Hydroelectricity): A Modelling Approach. *Eng. Geol.* **2017**, *217*, 71–80. [CrossRef]
23. Menéndez, J.; Ordóñez, A.; Álvarez, R.; Loredo, J. Energy from Closed Mines: Underground Energy Storage and Geothermal Applications. *Renew. Sustain. Energy Rev.* **2019**, *108*, 498–512. [CrossRef]
24. Références et projets | BRGM. Available online: https://www.brgm.fr/fr/resultats-donnees/references-projets (accessed on 18 January 2021).
25. Poty, E.; Chevalier, E. *Wallonie L'activité Extractive En Wallonie: Situation Actuelle et Perspectives*; Laboratoire d'analyses litho et zoostratigraphiques: Liège, Belgium, 2004.
26. Gombert, P.; Poulain, A.; Goderniaux, P.; Orban, P.; Pujades, E.; Dassargues, A. Potentiel de valorisation de sites miniers et carriers en step en France et en Belgique. *LHB* **2020**, *4*, 33–42. [CrossRef]
27. Poulain, A. Etude de l'impact Sollicitations Cycliques Au Sein d'aquifères Non Confinés, à Différentes Échelles. Applications à Des Systèmes de Pompage–Turbinage En Carrières, Université de Mons. 2018. Available online: https://web.umons.ac.be/fpms/fr/evenements/etude-de-limpact-de-sollicitations-cycliques-au-sein-daquiferes-non-confines-a-differentes-echelles-applications-a-des-systemes-de-pompage-turbinage-en-carrieres-par/ (accessed on 18 January 2021).
28. Poulain, A.; Goderniaux, P.; de dreuzy, J.-R. Study of Groundwater-Quarry Interactions in the Context of Energy Storage Systems. **2016**, *18*, EPSC2016-9055. Available online: https://orbi.uliege.be/handle/2268/217789 (accessed on 18 January 2021).
29. Pujades, E.; Jurado, A.; Orban, P.; Ayora, C.; Poulain, A.; Goderniaux, P.; Brouyère, S.; Dassargues, A. Hydrochemical Changes Induced by Underground Pumped Storage Hydropower and Their Associated Impacts. *J. Hydrol.* **2018**, *563*, 927–941. [CrossRef]
30. Pujades, E.; Jurado, A.; Orban, P.; Dassargues, A. Parametric Assessment of Hydrochemical Changes Associated to Underground Pumped Hydropower Storage. *Sci. Total Environ.* **2019**, *659*, 599–611. [CrossRef]
31. Pujades, E.; Orban, P.; Jurado, A.; Ayora, C.; Brouyère, S.; Dassargues, A. Water Chemical Evolution in Underground Pumped Storage Hydropower Plants and Induced Consequences. *Energy Procedia* **2017**, *125*, 504–510. [CrossRef]
32. Pujades, E.; Orban, P.; Bodeux, S.; Archambeau, P.; Erpicum, S.; Dassargues, A. Underground Pumped Storage Hydropower Plants Using Open Pit Mines: How Do Groundwater Exchanges Influence the Efficiency? *Appl. Energy* **2017**, *190*, 135–146. [CrossRef]
33. Pujades, E.; Orban, P.; Archambeau, P.; Kitsikoudis, V.; Erpicum, S.; Dassargues, A. Underground Pumped-Storage Hydropower (UPSH) at the Martelange Mine (Belgium): Interactions with Groundwater Flow. *Energies* **2020**, *13*, 2353. [CrossRef]
34. Water Framework Directive. Available online: https://eur-lex.europa.eu/eli/dir/2000/60/oj (accessed on 4 February 2021).
35. Anciaux, V. The Ardennes–Slate Throughout. *de Leekëppert, Bulletin de Liaison* **2019**, 5. Available online: http://www.ardoise.lu/wp/wp-content/uploads/2019/11/2019-ARDENNES_1_AnciauxThilmany-VERSION-ANGLAISE-FINAL.pdf (accessed on 18 January 2021).
36. Bouezmarni, M.; Denne, P.; Debbaut, V. Carte hydrogéologique de Wallonie 1/25000. 65/3-4, Bastogne–Wardin et 65/7-8 Fauvillers–Romeldange. 2006. Available online: http://hdl.handle.net/2268/98657 (accessed on 18 January 2021).
37. Bear, J.; Cheng, A.H.-D. *Modeling Groundwater Flow and Contaminant Transport*; Theory and Applications of Transport in Porous Media; Springer: Dordrecht, Netherlands, 2010.
38. Erpicum, S.; Archambeau, P.; Dewals, B.; Pirotton, M.; Pujades, E.; Orban, P.; Dassargues, A.; Cerfontaine, B.; Charlier, R.; Poulain, A.; et al. Underground Pumped Hydro-Energy Storage in Wallonia (Belgium) Using Old Mines–Potential and Challenges. **2017**, 2984–2991. Available online: http://hdl.handle.net/2268/214205 (accessed on 18 January 2021).
39. Brouyère, S.; Orban, P.; Wildemeersch, S.; Couturier, J.; Gardin, N.; Dassargues, A. The Hybrid Finite Element Mixing Cell Method: A New Flexible Method for Modelling Mine Ground Water Problems. *Mine Water Environ.* **2009**, *28*, 102–114. [CrossRef]
40. Wildemeersch, S.; Brouyère, S.; Orban, P.; Couturier, J.; Dingelstadt, C.; Veschkens, M.; Dassargues, A. Application of the Hybrid Finite Element Mixing Cell Method to an Abandoned Coalfield in Belgium. *J. Hydrol.* **2010**, *392*, 188–200. [CrossRef]

41. Celia, M.A.; Bouloutas, E.T.; Zarba, R.L. A General Mass-Conservative Numerical Solution for the Unsaturated Flow Equation. *Water Resour. Res.* **1990**, *26*, 1483–1496. [CrossRef]
42. Dassargues, A. *Hydrogeology: Groundwater Science and Engineering*; CRC Press: Raton, FL, USA, 2018; ISBN 0-429-89440-6.
43. Yeh, G.T.; Cheng, J.R.; Cheng, H.P. *3DFEMFAT: A 3-Dimensional Finite Element Model of Density-Dependent Flow and Transport through Saturated-Unsaturated Media*; Department of Civil and Environmental Engineering, Pennsylvania State University: University Park, PA, USA, 1994.
44. Brouyère, S. Etude et Modélisation du Transport et du Piégeage des Solutés en Milieu Souterrain Variablement Saturé. Evaluation des Paramètres Hydrodispersifs par la Réalisation et L'interprétation D'essais de Traçage in Situ. Doctoral Thesis, Université de Liège, Sart Tilman, Belgique, 18 December 2001.
45. Brouyère, S.; Carabin, G.; Dassargues, A. Climate Change Impacts on Groundwater Resources: Modelled Deficits in a Chalky Aquifer, Geer Basin, Belgium. *Hydrogeol. J.* **2004**, *12*, 123–134. [CrossRef]
46. Carabin, G.; Dassargues, A. Modeling Groundwater with Ocean and River Interaction. *Water Resour. Res.* **1999**, *35*, 2347–2358. [CrossRef]
47. Orban, P.; Brouyère, S. *Groundwater Flow and Transport Delivered for Groundwater Quality Trend Forecasting by TREND T2*. 2006. Available online: http://hdl.handle.net/2268/78481 (accessed on 18 January 2021).
48. Scanlon, B.R.; Mace, R.E.; Barrett, M.E.; Smith, B. Can We Simulate Regional Groundwater Flow in a Karst System Using Equivalent Porous Media Models? Case Study, Barton Springs Edwards Aquifer, USA. *J. Hydrol.* **2003**, *276*, 137–158. [CrossRef]
49. Hassan, S.M.T.; Lubczynski, M.W.; Niswonger, R.G.; Su, Z. Surface–Groundwater Interactions in Hard Rocks in Sardon Catchment of Western Spain: An Integrated Modeling Approach. *J. Hydrol.* **2014**, *517*, 390–410. [CrossRef]
50. Kitsikoudis, V.; Archambeau, P.; Dewals, B.; Pujades, E.; Orban, P.; Dassargues, A.; Pirotton, M.; Erpicum, S. Underground Pumped-Storage Hydropower (UPSH) at the Martelange Mine (Belgium): Underground Reservoir Hydraulics. *Energies* **2020**, *13*, 3512. [CrossRef]
51. Suwal, N.; Kuriqi, A.; Huang, X.; Delgado, J.; Młyński, D.; Walega, A. Environmental Flows Assessment in Nepal: The Case of Kaligandaki River. *Sustainability* **2020**, *12*, 8766. [CrossRef]

 *applied sciences*

Article

Monitoring Scheme for the Detection of Hydrogen Leakage from a Deep Underground Storage. Part 1: On-Site Validation of an Experimental Protocol via the Combined Injection of Helium and Tracers into an Aquifer

Stéphane Lafortune [1,*], Philippe Gombert [1], Zbigniew Pokryszka [1], Elodie Lacroix [1,2], Philippe de Donato [2] and Nevila Jozja [3]

[1] Ineris, Parc Technologique Alata, 60550 Verneuil-en-Halatte, France; philippe.gombert@ineris.fr (P.G.); zbigniew.pokryszka@ineris.fr (Z.P.); elodie.lacroix@ineris.fr (E.L.)

[2] Faculté des Sciences et Technologies, Laboratoire GeoRessources, Campus des Aiguillettes, Université de Lorraine, 3B Rue Jacques Callot, 54500 Vandoeuvre-lès-Nancy, France; philippe.de-donato@univ-lorraine.fr

[3] Ecole Polytechnique, Laboratoire CETRAHE, Université d'Orléans, 8 rue Léonard de Vinci, 45100 Orléans, France; nevila.jozja@univ-orleans.fr

* Correspondence: stephane.lafortune@ineris.fr; Tel.: +33-(0)-344-556-791

Received: 1 August 2020; Accepted: 28 August 2020; Published: 1 September 2020

Abstract: Massive underground storage of hydrogen could be a way that excess energy is produced in the future, provided that the risks of leakage of this highly flammable gas are managed. The ROSTOCK-H research project plans to simulate a sudden hydrogen leak into an aquifer and to design suitable monitoring, by injecting dissolved hydrogen in the saturated zone of an experimental site. Prior to this, an injection test of tracers and helium-saturated water was carried out to validate the future protocol related to hydrogen. Helium exhibits a comparable physical behavior but is a non-flammable gas which is preferable for a protocol optimization test. The main questions covered the gas saturation conditions of the water, the injection protocol of 5 m^3 of gas saturated water, and the monitoring protocol. Due to the low solubility of both helium and hydrogen, it appears that plume dilution will be more important further than 20 m downstream of the injection well and that monitoring must be done close to the well. In the piezometer located 5 m downstream the injection well, the plume peak is intended to arrive about 1 h after injection with a concentration around 1.5 mg·L^{-1}. Taking these results into account should make it possible to complete the next injection of hydrogen.

Keywords: hydrogen; underground storage; leakage; monitoring; protocol; helium; aquifer

1. Introduction

1.1. General Information Regarding the Underground Storage of Hydrogen

To contribute more effectively to the fight against climate change and the preservation of the environment, as well as reinforcing their energy independence, France published the Energy transition law for green growth in 2015 [1]. This law aims to increase the share of renewable energies to 23% of gross final energy consumption in 2020 and 32% in 2030, compared to 16% currently [2]. The development of these renewable energies will come up against the need to manage the fluctuating or intermittent nature of some of them. This will involve storing the energy produced in excess or not consumed so that this energy can be re-used later (directly as fuel or mixed with natural gas, or indirectly by converting it into heat or electricity). The underground environment has many advantages with regard to its potential for high capacity storage in the short or medium-term [3]. France already has 100 operational underground reservoirs of which 78 are salt cavities: these are very large underground cavities, of the order of a million m^3, formed by injecting freshwater into deep salt formations. Currently, the storage capacity of all of these salt cavities together totals around 14 million m^3 of liquid or liquefied hydrocarbons and 2 billion m^3 of natural gas [3]. Against a background of the gradual abandonment of fossil fuels, a number of research studies are looking into the possibility of storing hydrogen (H_2) in such deep salt cavities in the future.

It is within this context that the ROSTOCK-H project (Risks and Opportunities of the Geological Storage of Hydrogen in Salt Caverns in France and Europe) has been financed by GEODENERGIES the French Scientific Interest Group. This project started in 2017 and will end in 2021. One of its objectives is to define monitoring methods for the detection of sub-surface hydrogen leakage, with the dual aim of (i) sizing a measurement scheme capable of detecting diffuse hydrogen leaks, and (ii) studying the diffusive process and the chemico-physical impacts of hydrogen in a shallow aquifer. The approach is centered on two experimental simulations separated in time on the same experimental site. Simulation 1 consists of analyzing the migration in groundwater of a plume of water saturated with neutral gas (helium) and containing various tracers. The objective is to test the operation protocol envisaged for the future injection of hydrogen and to optimize the associated monitoring systems. Simulation 2 consists of creating a plume of dissolved hydrogen in groundwater, according to the same protocol used with helium, to simulate a sudden and brief leak from a deep geological hydrogen storage site towards a shallow aquifer. The evolution of the plumes thus created in the saturated zone, and any potential outgassing to non-saturated zone and the surface will then be followed. All of these simulations will take place at the Catenoy (Northern France) experimental site, which has already been used in the context of similar experiments that studied the behavior of CO_2 for the purpose of Carbon Capture and Storage, or CCS [4,5].

The first injection simulation, which is the subject of this article, therefore involves helium and aims to size the entire leak simulation system and to adapt its protocol and monitoring for the simulation of a hydrogen leak which will subsequently be carried out on the same experimental site. Helium, the gas is chosen for this test, exhibits a physical behavior similar to that of hydrogen, in particular a very low solubility and a high diffusion coefficient in water. At the same time, it is a non-flammable gas, as opposed to the highly flammable hydrogen. This fact makes the organization of this pre-test less complicated from a safety point of view while respecting the similarities with the future hydrogen experiment.

The test site is located in the chalk layer within the Paris Basin. The protocol adopted consists of extracting water from the shallow aquifer, saturating it with gas (helium), and then reinjecting it into the aquifer with tracers to follow the propagation of the dissolved gas plume. This test aims to improve the experimental protocol to be used for the subsequent experiment involving injecting dissolved hydrogen into the aquifer (simulation no. 2).

1.2. Risks Associated with Underground Hydrogen Storage

If there is a leak coming from a deep geological reservoir, the gas will migrate to the surface. In most cases, it will encounter at least one aquifer before reaching the surface [6]. If the leakage rate exceeds the dissolution potential of hydrogen in the groundwater, which is of the order of 2 mg·L^{-1} at surface conditions, which is low compared to other gases (11 mg·L^{-1} for dioxygen, 24 mg·L^{-1} for dinitrogen, 2500 mg·L^{-1} for carbon dioxide), part of the hydrogen will continue its migration to the surface. Hydrogen is then likely to accumulate in a confined underground area near the reservoir (cellar, underground car park, urban underground network, tunnel, etc.) where it will become a risk factor for explosion, fire, or asphyxiation. Indeed, hydrogen is a highly flammable gas with a very wide explosive range of between 4% and 75% at ambient pressure and temperature [7].

In the event of a potential hydrogen leak, the aquifer, therefore, represents the last warning barrier on the path of migration to the surface [6]. By transporting information from upstream to downstream, the aquifer constitutes a very favorable environment for the implementation of an integrated monitoring system immediately downstream of a deep storage site. As dissolved hydrogen is not normally present in water, detecting it within an aquifer will indicate a potential leak. This could be manifested as a direct detection (H_2 dissolved in water) or indirectly by means of the effects caused by this strongly reducing gas: decrease of the oxidation-reduction potential, decrease in the content of other dissolved gases in the water (mainly N_2, O_2, and CO_2), and oxidation-reduction reactions, for example [8–14]:

- reduction of nitrates (NO_3^-) to nitrites (NO_2^-), or even to ammonium (NH_4^+), and then to gaseous nitrogen (N_2);
- reduction of sulfates (SO_4^{2-}) to sulfides (SO_3^-), or even to hydrogen sulfide (H_2S);
- reduction of iron III to iron II;
- dissolving of metallic trace elements, if they are present in the aquifer rock, following the lowering of the oxidation-reduction potential.

The literature shows that, under normal pressure and temperature conditions, the reduction of nitrates and sulfates cannot take place except in the presence of a catalyst such as iron, copper, nickel, or platinum [8–14]. However, the frequent use of stainless steel, which contains iron and a significant amount of nickel (up to 20%), in the metal casings of a large number of water boreholes (for drinking water, mineral water, etc.) and hydrocarbon wells inevitably brings some of these catalysts into contact with the groundwater.

2. Materials and Methods

2.1. Presenting the Catenoy Site

The Catenoy (Oise) experimental site is located about 50 km north of Paris in the Paris sedimentary basin (Figure 1). The coordinates of the center of the site are as follows: latitude 49°22′05″ N, longitude 2°30′26″ E, altitude ~60 m asl. The geology corresponds to a few meters of Quaternary deposits (colluviums, loess) and Tertiary formations, lying over a hundred meters of Senonian chalk that is only visible in the thalwegs (see also Figure 2b). Under the site, the underground geology of the first 25 m is 3 m of colluvium, 4 m of Thanetian sands, and 18 m of chalk. At a few hundred meters on the south of the site, Ypresian and Lutetian layers form a hill. There are no major tectonic accidents nearby and the bedding of the chalk formation is horizontal. This chalk encloses an aquifer with a static level at a depth of 13 m which flows in the WSW-ENE direction [4].

Located in a former agricultural field that has been fallow for more than a decade, the site is equipped with 8 piezometers aligned in the direction of flow of the aquifer over a distance of 80 m, referenced PZ1 to PZ6 (including supplementary piezometers PZ2BIS and PZ2TER), plus a technical shed housing the monitoring material (Figure 2). The piezometers are about 25 m depth and are screened in the aquifer starting from 13 m depth.

The hydrodynamic characteristics of the chalk aquifer at the site were determined during a pumping test carried out in 2013 [4]. Depending on the piezometer considered, the following values were reported:

- Storage coefficient (porosity): 1.1×10^{-2} to 6.5×10^{-2}
- Hydraulic conductivity (permeability): 6.4×10^{-4} to 1.4×10^{-3} m·s^{-1}

In addition, a previous tracing test made it possible to estimate the flow rate of the aquifer at 3 m·d^{-1} at PZ3 and PZ5, and 10 m·d^{-1} at PZ4, the latter being situated in a preferential flow path (fissured area). This test, therefore, demonstrates the double porosity of the aquifer studied.

There is also a meteorological station on-site to measure the following parameters at hourly time intervals: atmospheric temperature and pressure, rainfall.

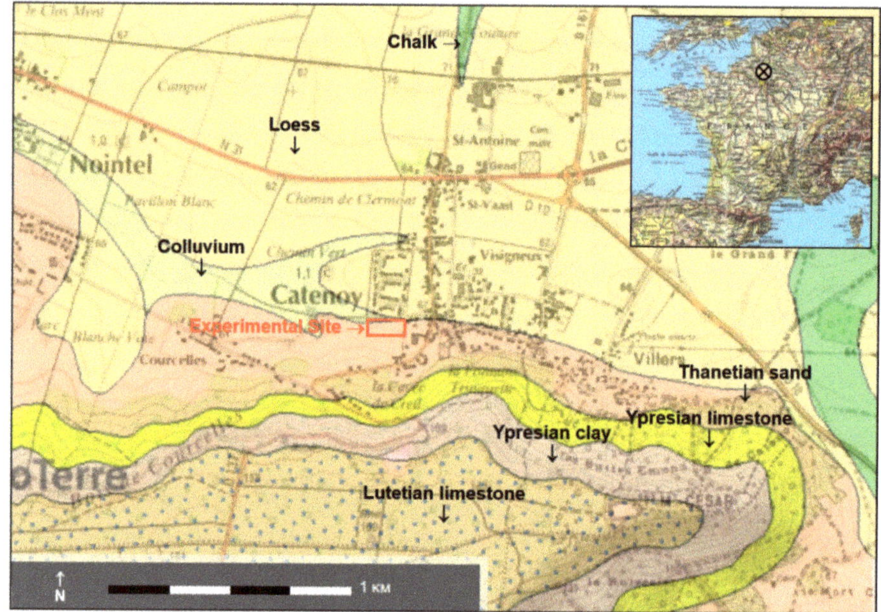

Figure 1. Location and geological context of the experimental site.

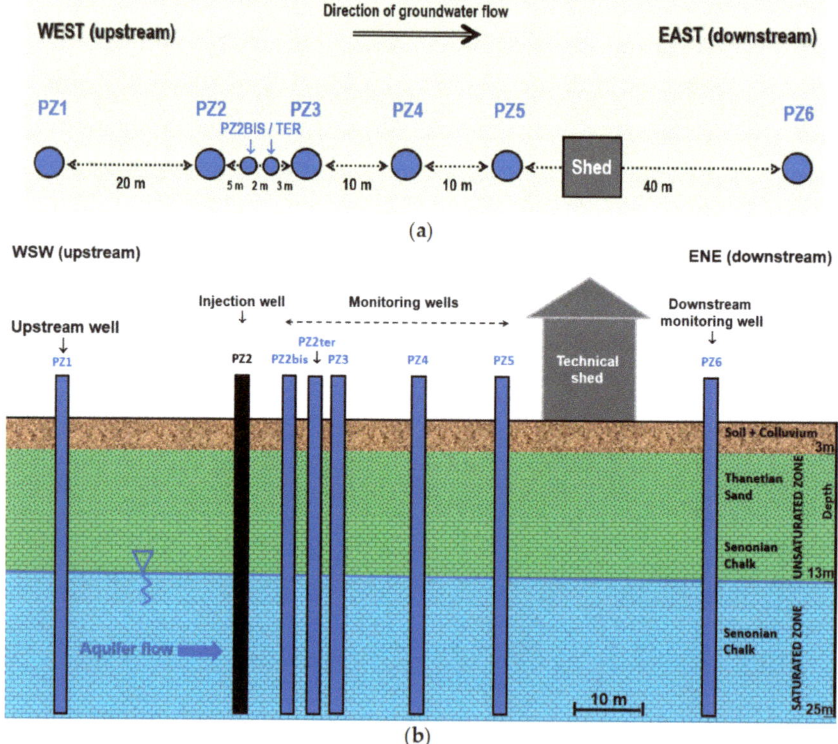

Figure 2. Diagram of the Catenoy experimental site: (**a**) planimetry; (**b**) cross-section.

2.2. Establishing the Baseline

Hydrogen is a very mobile gas that can leak towards the surface and accumulate in the groundwater and the soil. A complete monitoring protocol could interest the saturated zone, the unsaturated zone, the soil, and the surface because hydrogen can be detected in all these compartments as it can be seen in natural hydrogen emission areas [15]. However, the leakage simulation protocol (see under) is based on the injection of water saturated with dissolved gas (helium or hydrogen) directly into the aquifer. The water will be previously saturated at atmospheric conditions, i.e., at a pressure of 0.10 MPa, and then injected from 2 m to 11 m under the water table, where the hydrostatic pressure is between 0.12 and 0.21 MPa. Thus, the water will always remain undersaturated and nor helium nor hydrogen will degas. In these conditions, the only way for the dissolved gas to propagate is to follow the groundwater flow. In this study, the monitoring system has thus be designed for the saturated zone, with a light control of eventual weak degassing in the internal atmosphere of the piezometers but only for safety purposes.

Prior to setting up a monitoring system for the hydrogen injection test, a baseline of the initial piezometric, chemico-physical and hydrogeochemical values of the aquifer was established over 388 days starting on 27 October 2018. On all of the 7 main piezometers of the site (PZ1, PZ2, PZ2BIS, PZ3, PZ4, PZ5, and PZ6), the baseline corresponds to more than 200 measurements of each of the main chemico-physical parameters of the water: pH, temperature, electrical conductivity (EC), oxidation-reduction potential (ORP) and dissolved O_2 (Table 1). The water has a neutral pH (pH = 7.25), is moderately mineralized (EC = 562 $\mu S \cdot cm^{-1}$), and oxygenated (O_2 = 5.44 $mg \cdot L^{-1}$) due to its proximity to the soil surface, and is thus globally oxidative (ORP = +103 mV).

Table 1. Baseline of the chemico-physical parameters.

Parameter	O_2	pH	T	EC	ORP
Unit	$(mg.L^{-1})$	-	(°C)	$(\mu S.cm^{-1})$	(mV)
Number	208	223	224	223	221
Average	5.4	7.3	12.1	562	103
SD	1.7	0.3	0.6	66	89

Legend: O_2 = dissolved oxygen; T = Temperature; EC = Electric Conductivity; ORP = Oxidation-Reduction Potential; SD = Standard Deviation.

These chemico-physical parameters are quite stable over space and time. Figure 3 represents the boxplots of the dissolved oxygen and the oxidation-reduction potential at each piezometer during the baseline. Figure 4 represents the evolution of these main chemico-physical parameters over time and seems to show a certain sensitivity to the depth of the aquifer, which varies from 13.06 m to 13.94 m (Figure 3).

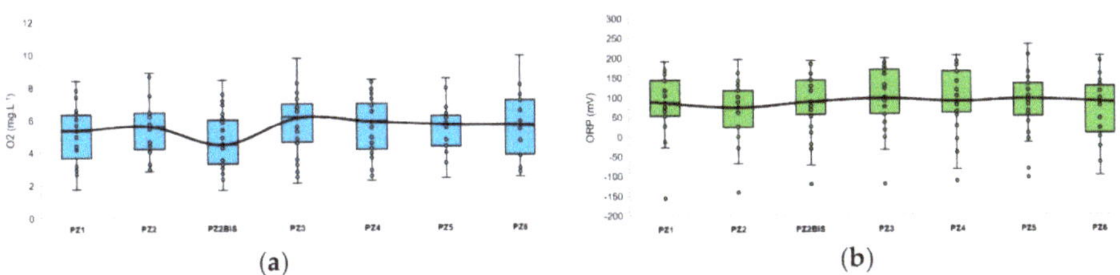

Figure 3. Boxplots of dissolved oxygen concentration and oxidation-reduction potential along the experimental site during the baseline: (**a**) Dissolved oxygen concentration; (**b**) oxidation-reduction potential. The colored boxes represent the 1st and 3rd quartiles (respectively Q1 and Q3), the black line is the median line, the dots are the measured values, the whiskers are the upper and lower extreme limits calculated according to the Tukey's formula: Q1 + 1.5 (Q3 − Q1) and Q3 − 1.5 (Q3−Q1).

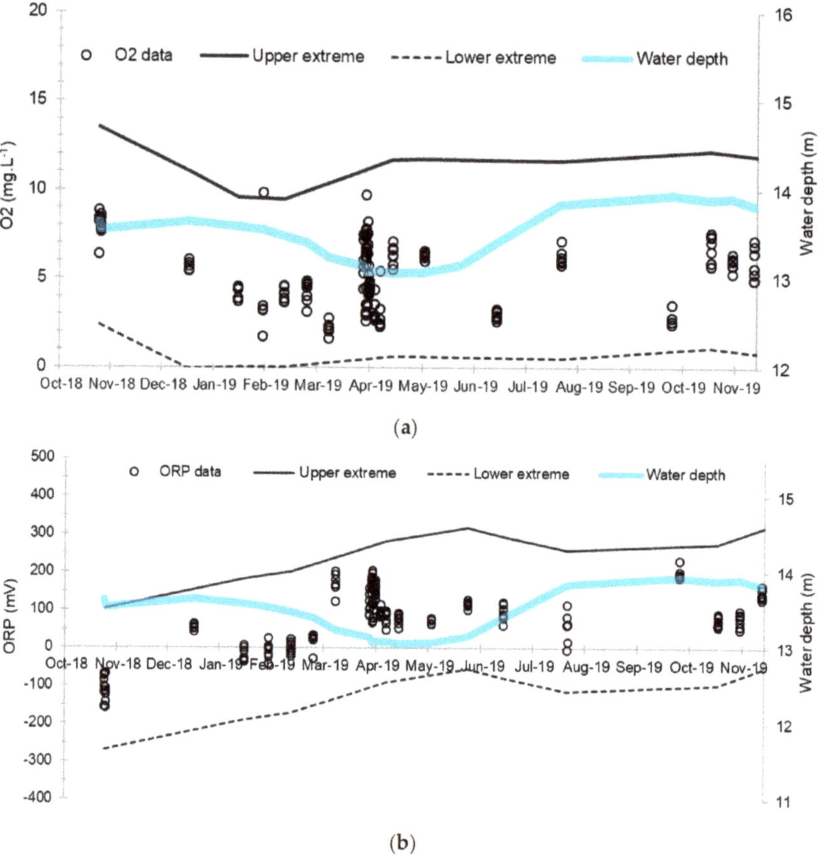

Figure 4. Evolution of dissolved oxygen concentration and oxidation-reduction potential and their extreme limits during the establishment of the baseline: (**a**) Dissolved oxygen concentration; (**b**) Oxidation-reduction potential. The solid and dashed lines correspond respectively to the upper and lower extreme limits calculated according to the Tukey's formula: $Q1 + 1.5 (Q3 - Q1)$ and $Q3 - 1.5 (Q3 - Q1)$ where Q1 and Q3 are the 1st and 3rd quartiles.

During the acquisition of the baseline data, 94 water samples were taken to analyze the major ions (Ca^{2+}, Mg^{2+}, Na^+, K^+, HCO_3^-, Cl^-, SO_4^{2-}, NO_3^-) and the main minor ions liable to be modified by a hydrogen-water-rock interaction (NO_2^-, NH_4^+, SO_3^{2-}, S_2^-, Fe, Mn). To comply with the storage conditions for all the elements, the analyses were carried out within 24 h of each sample being taken, using the methods presented in Table 2.

Table 2. Analytical methods and detection thresholds for the analyzed elements ($mg·L^{-1}$).

Parameter	HCO_3^-	Ca^{2+}	Mg^{2+}	Na^+	K^+	Cl^-	SO_4^{2-}	NO_3^-	NO_2^-	NH_4^+	SO_3^{2-}	S^{2-}	Fe	Mn
Method	Titration					Ionic Chromatography							ICP-MS	
DL ($mg·L^{-1}$)	0.04		0.05				0.01	0.02	0.02		0.01		0.001	

Legend: ICP-MS = Inductively Coupled Plasma-Mass Spectrometry; IC = Ionic Chromatography; DL = Detection Limit.

Regarding the major elements, Table 3, Figures 5 and 6 show that their behavior is also very stable throughout the baselining. The water generally exhibits bicarbonate-calcium facies, characteristic of chalk waters. This dominant hydrochemical facies is slightly altered by the presence of nitrate ions from agricultural inputs.

Table 3. Main characteristics of the major ions analyzed during baselining (mg·L^{-1}).

Parameter	Ca^{2+}	Mg^{2+}	Na$^+$	K$^+$	HCO$_3^-$	Cl$^-$	SO$_4^{2-}$	NO$_3^-$	Total
Average	97.1	11.5	12.6	4.69	298.8	23.6	27.9	33.4	97.1
SD	7.4	0.8	1.0	0.23	10	2.3	2.8	2.5	7.4

Legend: SD = Standard Deviation.

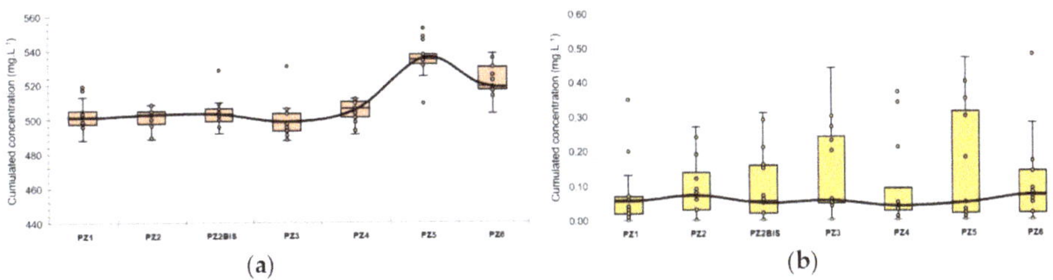

(a)　　　　(b)

Figure 5. Boxplots of major and minor ions concentration along the experimental site during the baseline: (**a**) major elements; (**b**) minor elements. The colored boxes represent the 1st and 3rd quartiles (respectively Q1 and Q3), the black line is the median line, the dots are the measured values, the whiskers are the upper and lower extreme limits calculated according to the Tukey's formula: Q1 + 1.5 (Q3 − Q1) and Q3 − 1.5 (Q3 − Q1).

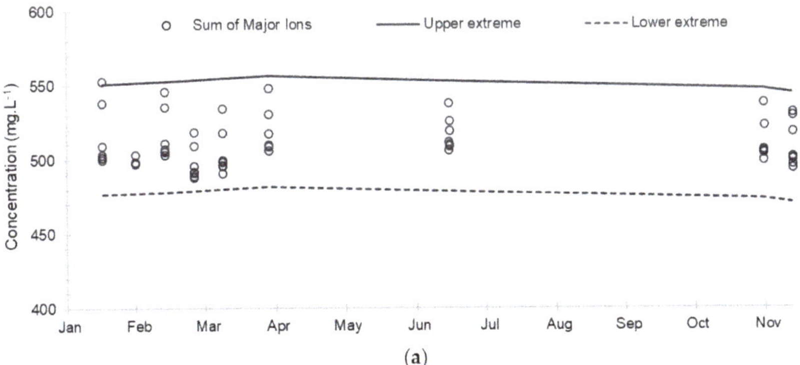

(a)

Figure 6. *Cont.*

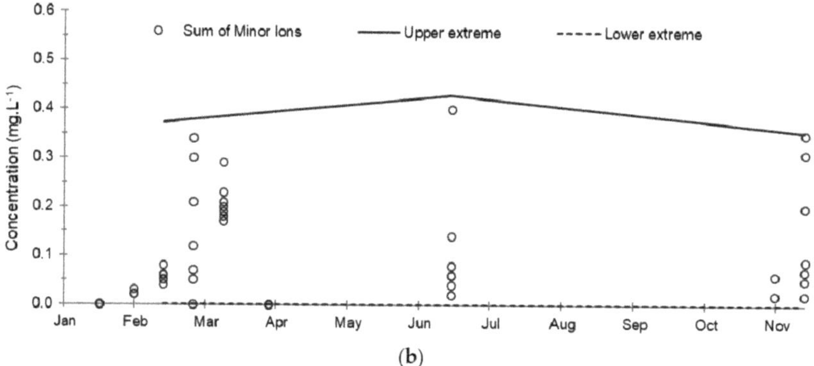

Figure 6. Evolution of the cumulative concentrations of major and minor ions and their extreme limits during the establishment of the baseline in 2019: (**a**) Sum of major ions; (**b**) Sum of minor ions. The solid and dashed lines correspond respectively to the upper and lower extreme limits calculated according to the Tukey's formula: Q1 − 1.5 (Q3 − Q1) and Q3 + 1.5 (Q3 − Q1) where Q1 and Q3 are the 1st and 3rd quartiles.

Regarding the minor elements analyzed, Table 4 shows the absence of nitrite and sulfide ions above the detection thresholds, as well the absence of sulfide ions except in five samples taken at PZ2 in the first half of 2019 where the concentrations ranged from 0.03 to 0.11 mg·L^{-1}. Ammonium ions are present in 74% of the samples, probably related to the application of nitrogenous fertilizers nearby, with a fairly fluctuating concentration with an average of 0.10 mg·L^{-1}. The totals of dissolved iron and manganese were also analyzed, but they were only minutely present in the water due to the mineralogical composition of the aquifer rock, which is made up of more than 95% calcite [4]: their ionized forms were therefore not researched. Ultimately, the evolution in the total of these minor ions varied little during baseline monitoring, the fluctuations observed being mainly due to the ammonium ions (Table 4).

Table 4. Main characteristics of minor ions analyzed during baselining.

Parameter	NO$_2^-$	NH$_4^+$	SO$_3^{2-}$	S^{2-}	Total Fe	Total Mn	Total (N + S)
Average	<DL	0.10	0.01	<DL	0.99	0.11	0.11
SD	-	0.13	0.03	-	1.58	0.22	0.13
CV	-	123%	(490%)	-	160%	193%	114%

Legend: DL = Detection Limit; SD = Standard Deviation; CV = Coefficient of Variation; Total (N+S) = Total of sulfur and nitrogen ions.

2.3. Preparing the Test

The helium was injected with the aim of testing and optimizing a future hydrogen injection device using an inert gas, and to configure the monitoring protocol (types of measurement and time intervals) depending on the piezometer being monitored. The objective of this test is to create a plume of dissolved helium in the aquifer, comparable to the future plume of dissolved hydrogen, and to monitor its propagation in the saturated zone.

Before this test, the propagation of the dissolved He plume was modeled in 1D using PHREEQC. Modeling parameters were determined using the results of previous CO$_2$ injection tests [4,5]. The result is shown in Figure 7 and shows a maximum dissolved helium concentration between 1.46 mg·L^{-1} and 8×10^{-21} mg·L^{-1} from PZ2BIS to PZ6, and a peak arrival time between 100 min and 23 days. Peak values at PZ5 and PZ6 are expected to be below 1 µg·L^{-1} and thus it will not be possible to detect helium in these two piezometers.

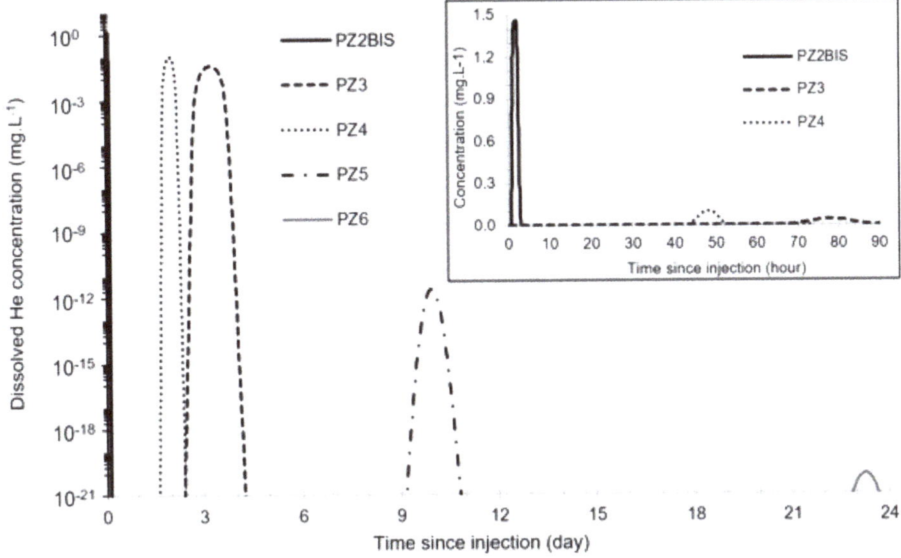

Figure 7. Model of the propagation of the dissolved He plume using PHREEQC.

Then, the water from the aquifer was extracted beforehand by pumping in the PZ2 piezometer (future injection well) to fill two HDPE tanks (Figure 8a): a first 1 m^3 tank which contains the tracers to help determine the arrival of the plume of dissolved gas and precisely quantify its kinetics, and a second 5 m^3 tank in which the water will be saturated with helium by bubbling it. It was decided not to incorporate the tracers in the tank of water saturated with helium to avoid the risk of reducing the solubility of this gas.

(a)   (b)

Figure 8. Views of the two tanks and the helium saturation device: (a) View of the 1 m^3 tracers tank [A], the 5 m^3 He saturated tank [B] and the compressed He cylinder [C]; (b) View from the manhole of the bubbling device in the 5 m^3 He tank (20 m of PVC pipe pierced with 200 needles of 0.5 mm diameter).

In the first 1 m^3 tank, two types of tracers were used:

- fluorescent organic tracers that exhibit no analytical interference between them, to allow in situ detection in real-time of the arrival of the injected plume thanks to the installation of a GGUN FL-30 field fluorimeter; these are uranine or fluorescein sodium (green dye), sulforhodamine B (red dye) and Amino G Acid (a colorless tracer emitting in blue); however, previous experiments with these types of organic tracers with a long carbonaceous molecule (C_{20} to C_{40}) have shown that not all of them were conservative when transferred to an aquifer composed of chalk with finely porous matrix permeability, as is the case in Catenoy [5];

- inorganic ionic tracers, which are highly conservative but colorless; they are analyzed a posteriori in the laboratory, from a water sample to precisely quantify the kinetics of the plume; these are lithium (as lithium chloride, LiCl) and bromide (as potassium bromide, KBr)

For the final hydrogen injection experiment, only the most efficient fluorescent tracer and ionic tracer in terms of their recovery will be used. The objective of this first test is, therefore, both to select these two tracers from the five tested, and to validate the principle of a prior injection of tracers to predict the arrival of the dissolved gas plume and, as a result, to improve the monitoring system. A quantity of 1 g of each tracer was diluted in the 1 m^3 tank: it will be noted that, for ionic tracers, this is 1 g of tracer ion (Li^+ and Br^-), which corresponds to 6.14 g of LiCl and 1.49 g of KBr.

To saturate the water with helium, a bubbling device was installed on the interior floor of the second tank to create a curtain of bubbles facilitating the dissolution of the gas. It is a PVC pipe of 20 m length, pierced with 200 holes (Figure 8b). This device is connected in a loop to a rotameter to regulate the flow of gas injected from a compressed gas cylinder (Figure 8a).

The two tanks were then emptied successively by gravity into the PZ2 borehole and the water then entered the aquifer.

The injection device consists of a reinforced PVC pipe 70 mm in diameter. This pipe is directly connected, by means of a tee fitting, to the outlet located at the base of the two tanks, each being isolated by a valve (Figure 9a). The injection pipe is plugged at its lower end and ballasted to ensure that it descends to the bottom of the well, at a depth of 25 m. To better distribute the injected fluid over the entire screened height in the injection well, the submerged part of the injection pipe is drilled with 46 holes that are distributed two by two from 12 to 23 m deep, and four by four from 23 to 25 m deep (Figure 9b). The shallowest holes are located 0.2 m below the water table to ensure that the dissolved helium is injected under a slight hydrostatic overpressure, and therefore cannot degas.

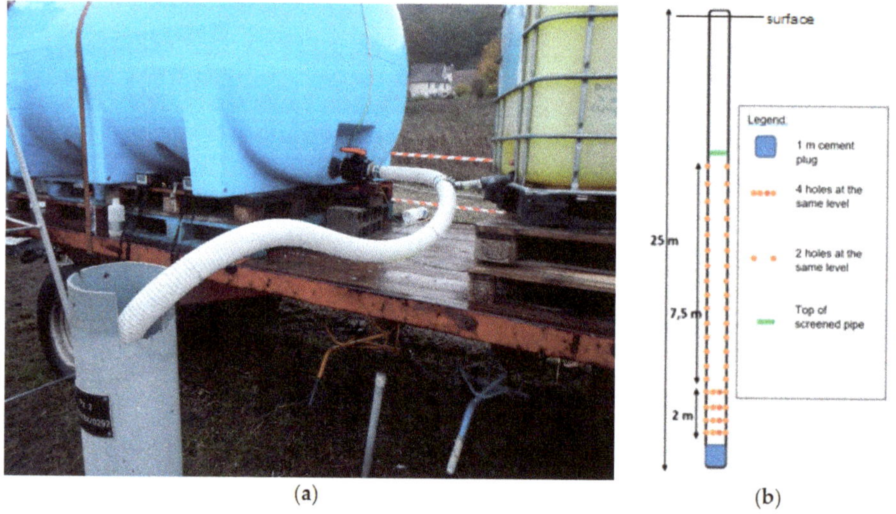

Figure 9. View of injection installation: (**a**) injection pipe and its connection to the tanks; (**b**) diagram of the arrangement of the holes along the injection pipe.

2.4. Conducting the Test

The tanks were filled with groundwater on the morning of 1 April 2019 using a submerged electric pump installed in PZ2 piezometer, the future injection well:

- the first 1 m^3 tank was filled with groundwater to 0.8 m^3, then the tracers, some of which are photosensitive, were added after sunset to avoid degradation by light; the volume of water was then increased to 1 m^3 for better mixing;
- the second 5 m^3 tank was filled with groundwater, then helium gas was continuously sent into the bubbling circuit until the following day at 12:00; the total time allowed for the helium to dissolve in the water was approximately 20 h; based on similar experiments carried out in the past, this was more than sufficient to ensure helium saturation of the water in the tank.

The injection of the water and tracers from the first tank (1 m^3) was performed by gravity on the 2nd of April 2019 from 10:35 to 11:10, which represents an injection flowrate of 1.7 m$^3 \cdot$h^{-1}. The helium-saturated water from the second tank (5 m^3) was injected again by gravity immediately after, i.e., from 11:12 to 12:47. This injection lasted 1 h 35 min, which corresponds to a flow rate of 3.2 m$^3 \cdot$h^{-1}.

2.5. Monitoring the Saturated Zone

The equipment installed to monitor the saturated zone during the helium injection test was as follows:

- Two physicochemical sensors for measuring temperature, pH, electrical conductivity, oxidation-reduction potential, and dissolved oxygen; one was permanently installed (for the duration of the test) in the PZ2BIS piezometer located 5 m downstream of the injection borehole while the other one was mobile to take measurements in all other piezometers;
- A GGUN-FL30 field fluorimeter which provides live analysis of the fluorescence of the water extracted from the piezometers; it is a multichannel device that can successively analyze the three fluorescent tracers used;
- A GRUNDFOS-MP1 submersible pump which was first used to fill the tanks from the PZ2 and moved to the PZ2BIS shortly before the start of the injection; it made it possible to regularly sample the groundwater to carry out laboratory analyses of the tracers, dissolved gases (helium) and major elements (calcium, magnesium, sodium, potassium, bicarbonates, chlorides, sulfates, nitrates);
- A Raman and Infrared (IR) spectrometer, installed in the PZ2TER piezometer located 7.5 m downstream of the injection well [16,17]; it makes possible to analyze the concentration of mononuclear diatomic molecules (H$_2$, O$_2$, N$_2$) as well as polar molecules (CO$_2$ and CH$_4$) in the water; since the detection of a monoatomic gas such as He is not possible with this type of sensor, only the indirect effect on the concentration of other dissolved gases can be detected;
- A device for water pumping from the aquifer and degassing by mechanical agitation; it is combined with an ALCATEL ASM 122D transportable mass spectrometer to measure the helium concentration in the degassed gas mixture.

3. Results

3.1. Tracers

The tracers were analyzed using spectrofluorimetry in the CETRAHE lab at the University of Orléans (Table 5). The assay of each of these tracers was performed using a calibration curve established with the same tracer used for the test. It should be noted that the spectral analysis technique, via the characteristic excitation and emission spectra, makes it possible to confirm the presence of these fluorescent tracers even when the concentration is low and close to the detection limit, to avoid any confusion with the natural fluorescence of water. The maximum net concentrations obtained at each piezometer are summarised in Table 6.

Table 5. Analytical methods and detection thresholds for the analyzed tracers.

Tracer	Uranine	Sulforhodamine B	Amino G Acid	Li	Br
Method	ICP-MS	ICP-MS	ICP-MS	IC	IC
DL ($\mu g \cdot L^{-1}$)	0.001	0.05	0.05	1	1

Legend: ICP-MS = Inductively Coupled Plasma-Mass Spectrometry; IC = Ionic chromatography; DL = Detection Limit).

Table 6. Maximum net tracer concentration per piezometer (in $\mu g \cdot L^{-1}$).

Piezometer	Distance * (m)	Uranine (DL = 0.001)	Sulforhodamine (DL = 0.050)	AGA (DL = 0.100)	Lithium (DL = 0.5)	Bromide (DL = 0.5)
PZ1	−20	0.0	0.0	0.0	2.0 *	2.8 *
PZ2	0	10.2	4.1	23.9	22.0	3.1
PZ2BIS	+5	19.8	21.1	30.0	148.1	45.7
PZ3	+10	1.2	0.3	0.0	3.4	5.0
PZ4	+20	0.5	0.1	0.0	2.7	1.1 *
PZ5	+30	0.3	0.0	0.0	2.0 *	0.5 *
PZ6	+60	0.0	0.0	0.0	1.8*	0.9 *

Legend: Distance = Distance from injection well in the downstream direction (positive values) and upstream direction (negative value), DL = Detection Limit, AGA = Amino G Acid, * background noise.

It thus appears that the PZ1 piezometer (upstream of the injection well) was not reached by the tracer plume and that the PZ2BIS piezometer (5 m downstream) is the only piezometer where the presence of all the tracers was proven. Starting from PZ3, located 10 m downstream of the injection well, some tracers such as the Amino G Acid and the bromide were not detected. Starting from PZ5, located 30 m downstream of the injection well, sulforhodamine B was also no longer detected. At PZ6, the piezometer that is most removed (60 m downstream of the injection well), no tracer was detected in significant concentrations by the end of the monitoring period. Uranine and lithium are the only tracers that were detected in all the piezometers located downstream of the injection point, except at PZ6. These are thus the best-suited tracers for this hydrogeological context, the first one because it is easily detectable in situ (using a field fluorimeter) including at low concentrations (0.1 $\mu g \cdot L^{-1}$) and the second one because it proved to be more conservative.

For uranine and lithium, the results were also interpreted using TRAC software [18] considering Fried's analytical solution [19] for the brief injection of a mass of tracer into an infinite volume in flow (Equation (1)):

$$C_{(x,y,t)} = \frac{m}{4\,b\,\pi\,\omega\,t\,\sqrt{D_L\,D_T}} \cdot \exp\left[-\frac{(x - u\,t)^2}{4\,D_L\,t} - \frac{y^2}{4\,D_T\,t} \right] \tag{1}$$

where $C_{(x,y,t)}$ is the concentration of tracer (kg·m^{-3}) at the point with coordinates (x, y) (m) and at time t (s), m is the mass of injected tracer (kg), b the thickness of the aquifer (m), ω the cinematic porosity (−), D_L the longitudinal dispersion (m$^2 \cdot$s^{-1}), D_T the transversal dispersion (m$^2 \cdot$s^{-1}), and u the real flow speed (m·s^{-1}). In the context of this test, the fixed parameters are $m = 10^{-3}$ kg, $b = 14$ m and x which corresponds to the distance of each piezometer from the injection well. Note that the y coordinate has been left free, which makes it possible to check whether the piezometers are properly aligned in the main flow axis of the aquifer with respect to the injection well.

At PZ2BIS, 5 m downstream of the injection well, the concentration peak was achieved on the day of injection itself at 11:44 for lithium and at 15:28 for uranine, that is, 1.15 and 4.83 h respectively after the start of injection (Figure 10). Despite this difference in transit time, the hydrodynamic parameters used for the calibration of the breakthrough curves are the same for the two tracers: only the retardation factor varies, being fixed at 1.0 for lithium and 1.6 for uranine. The cinematic porosity is thus equal to 1.40×10^{-2} and the permeability 1.10×10^{-3} m·s^{-1}, values in accordance with those previously obtained in test pumping.

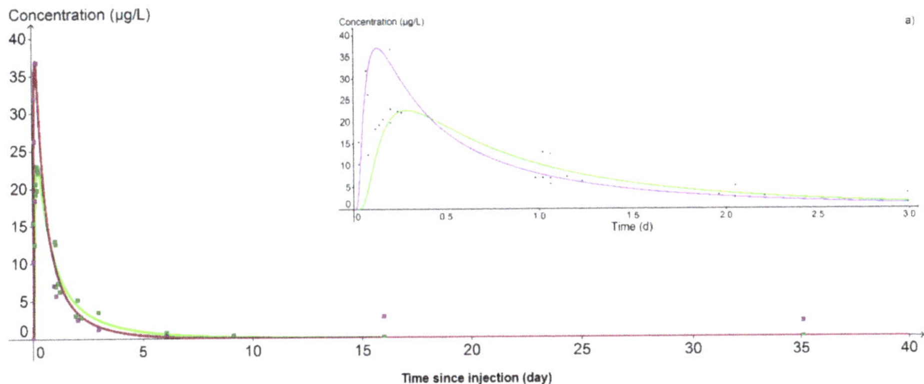

Figure 10. Evolution of uranine (green curve) and lithium (purple curve) concentrations at the nearest downstream piezometer (PZ2BIS). (**a**) Detail of the 3 first days. The peak at 148.1 μg·L^{-1} obtained with the lithium is not shown because it could not be simulated.

From the PZ3 piezometer, 10 m downstream of the injection well, it is no longer possible to fit the data on a single curve because the recovery is bimodal. The first peak reflects a rapid arrival of the tracer by a preferential path (fissured zone?) while the second peak corresponds to a slower propagation within the matrix aquifer. In Figure 11, two distinct fits were therefore applied to each first peak (dashed curves) and second peak (solid curves). Table 7 shows that the porosity obtained is fairly uniform around the mean value of 5.97×10^{-2} m·s^{-1} regardless of the piezometer or tracer studied, but that the permeability varies more strongly around the mean value of 4.11×10^{-3} m·s^{-1} depending on the adjustment made. These values are also significantly higher than those obtained at PZ2BIS, which is interpreted as resulting from an environment with multiple porosity, of both matrix and fissure type, once a larger aquifer volume is involved. As before, we observe a faster propagation of the plume at PZ4 (3.9 m·d^{-1}) than at PZ3 (1.6 m·d^{-1}): however, these speeds are 2 to 3 times lower than during the tracing test carried out in 2012 (10 m·d^{-1} and 3 m·d^{-1}, respectively), which seems to be due to a low groundwater table which started exceptionally early this year. It should also be noted that this speed artificially reached 104 m·d^{-1} during the injection, at the PZ2BIS which is a piezometer directly influenced by the injection conditions.

Table 7. Average hydrodynamic characteristics resulting from the calibration of the recovery curves.

Peak	First Peak (With PZ2BIS)		Second Peak (Without PZ2BIS)		Average
Tracer	Lithium	Uranine	Lithium	Uranine	
Porosity (−)	4.85×10^{-2}	6.35×10^{-2}	5.67×10^{-2}	7.00×10^{-2}	5.97×10^{-2}
Permeability (m·s^{-1})	4.47×10^{-3}	1.11×10^{-2}	3.03×10^{-4}	5.86×10^{-3}	4.11×10^{-3}

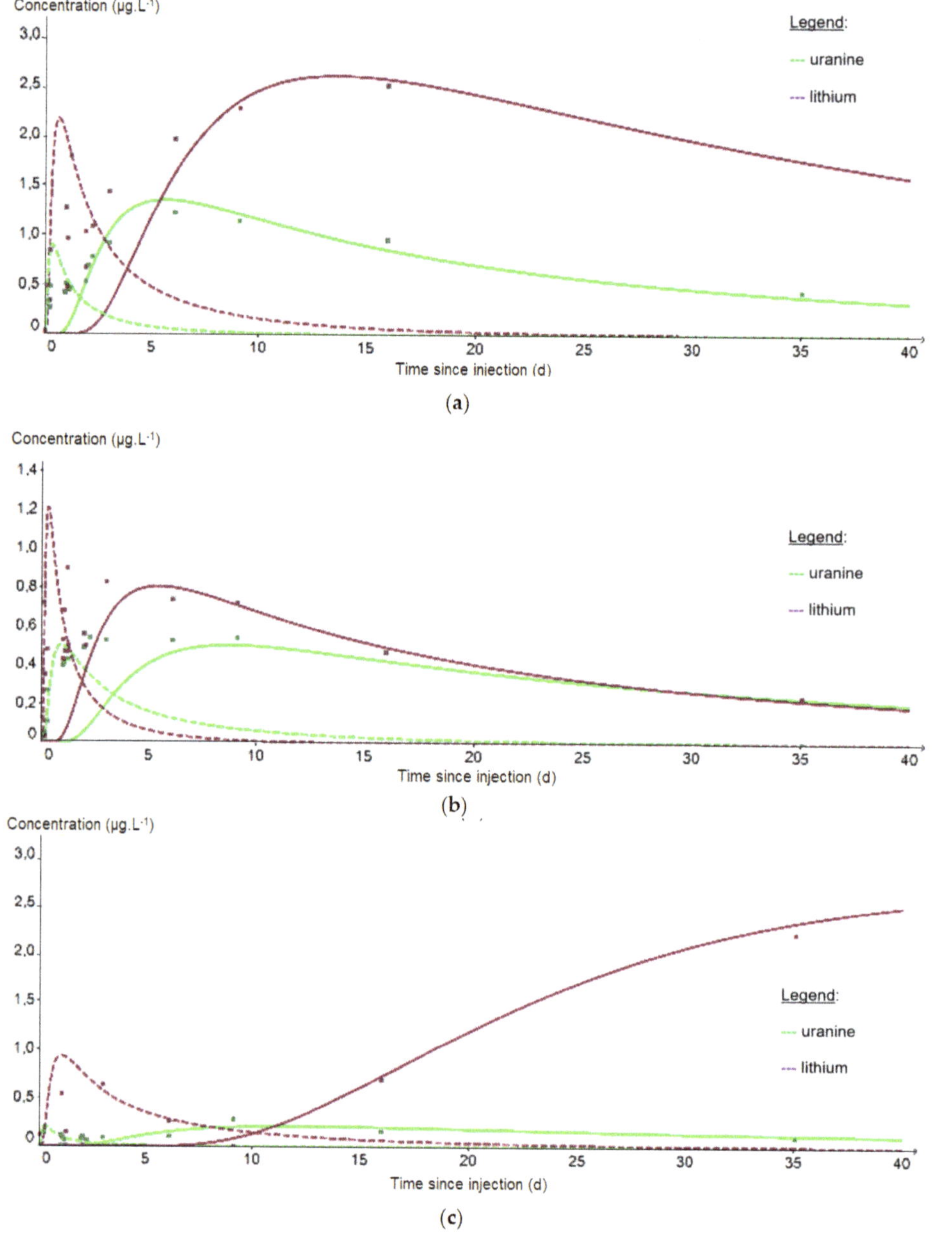

Figure 11. Evolution of the tracer concentrations at the downstream piezometers: (**a**) PZ3; (**b**) PZ4; (**c**) PZ5. The curves correspond to the fit of the first peak for the dashed curve and of the second peak for the solid curve; the purple curve represents lithium and the green curve uranine.

3.2. Dissolved Helium Concentration

The dissolved helium was extracted from the water sampling vessels by partial degassing by mechanical agitation, after which the extracted gaseous mixture was directly analyzed on site using an ALCATEL ASM 122D mass spectrometer. The results obtained during baseline measurements at the two reference piezometers, located upstream (PZ1) and far downstream (PZ6), indicate that the groundwater does not contain a significant amount of helium. The measured values of dissolved gas are less than the equilibrium concentration with the surface-atmosphere (helium content about 5 ppm).

During the experiment, the arrival of the helium plume at the PZ2BIS piezometer, 5 m downstream of the injection well, occurred very quickly after injection (Figure 12): the maximum concentration of 1.47 mg·L^{-1} was measured 30 min after injection. After this, the helium concentration in water decreases. It has almost returned to its initial state at this piezometer 40 days after injection. In PZ3 and PZ4 piezometers, located respectively 10 m and 20 m downstream, the dissolved helium concentrations were significantly lower as at PZ2BIS. The arrival of the helium was detected 3 h after the injection at PZ3 piezometer; and a little more than 5 h after at PZ4. At these two piezometers, the maximum helium concentrations were about 3 and 8 µg·L^{-1}, respectively, and were recorded 9 days after injection. No significant trace of dissolved helium was measured at the other piezometers located further downstream (PZ5 at 30 m and PZ6 at 60 m).

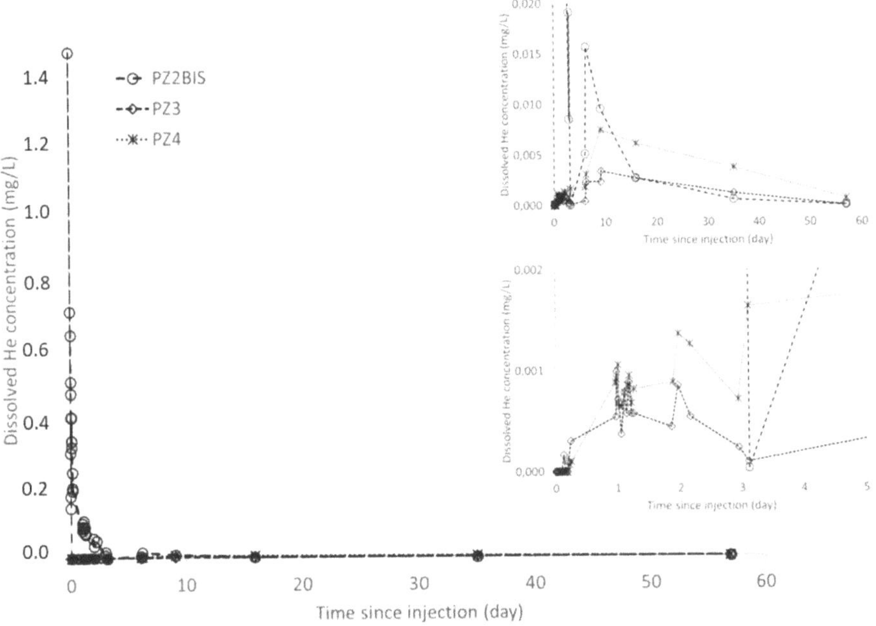

Figure 12. Dissolved helium concentrations in the samples taken at the downstream piezometers PZ2BIS, PZ3, and PZ4.

4. Discussion

Following this test, we can see that the experimental protocol for saturating water with gas and then injecting it into the groundwater is operational, as is the way of monitoring the saturated zone. However, the results obtained lead us to propose a certain number of improvements for the hydrogen injection experiment.

Concerning the gas saturation of the water in the 5 m^3 tank, it will not be possible—for safety reasons—to allow hydrogen to bubble all night to obtain maximal saturation at the time of injection, as was done with the helium. Hydrogen is an easily flammable gas requiring the establishment of an ATEX zone and suitable control measures. These measures are difficult to ensure overnight. As a result, the hydrogen bubbling will have to be interrupted in the evening to resume the next morning. This can lead to a delay of several hours in reaching an optimum level of hydrogen saturation in the water in the tank and, consequently, the same delay in all the subsequent operations (injection, monitoring measurements, etc.). To increase saturation kinetics, the number of gas outlets at the bottom of the 5 m^3 tank will be doubled, from 200 to 400.

The first 1 m^3 tank will hold the fluorescent and ionic tracers that have been shown to provide the best performances: uranine and lithium. Considering the weakness of the signal obtained during this test, owing to a strong dilution of the tracers in the groundwater, they will be used at a concentration higher than an order of magnitude: 10 g·L^{-1} instead of 1 g·L^{-1}. In addition, the water in this tank will also be saturated with helium, to be used as an inert tracer gas to compare its behavior with that of hydrogen, a potentially reactive gas in this aquifer context.

The second 5 m^3 tank will be saturated with hydrogen by means of bubbling in a gaseous state during the first day of preparing the materials, as well as during the following morning. The injection of the hydrogen-saturated water will therefore take place at the start of the afternoon. In the test conducted with helium, the two tanks were drained successively, but owing to their respective geometries, the injection rate of the second tank was found to be significantly higher (3.2 m^3·h^{-1}) than that of the first (1.7 m^3·h^{-1}). Following this, the two plumes probably coalesced, which hampered the interpretation of the tracer breakthrough curves, and probably diluted the plume of dissolved gas. To avoid this, we will apply a latency time of $\frac{1}{2}$ h between the two injections, and the emptying rate of the second tank containing dissolved hydrogen will be retained at less than or equal to that of the first tank containing the tracers.

The PZ2BIS piezometer placed directly downstream of the point of injection provided the best recovery curves and will thus be considered to be the principal monitoring piezometer. As such, given the speed of the response obtained (1.15 h), it must be equipped with a specific monitoring device to provide continuous data acquisition: dissolved hydrogen measurement probe, physicochemical measurement probe, and borehole fluorimeter. This equipment must be available in duplicate to be able to monitor the other piezometers manually. As soon as the tracer signal has disappeared from the PZ2BIS, the continuously recording borehole fluorimeter will be moved to the piezometers located further downstream (PZ3, PZ4, and PZ5).

The piezometers must not all be sampled at the same frequency, but at specific time intervals in function to their distance from the injection well, and the current hydrogeological conditions, by taking particular account of the propagation speed of the fluorescent tracer. During the current test, the duration of monitoring (40 days) did not permit a satisfactory sampling of the most distant piezometers, namely PZ5 (30 m downstream) and PZ6 (60 m downstream). This duration will therefore be significantly increased, but at a rate of only one sample per week from the 5th week of monitoring: the total duration of the monitoring may vary from 60 to 80 days depending on the hydrogeological conditions at the time (high or low water). It is, however, suggested that a period of high water is favored to reduce the monitoring time. In all cases, only the PZ2TER will operate continuously for the measurement of dissolved gases using Raman and IR spectrometers (O_2, N_2, H_2, CO_2, and CH_4).

Finally, the piezometry of the aquifer will be measured twice a day at all piezometers during the week of injection, and then once a day thereafter, to detect any variation in the speed or flowing direction of the aquifer. An automatic water depth measurement probe will also be placed at the bottom of PZ2 to measure the amplitude of the piezometric dome induced by the injection.

5. Conclusions

A test of the combined injection of tracers (organic and ionic) and helium-saturated water was done in April 2019 to assess and optimize the concept of injecting water saturated with hydrogen, planned for later, and monitoring its physicochemical properties.

The test has confirmed the technical feasibility, under field conditions, of saturating a significant quantity of water with a low-solubility gas and injecting it in a controlled manner into a shallow aquifer.

It was possible to properly monitor the propagation of the dissolved gas plume in the aquifer with the means of analysis used. Helium could be detected up to 20 m downstream of the injection well by mass spectrometer analysis of the gas mixture obtained through partial degassing of water samples by mechanical agitation.

Among the five tracers used, uranine and lithium were shown to be the most effective. The first is a colored fluorescent organic tracer, easily and continuously detectable in situ but affected in this specific hydrogeological context (fine matrix and fissure porosity) by a certain retardation factor with respect to the propagation of the water. The second is a colorless ionic tracer, not affected by such a retardation factor, but not detectable in situ. To obtain a cleaner signal, the tracer mass used will be ten times higher when injecting the water saturated with hydrogen.

The temporal modeling of the post-injection evolution of these two tracers reveals, as a function of the distance from the injection well, two distinct hydrodynamic regimes linked to the existence of a multiple porosity, of both matrix and fissure type. These elements will be essential in understanding the transport of the hydrogen plume during the next injection simulation.

The preparation and the conditions of injection of the tracer tank and hydrogen-saturated water tank have been modified to take the results obtained into account: doubling of the number of bubbling outlets in the 5 m^3 tank bubbling device, establishment of a latency period between the two injections, reduction of the flow rate of the second tank.

Finally, the protocol of monitoring has also been modified: the establishment of specific monitoring of the PZ2BIS piezometer with continuous in situ recordings of a maximum of data, adaptation of the sampling schedule to the specificity of each piezometer and increase in the overall monitoring time.

Thus, the adoption of all of these improvements will permit proper execution of the main experiment of injecting hydrogen-saturated water and carrying out the associated monitoring, which will also be preferentially done during periods of high water.

Author Contributions: Conceptualization, methodology, and validation, P.G., E.L., S.L., and Z.P.; analysis, S.L., P.G., Z.P., E.L., P.d.D., and N.J.; writing—original draft preparation, S.L., P.G., and Z.P.; writing—review and editing, S.L., P.G., Z.P., E.L., P.d.D., and N.J.; supervision, P.G. All authors have read and agreed to the published version of the manuscript.

Funding: This research was funded by the French Scientific Interest Group GEODENERGIES in the framework of the ROSTOCK-H project (Risks and Opportunities of the Geological Storage of Hydrogen in Salt Caverns in France and Europe).

Conflicts of Interest: The authors declare no conflict of interest. The funders had no role in the design of the study; in the collection, analyses, or interpretation of data; in the writing of the manuscript, or in the decision to publish the results.

References

1. Légifrance. Loi n° 2015-992 du 17 août 2015 Relative à la Transition Energétique Pour la Croissance Verte. Available online: https://www.legifrance.gouv.fr/eli/loi/2015/8/17/DEVX1413992L/jo/texte (accessed on 9 December 2019).
2. Coltier, Y.; Plouhinec, C. Chiffres Clés des Energies Renouvelables. Edition 2019. Available online: https://www.statistiques.developpement-durable.gouv.fr/sites/default/files/2019-05/datalab-53-chiffres-cles-des-energies-renouvelables-edition-2019-mai2019.pdf (accessed on 9 December 2019).
3. Ineris. Le Stockage Souterrain Dans le Contexte de la Transition Energétique. Maîtrise des Risques et Impacts. Collection Ineris Références. 2016. Available online: https://www.ineris.fr/sites/ineris.fr/files/contribution/Documents/ineris-dossier-ref-stockage-souterrain.pdf (accessed on 13 December 2019).
4. Gombert, P.; Pokryszka, Z.; Lafortune, S.; Lions, J.; Grellier, S.; Prevot, F.; Squarcioni, P. Selection, instrumentation and characterization of a pilot site for CO_2 leakage experimentation in a superficial aquifer. *Energy Procedia* **2014**, *63*, 3172–3181. [CrossRef]
5. Gal, F.; Lions, J.; Pokryszka, Z.; Gombert, P.; Grellier, S.; Prévot, F.; Squarcioni, P. CO_2 leakage in a shallow aquifer—Observed changes in case of small release. *Energy Procedia* **2014**, *63*, 4112–4122. [CrossRef]

6. Lions, J.; Devau, N.; de Lary, L.; Dupraz, S.; Parmentier, M.; Gombert, P.; Dictor, M.-C. Potential impacts of leakage from CO_2 geological storage on geochemical processes controlling fresh groundwater quality: A review. *Int. J. Greenh. Gas Control* **2014**, *22*, 165–175. [CrossRef]

7. AFHYPAC. Inflammabilité et Explosivité de L'hydrogène. FICHE 7.1, Février 2015, 5 p. Available online: http://www.afhypac.org/documents/tout-savoir/fiche_7.1_inflammabilite_explosivite_rev._fev_2015_ineris_bwe_pm.pdf (accessed on 9 December 2019).

8. Siantar, D.P.; Schreier, C.G.; Chou, C.S.; Reinhard, M. Treatment of 1,2-dibromo-chloropropane and nitrate-contamined water with zero-valent iron or hydrogen/palladium catalysts. *Water Res.* **1996**, *30*, 2315–2322. [CrossRef]

9. Smirnov, A.; Hausner, D.; Laffers, R.; Strongin, D.R.; Schoonen, M.A.A. Abiotic ammonium formation in the presence of Ni-Fe metals and alloys and its implication for the Hadean nitrogen cycle. *Geochem. Trans.* **2008**, *9*, 5. [CrossRef] [PubMed]

10. Truche, L.; Berger, G.; Destrigneville, C.; Pages, A.; Guillaume, D.; Giffaut, E.; Jacquot, E. Experimental reduction of aqueous sulphate by hydrogen under hydrothermal conditions: Implication for the nuclear waste storage. *Geochim. Cosmochim. Acta* **2009**, *73*, 4824–4835. [CrossRef]

11. Truche, L.; Berger, G.; Albrecht, A.; Domergue, L. Abiotic nitrate reduction induced by carbon steel and hydrogen: Implications for environmental processes in waste repositories. *Appl. Geochem.* **2013**, *28*, 155–163. [CrossRef]

12. Truche, L.; Berger, G.; Albrecht, A.; Domergue, L. Engineered materials as potential geocatalysts in deep geological nuclear waste repositories: A case study of the stainless steel catalytic effect on nitrate reduction by hydrogen. *Appl. Geochem.* **2013**, *35*, 279–288. [CrossRef]

13. Berta, M.; Dethlefsen, F.; Ebert, M.; Schäfer, D.; Dahmke, A. Geochemical Effects of Millimolar Hydrogen Concentrations in Groundwater: An Experimental Study in the Context of Subsurface Hydrogen Storage. *Environ. Sci. Technol.* **2018**, *52*, 4937–4949. [CrossRef] [PubMed]

14. Lagmöller, L.; Dahmke, A.; Ebert, M.; Metzgen, A.; Schäfer, D.; Dethlefsen, F. Geochemical effects of hydrogen intrusions into shallow groundwater—An incidence scenario from Underground Gas Storage. *Groundw. Qual.* **2019**, Presentation nr. 90. Available online: https://www.uee.uliege.be/upload/docs/application/pdf/2019-10/90._vortrag_gq_2019_lagmoller_05.09.2019_fd5_ds_am_me_ad_ll12.pdf (accessed on 9 December 2019).

15. Prinzhofer, A.; Moretti, I.; Françolin, J.; Pacheco, C.; D'Agostino, A.; Werly, J.; Rupin, F. Natural hydrogen continuous emission from sedimentary basins: The example of a Brazilian H2-emitting structure. *Int. J. Hydrog. Energy* **2019**, *44*, 5676–5685. [CrossRef]

16. Labat, N.; Lescanne, M.; Hy-Billiot, J.; de Donato, P.; Cosson, M.; Luzzato, T.; Legay, P.; Mora, F.; Pichon, C.; Cordier, J.; et al. Carbon capture and storage: The Lacq pilot (project and injection period 2006–2013). *Chapter 7 Environ. Monit. Model.* **2015**, 182–188. Available online: https://www.globalccsinstitute.com/archive/hub/publications/194253/carbon-capture-storage-lacq-pilot.pdf (accessed on 13 December 2019).

17. Lacroix, E.; de Donato, P.; Lafortune, S.; Caumon, M.-C.; Barres, O.; Derrien, M.; Piedevache, M. Metrological development based on in situ and continuous monitoring of dissolved gases in an aquifer: Application to the geochemical baseline definition for hydrogen leakage survey. *Anal. Methods* **2020**. under review.

18. Gutierrez, A.; Klinka, T.; Thiery, D.; Buscarlet, E.; Binet, S.; Jozja, N.; Defarge, C.; Leclerc, B.; Fecamp, C.; Ahumada, Y.; et al. TRAC, a collaborative computer tool for tracer-test interpretation. In *EPJ Web of Conferences*; EDP Sciences: Les Ulis, France, 2013. [CrossRef]

19. Fried, J.J. *Groundwater Pollution*; Elsevier: New York, NY, USA, 1975.

Article

Empirical Analysis and Geomechanical Modelling of an Underground Water Reservoir for Hydroelectric Power Plants

Javier Menéndez [1],[*], Falko Schmidt [2], Heinz Konietzky [3], Antonio Bernardo Sánchez [4] and Jorge Loredo [5]

[1] Renewable Energy Department, Hunaser Energy, Avda. Galicia 44, 33005 Oviedo, Spain
[2] Geomechanical Department, Geotechnical Engineer, 39011 Santander, Spain; falko@geomecanica.es
[3] Geotechnical Institute, TU Bergakademie Freiberg, 09599 Freiberg, Germany; Heinz.Konietzky@ifgt.tu-freiberg.de
[4] Department of Mining Technology, Topography and Structures, University of León, 24071 León, Spain; abers@unileon.es
[5] Department of Mining Exploitation, University of Oviedo, 33004 Oviedo, Spain; jloredo@uniovi.es
[*] Correspondence: jmenendezr@hunaser-energia.es; Tel.: +34-985107300

Received: 28 July 2020; Accepted: 20 August 2020; Published: 24 August 2020

Abstract: The European Union policy of encouraging renewable energy sources and a sustainable and safe low-carbon economy requires flexible energy storage systems (FESSs), such as pumped-storage hydropower (PSH) systems. Energy storage systems are the key to facilitate a high penetration of the renewable energy sources in the electrical grids. Disused mining structures in closed underground coal mines in NW Spain have been selected as a case study to analyze the construction of underground pumped-storage hydropower (UPSH) plants. Mine water, depth and subsurface space in closured coal mines may be used for the construction of FESSs with reduced environmental impacts. This paper analyzes the stability of a network of tunnels used as a lower water reservoir at 450 m depth in sandstone and shale formations. Empirical methods based on rock mass classification systems are employed to preliminarily design the support systems and to determinate the rock mass properties. In addition, 3D numerical modelling has been conducted in order to verify the stability of the underground excavations. The deformations and thickness of the excavation damage zones (EDZs) around the excavations have been evaluated in the simulations without considering a support system and considering systematic grouted rock bolts and a layer of reinforced shotcrete as support system. The results obtained show that the excavation of the network of tunnels is technically feasible with the support system that has been designed.

Keywords: mining structures; underground reservoir; empirical analysis; numerical modelling; energy storage; hydropower plants

---

1. Introduction

In the current evolving energy context, characterized by an increasing of variable renewable energy (VRE) in the electricity mix, the development of flexible energy storage systems (FESSs), such as pumped storage hydropower (PSH) or compressed air energy storage (CAES) plants are required. PSH is the most mature technology to provide ancillary services to the electrical grid [1]. PSH systems accounted for 150 GW worldwide in 2016 (40 GW in the European Union) [2] and the capacity could be 325 GW in 2030 [3].

Underground pumped storage hydropower (UPSH) is an alternative to store large amounts of electrical energy with low environmental impacts [4,5]. Other energy storage systems such as

Li-ion batteries are more efficient, but more expensive to install [6,7]. Geomechanical studies of the underground infrastructure are required to assess the technical feasibility of subsurface energy storage plants. Menendez et al. [8] carried out a stability analysis of the underground infrastructure of UPSH plants in closed mines. The stability of the powerhouse cavern and the effect of air pressure on the excavations (tunnels and air shafts) during the operation time were analyzed. Uddin and Asce [9] studied the behavior of a limestone mine as UPSH plant, at 671 m depth with a volume of 9.6 million m^3. Carneiro et al. [10] presented the opportunities for large-scale energy storage in geological formations in mainland Portugal. UPSH, CAES and gas storage systems (hydrogen and methane) were analyzed. Khaledi et al. [11] analyzed compressed air storage caverns in rock salt considering thermo-mechanical cyclic loading. The study concludes that the stability is affected by the operating pressure (10 MPa) and the increased creep rate accelerates the volume convergence. Liu et al. [12] applied empirical analysis and numerical methods to provide a support design for an underground water-sealed oil storage cavern. Chen et al. [13] developed a numerical model to analyze the stability of a small-spacing two-well salt cavern used as gas storage. Rutqvist et al. [14] investigated the thermodynamic and geomechanical performance of CAES systems in concrete-lined rock caverns. Zhu et al. [15] developed an equation for prediction of displacement at the key point on the high sidewalls of powerhouse caverns, considering four basic factors (the rock deformation modulus, the overburden depth, the height of the powerhouse and the lateral pressure coefficient of the initial stress). Harza [16] suggested the idea to use a closed underground mine as a lower reservoir and develop an UPSH plant in 1960. At the end of the 1960s, Swedish engineers proposed the exploitation of a surface upper reservoir and the construction of a new lower water reservoir in an underground rock cavern [17]. Sorensen suggested an optimistic future for the development of UPSH plants [18]. The Mount Hope project, located in northern New Jersey (USA) was proposed in 1975 [19]. It intended to use the facilities of a closed iron mine as a lower water reservoir, but it was never developed. In 1978, an UPSH project was presented with a lower reservoir formed by a network of 15 × 25 m elliptical tunnels, at a depth of 1000 m [20]. During the 1980s, a project to install an UPSH plant was proposed in the Netherlands [21], but the project was finally not developed due to the poor quality of the rock mass. Wong assessed the possibility of constructing UPSH plants in the Bukit Timah granite of Singapore [4]. Coal mining structures in closed underground coal mines in the Asturian Central Coal Basin (ACCB), NW Spain, have been proposed as a lower water reservoir of UPSH plants [22,23]. Recently, several studies have been also carried out in Germany to assess the possibilities to develop UPSH plants on closed underground coal mines in the Harz and Ruhr regions [24–26]. Pujades et al. [27] and Kitsikoudis et al. [28] proposed a closed slate mine in Belgium as a lower reservoir of an UPSH plant. The slate mine consists of nine large caverns with an available volume of 550,000 m^3.

This paper analyzes the stability of a network of tunnels as a lower water reservoir at 450 m depth in sandstone and shale rock formations in closed mines. The rock mass was characterized according to the rock mass quality (Q-System) and the rock mass properties were estimated. In addition, 3D numerical analysis using FLAC 3D have been performed to verify the stability of the excavations. The maximum thickness of the excavation damage zone (EDZ) and the vertical and horizontal displacements when the support system is applied have been compared to the unsupported case in central and transversal tunnels. The axial force, bending moment and shear force in the shotcrete layer have also analyzed. The results obtained show that the excavation of the network of tunnels is technically feasible with the support system that has been designed.

2. Methodology

2.1. Preliminary Energy Balance

The storable amount of electricity in UPSH systems depends on the reservoirs' capacity and the hydraulic net head [26]. Figure 1 shows the stored energy per cycle in a closed coal mine considering water masses between 0.1–0.5 Hm^3 and net heads between 100–600 mH$_2$O. A Francis turbine efficiency of 90% in turbine and pump modes has been considered [29]. Finally, the power output and the power input of the turbines depends on the time at full load. The energy storage reaches 480 MWh cycle^{-1} considering a gross head of 450 mH$_2$O and a useful water capacity of 450,000 m^3.

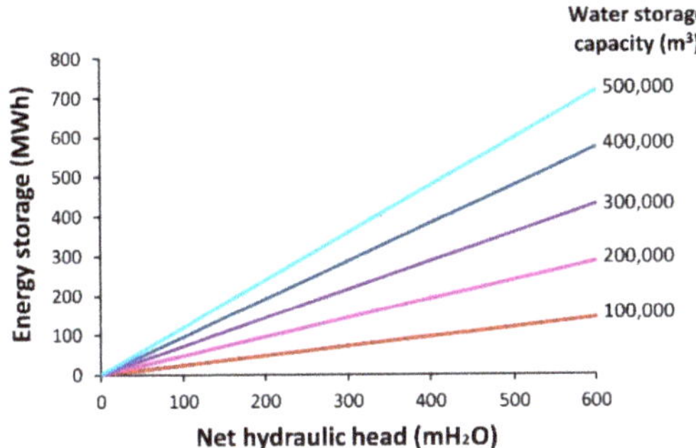

Figure 1. Storable amount of energy of an UPSH plant, considering a turbine efficiency of 90%.

2.2. Geology

The Asturian Central Coal Basin (ACCB) is a coal mining area located in NW Spain. It is precisely in the ACCB where the terrigenous carboniferous sediments are best represented, reaching a thickness close to 6000 m between the Namurian B and Westfalian D. The succession is divided at large scale into two sectors: one lower denominated Lena Group, characterized by limestones and thin seams of coal, and another superior called Sama Group, in which limestones are less abundant and there are levels of sandstones, shales and seams of exploitable bituminous and hard coal [8].

2.3. Underground Hydroelectric Power Plant

UPSH systems consist of two water reservoirs; the upper reservoir is located above ground, while the lower reservoir is underground. The scheme of an underground hydroelectric power plant and lower water reservoir is shown in Figure 2a. In the present work, a subsurface water reservoir conformed by a new network of tunnels with an arched roof and straight walls cross-section of 30 m^2 (Figure 2b) is considered. The water reservoir consists of a central tunnel connected to the ventilation shaft, and 200 m length of transversal tunnels with a distance between them of 20 m. To avoid water leakage, the tunnels' surface is covered with an impermeable high-strength membrane.

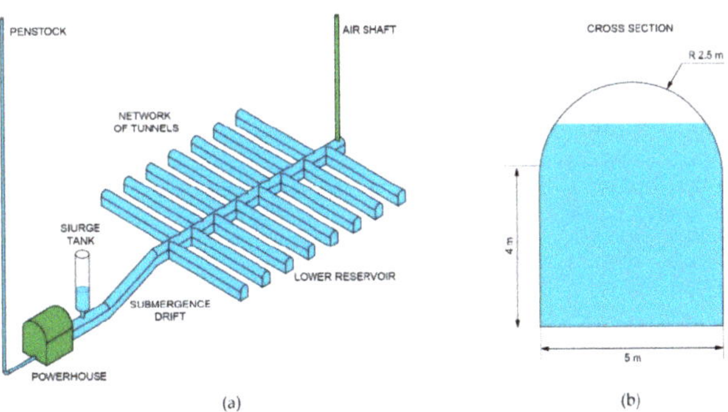

Figure 2. (**a**) Scheme of an UPSH plant with a lower water reservoir and an air shaft in a closed underground coal mine; (**b**) cross section of tunnels.

2.4. Empirical Analysis

The Q-system method was developed by Barton et al. [30] in the Norwegian Geotechnical Institute (NGI) to classify rock masses. The method is based on about 200 case histories of caverns and tunnels. The Q-system has been updated and it is now based on 1260 case records [31]. Barton carried out some changes and adapted it to give support recommendations due to the increasing use of steel fibre reinforced shotcrete (SFR) in underground excavation support [32]. The quality of rock masses (Q) is obtained by applying Equation (1).

$$Q = \frac{RQD}{J_n} \frac{J_r}{J_a} \frac{J_w}{SRF} \tag{1}$$

where RQD is the rock quality designation, J_n is the joint set number, J_r is the joint roughness, J_a is the joint alteration, J_w is the water reduction factor and SRF is the stress reduction factor. The Q value varies from 0.001 for exceptionally poor quality to 1000 for an exceptionally good quality rock mass. A stress-free form Q_N was defined by Goel et al. [33], which is given by Equation (2).

$$Q_N = \frac{RQD}{J_n} \frac{J_r}{J_a} J_W \tag{2}$$

Barton defined a new parameter Q_C (Equation (3)) to improve the correlation among the engineering parameters, where σ_{ci} is the strength of intact rock in MPa [32].

$$Q_{CN} = Q \frac{\sigma_{ci}}{100} \tag{3}$$

The bolt length (L_b) may be calculated in terms of the excavation width B, by the Equation (4), proposed by Barton et al. [30].

$$L_b = 2 + (0.15\ B) \tag{4}$$

2.5. Material Properties

The physical and mechanical properties of the shale and sandstone were obtained from laboratory testing at the University of Oviedo Rock Mechanics Laboratory on intact rock samples following International Society Rock Mechanics (ISRM) methods [34]. Table 1 shows the properties of the intact rocks and the geological strength index (GSI). The rock mass properties were determined by means of empirical methods considering a blast damage factor $D = 0.8$. Table 2 shows the rock mass properties of the shale and sandstone formations that have been employed as input data in the numerical analysis.

Table 1. Properties of intact rocks.

Lithology	Unit Weight, γ (KN m^{-3})	Intact Modulus, E_i (MPa)	Compressive Strength, σ_{ci} (MPa)	Intact Rock Constant (m_i)	GSI
Shale	23.83	28,988	59.7	9.2	35
Sandstone	25.87	43,650	150.8	15.4	50

Table 2. Rock mass properties considered in the models.

Lithology	Young's Modulus (MPa)	Poisson's Ratio	Tensile Strength (MPa)	Cohesion (MPa)	Friction Angle (°)
Shale	3287	0.27	0.048	0.82	37.7
Sandstone	13,409	0.25	0.226	2.02	52.7

2.6. Numerical Modelling

To verify the results of the empirical analysis, a 3D numerical analysis was developed. FLAC 3D commercial software was applied to obtain the deformations and the failure states considering the excavation of the network of tunnels and the support system [9,15,35]. Figure 3 shows the geometry of the models of the lower reservoir in the form of a network of tunnels that have been selected to conduct the numerical analysis. The size of the transversal tunnels model (Figure 3a) is 82.5 m long, 12.5 m wide and 69 m high, and the central tunnel–junction zone model (Figure 3b) is 174 m long, 12.5 m wide and 80 m high. It is assumed that there are roller boundaries at the bottom and along the sides and there is an unconstrained boundary at the top of model for application of uniform vertical stress, in order to simulate the primary stress field. The mesh was refined close to the contour of the excavations and gradually was coarser at positions outwards, for increasing the accuracy of the calculations.

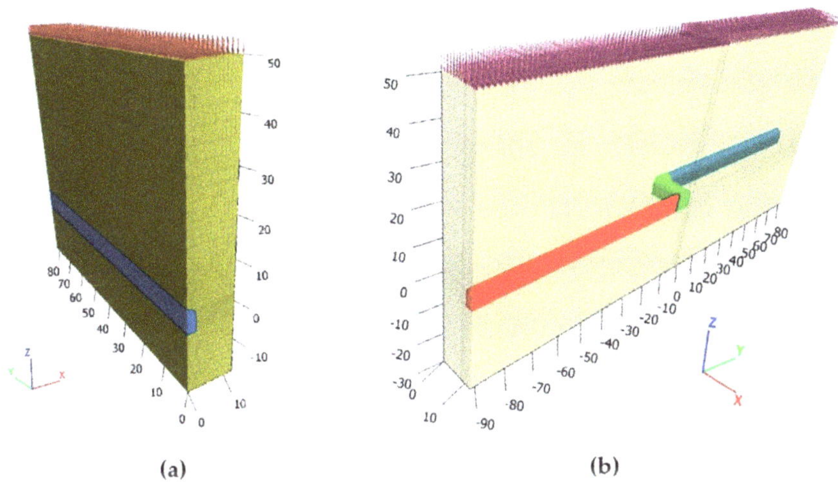

(a)   (b)

Figure 3. Geometry of the models of the network of tunnels: (**a**) transversal tunnels; (**b**) central tunnel-junction zone.

2.7. Simulation Procedure

For the models of the network of tunnels, the solution steps included: (1) establishment of the initial stress field by applying vertical and horizontal stresses and once equilibrium was reached deformations are reset, (2) excavation of the central and transversal tunnels, and (3) installation of the support system immediately after excavation. Failure states and total displacements caused by excavations and installation of the support system were analyzed. The rock mass is considered as homogenous and linearly elastic-perfectly plastic, according to the Mohr-Coulomb (M-C) field criterion. Plastic failure occurs when the shear stress on a certain plane reaches a limit called the shear yield stress. The M-C failure criterion is expressed in Equation (5).

$$\frac{1}{2}(\sigma_1 - \sigma_3) = c \cos\phi - \frac{1}{2}(\sigma_1 + \sigma_3) \sin\phi \tag{5}$$

where σ_1 and σ_3 are the maximum and minimum principal stresses, respectively, in MPa, c is the cohesion, in MPa, and ϕ is the friction angle, in degrees. The initial primary stress field at the tunnels' depth of 450 m is 10.35×10^6 Pa in the vertical direction and 5.17×10^6 Pa in the horizontal direction, considering a density of 23 KN m^{-3}. The in-situ horizontal stresses were obtained from the empirical equations from global data and a compilation of such data within the ACCB.

2.8. Numerical Modelling. Support System Design

Table 3 shows the design of the support system in the transversal tunnels and the central tunnel for the existing rock masses in the study area. The bolt lengths were calculated by Equation (4), as 2.75 m; therefore, the rock bolt length design was 3 m. In the shale formation grouted rock bolts were selected with the following specifications: 25 mm diameter, 245 KN (25 t) of load capacity, 3 m long, and spacing between bolts of 1.2 m. Furthermore, there was a layer of 150 mm thick reinforced shotcrete in the central tunnel and a layer that was 120 mm thick in transversal tunnels. In the sandstone formation, grouted rock bolts were selected with the following specifications: 25 mm in diameter, 245 kN of load capacity, 3 m long, and spacing between bolts of 1.5 m. A layer of 80 mm thick reinforced shotcrete in the central tunnel and 60 mm thick in transversal tunnels is also simulated.

Table 3. Support system design.

Shale Rock Mass	
Central tunnel	Systematic bolting 245 kN, ϕ = 25 mm, L = 3 m, Reinforced shotcrete 150 mm
Transversal tunnels	Systematic bolting 245 kN, ϕ = 25 mm, L = 3 m, Reinforced shotcrete 120 mm
Sandstone Rock Mass	
Central tunnel	Systematic bolting 245 kN, ϕ = 25 mm, L = 3 m, Reinforced shotcrete 80 mm
Transversal tunnels	Systematic bolting 245 kN ϕ = 25 mm, L = 3 m, Reinforced shotcrete 60 mm

3. Results and Discussion

3.1. Q-System Method

To estimate the support system it is necessary to establish an equivalent diameter for the tunnel, D_e, which is the width/height of excavation divided by the ESR (excavation support ratio), which for transversal tunnels can be established at 1.3, so D_e = 5.0. For the central tunnel, a value of ESR of 1.0 has been estimated, obtaining an equivalent diameter for the tunnel, D_e = 6.5. The evaluations of rock masses in sandstone and shale rock masses by using the Q-system and the GSI are summarized in Table 4. By applying Equation (1), in sandstone formation a value of Q of 1.27 is obtained, while in the shale rock mass the value of Q is 0.23.

Table 4. Evaluating results using the Q-system.

Lithology	RQD	J_n	J_r	J_a	J_w	SRF	Q	Q_N	GSI
Shale	31	15	1	2	0.4	1.8	0.23	0.41	35
Sandstone	48	12	1.4	2.2	0.5	1	1.27	1.27	50

The suggested support systems based on the Q-system for the central tunnel and the transversal tunnels are given in Table 5 following the recommendations proposed by Grimstad and Barton [31] using an equivalent diameter of 5.0 and 6.5 for transversal tunnel and central tunnel, respectively.

Table 5. Empirical analysis. Support categories for shale and sandstone formations.

Shale Rock Mass	
Central tunnel	Systematic grouted bolts spaced 1.4 m, L = 2.7 m, Fibre reinforced shotcrete 120 mm
Transversal tunnels	Systematic grouted bolts spaced 1.4 m, L = 2.7 m, Fibre reinforced shotcrete 100 mm
Sandstone Rock Mass	
Central tunnel	Systematic grouted bolts spaced 1.7 m, L = 2.4 m, Fibre reinforced shotcrete 80 mm
Transversal tunnels	Systematic grouted bolts spaced 1.7 m, L = 2.4 m, Fibre reinforced shotcrete 50 mm

3.2. Numerical Simulations

The results of the calculations that have been conducted in sandstone and shale rock masses, in transversal tunnels, the central tunnel and the junction zone between them are shown in this section. Table 3 shows the properties of the support system used in the numerical analysis (bolt diameter, length, spacing, load capacity, and thickness of the spayed concrete layer). Figure 4 shows the vertical and horizontal displacements in the transversal tunnels for shale rock mass according to the model indicated in Figure 3a. The maximum displacement is located at the tunnel walls (17.4 mm).

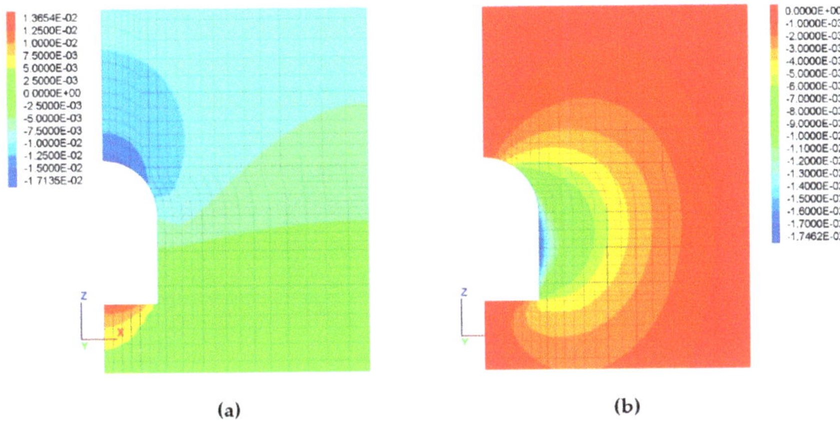

(a)   (b)

Figure 4. Transversal tunnels in shale formation: (**a**) vertical displacements [m]; (**b**) horizontal displacements [m].

Table 6 shows the summary of the numerical simulations results in transversal tunnels for shale and sandstone rock masses: vertical and horizontal displacements, thickness of the EDZ, axial load in rock bolts and axial force, bending moment and shear force in the shotcrete layer at the walls, roof and floor of tunnels for unsupported and supported cases. By installation of the support, the EDZ and total displacements (vertical and horizontal) notably decreased. Because of the lower quality of the rock mass, in shale formation the displacement values are higher than in sandstone rock mass. The maximum value of displacement reached 17.5 mm at the walls of the tunnels in shale rock mass. The maximum thickness of EDZ is located at the tunnel walls in shale formation, reaching 2.1 m. In shale formation, the vertical and horizontal displacements are reduced to 12.5 mm (27%) and 14.8 mm (15%), respectively. In sandstone formation, the vertical (roof) and horizontal (walls) displacements are reduced to 1.92 mm (34%) and 2.27 mm (22%), respectively. In addition, the thickness of the EDZ is also reduced by 38% in shale formation when the support system is applied.

Table 6. Numerical simulations results in transversal tunnels.

Variable		Unsupported Case		Supported Case	
		Shale	Sandstone	Shale	Sandstone
Vertical displacements (mm)		17.1	2.93	12.5	1.92
Horizontal displacements (mm)		17.4	2.91	14.8	2.27
Thickness of EDZ (m)		2.1	0.72	1.3	0.37
Axial load rock bolts (kN)				130.0	30.5
Shotcrete	Axial force (kN)			699.46	62.41
	Bending moment (KNm)			1.5	0.03
	Shear force (kN)			5.02	0.08

Figure 5a also shows the rock bolts force for sandstone rock mass. The maximum load in the grouted rock bolts reached 30.5 kN at the tunnel walls. Figure 5b shows the failure states for transversal tunnels in sandstone formation as the model reaches equilibrium. A combination of shear and tensile failure initiation mechanisms are observed at the floor and walls. The failure mode changes to only shear at the roof of the tunnels. The thickness of the EDZ is 0.72 m in sandstone formation. The simulation results indicate that the designed support systems can guarantee the tunnels stability.

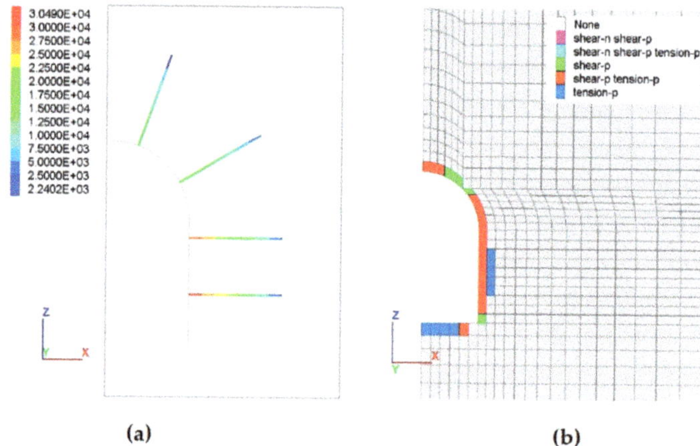

Figure 5. Transversal tunnels in sandstone formation: (**a**) rock bolts axial force [N]; (**b**) state of plasticity.

Figure 6a indicates the rock bolt axial force for transversal tunnels in shale formation. The maximum bolt force is located at the walls of the tunnels (130 kN). Figure 6b shows the plasticity zones for shale formation after reaching balance. Shear failure is observed at the roof, walls and floor of the transversal tunnels. The thickness of the EDZ is larger in shale rock mass, reaching a value of 2.1 m. After support installation the thickness of the EDZ is reduced to 1.3 m in shale formation and 0.37 m in sandstone formation.

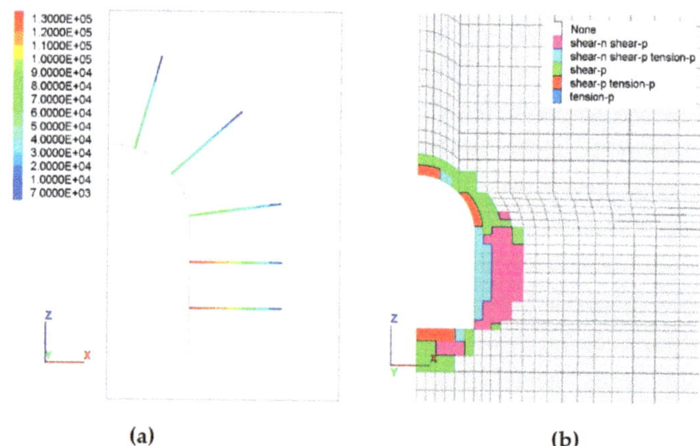

Figure 6. Transversal tunnels in shale formation: (**a**) rock bolts axial force [N]; (**b**) state of plasticity.

In addition to the stability analysis in the transversal tunnels, the state of plasticity and the total displacements have also been analyzed in the central tunnel. Figure 7 shows the axial force in the rock bolts and the state of plasticity in the central tunnel for shale formation. A combination of shear and tensile failure initiation mode is seen at the roof, walls and floor of the central tunnel. Table 7 shows the summary of the numerical simulations results in the central tunnel in shale and sandstone formations for unsupported and supported cases, with the support system indicated previously in Table 3.

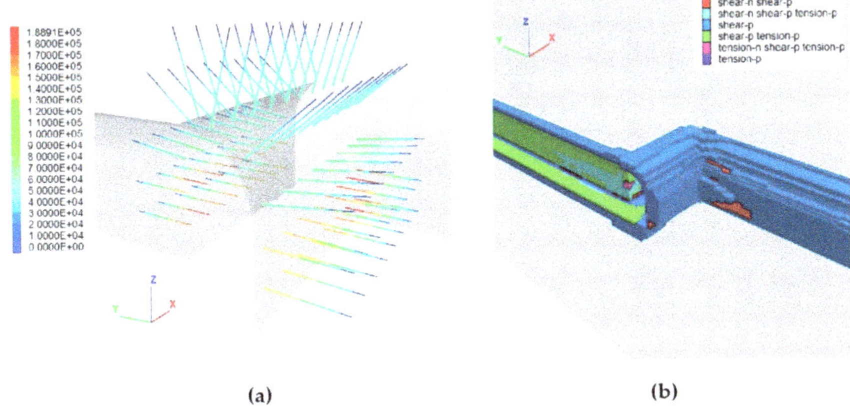

Figure 7. The central tunnel in shale formation: (**a**) rock bolts axial force [N]; (**b**) state of plasticity.

Table 7. Numerical simulation results in the central tunnel.

Variable	Unsupported Case		Supported Case	
	Shale	Sandstone	Shale	Sandstone
Vertical displacements (mm)	20.92	3.29	12.97	2.51
Horizontal displacements (mm)	18.32	2.72	11.62	1.54
Thickness of EDZ (m)	2.9	1.1	1.75	0.65
Axial load rock bolts (kN)			188.9	18.5
Shotcrete — Axial force (kN)			883.64	56.29
Shotcrete — Bending moment (KNm)			8.08	0.16
Shotcrete — Shear force (kN)			22.35	0.42

The maximum value of displacement reached is 20.92 mm at the roof of the tunnel in shale rock mass. The thickness of the EDZ reached is 2.9 m is shale formation for the unsupported case. The vertical and horizontal displacements in sandstone formation reach 3.29 and 2.72 mm, respectively. The area of the EDZ and the maximum displacements notably decreased when the support system is applied. In shale formation, the vertical (roof) and horizontal displacements decreased down to 12.97 mm (38%) and 11.62 mm (36.5%), respectively. The rock bolt load reaches a value of 188.9 kN at the walls, while the elastic capacity of the rock bolts is 245 kN (safety factor of 1.29). The axial force, bending moment and shear force in reinforced shotcrete layer have also been analyzed, reaching 883.64 kN, 8.08 kNm, and 22.35 kN, respectively, in shale rock mass.

Finally, the stability analysis was carried out in the junction zone between the central and transversal tunnels. Figure 8a shows the horizontal displacement and Figure 8b shows the shear force in the fibre reinforced shotcrete layer.

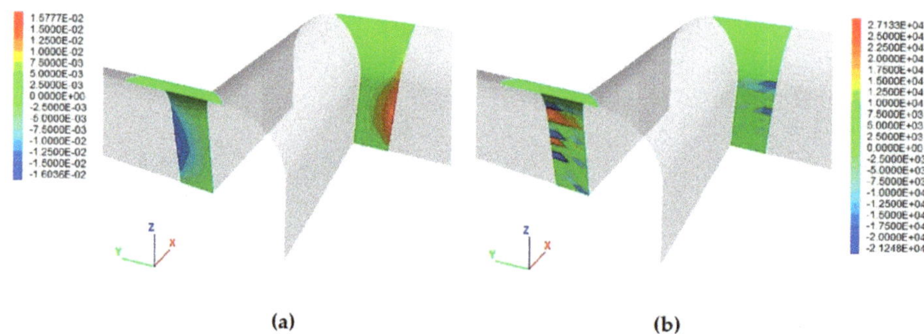

Figure 8. Junction zone in shale formation: (**a**) horizontal displacements; (**b**) shear force in shotcrete.

Table 8 shows the results for the junction zone between the central and transversal tunnels. The maximum value of displacement reached 18.20 mm at the roof of the tunnels in shale rock mass. By installation of support, the thickness of the EDZ and deformations are reduced. In shale formation, the vertical and horizontal displacements decreased down to 13.84 mm at the roof and 11.51 mm in the walls, when the support system is applied. The axial force, bending moment and shear force in reinforced shotcrete layer have also been analyzed in the junction zone, reaching 1570 kN, 9.31 kNm, and 27.13 kN, respectively, in shale rock mass. The shear force indicated in Figure 8b shows peaks due to the head of the rock bolts.

Table 8. Numerical simulation results in the junction zone.

Variable		Unsupported Case		Supported Case	
		Shale	Sandstone	Shale	Sandstone
Vertical displacements (mm)		18.20	2.96	13.84	1.76
Horizontal displacements (mm)		16.03	2.17	11.51	1.82
Thickness of EDZ (m)		2.9	1.2	1.8	0.7
Axial load rock bolts (kN)				160.4	29.6
Shotcrete	Axial force (kN)			1,570.0	168.72
	Bending moment (KNm)			9.31	0.21
	Shear force (kN)			27.13	0.59

Figure 9 shows the rock bolts force in the central tunnel and the transversal tunnels depending on the distance from the rock bolt head. Rock bolts located at the walls in shale and sandstone formations have been selected. The maximum axial load is reached in shale rock mass, at the central tunnel. It has been observed that in all scenarios the axial load reached is less than 245 kN (load capacity designed).

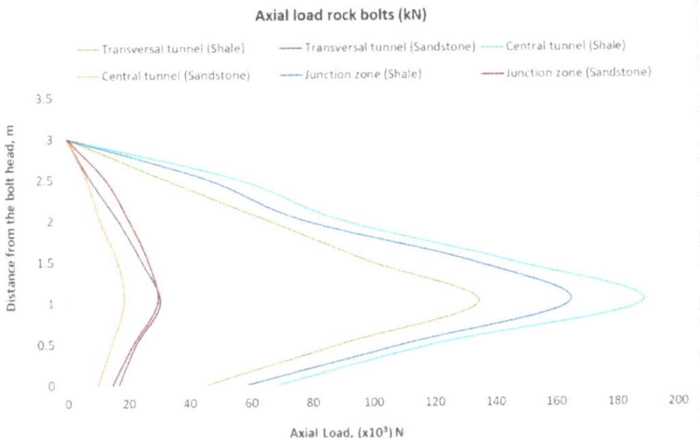

Figure 9. Axial load in rock bolts. Shale and sandstone rock masses.

4. Conclusions

Flexible energy storage systems allow increasing the generation of electricity by means of intermittent renewable energy sources. A closed coal mine in the Asturian Central Coal Basin (NW Spain) is proposed as a subsurface water reservoir of underground pumped storage hydropower plants. Underground pumped storage hydropower plants provide a large amount of electrical energy with rapid response and low environmental impacts. An underground reservoir conformed by a network of tunnels with an arched roof cross section of 30 m^2 has been designed. For the construction of the underground water reservoir it is necessary to analyze the stability of the excavation of the network of tunnels. Transversal tunnels, the central tunnel and junction zones have been analyzed.

Empirical analysis and three-dimensional numerical simulations have been carried out in this work. According to the empirical analysis, grouted rock bolts and a layer of fibre reinforced shotcrete were recommended as the support system. The rock mass properties used as input parameters for numerical modeling were obtained from laboratory tests and estimated from rock mass classification systems. In the numerical model, the state of plasticity, the axial load in rock bolts and the deformations around the tunnels were analyzed. In addition, the axial force, the bending moment and the shear force were also checked in the reinforced shotcrete layer. The maximum displacements and thickness of the excavation damage zone were reached in shale formation. As shown by the numerical simulations, the proposed support from empirical analysis was feasible. By applying the designed support system, the area of the excavation damage zone and the maximum displacements significantly decreased. The results of the numerical analysis show that no significant failure is expected. All this shows that a combination of empirical and numerical methods is appropriate to design a proper support system in underground infrastructures.

Based on further rock lab testing, a softening factor for the rock mass after failure should be applied in order to get more precise predictions for displacements and the extension of the excavation damage zone as well as the support loads.

Author Contributions: Conceptualization, J.M.; investigation, J.M., F.S., H.K., A.B.S. and J.L.; methodology, J.M. and F.S.; software, J.M., F.S.; validation, J.M, F.S., H.K. and A.B.S.; writing original draft, J.M. and F.S.; writing review and editing, J.M., A.B.S. and J.L.; supervision, F.S., H.K., A.B.S. and J.L. All authors have read and agreed to the published version of the manuscript.

Funding: This research received no external funding.

Conflicts of Interest: The authors declare no conflict of interest.

Nomenclature

3D	Three-dimensional
ACCB	Asturian Central Coal Basin
CAES	Compressed air energy storage
EDZ	Excavation damage zone
ESR	Excavation support ratio
FESS	Flexible energy storage system
GSI	Geological strength index
MC	Mohr-Coulomb
PSH	Pumped-storage hydropower
RQD	Rock quality designation,
SFR	Steel fibre reinforced shotcrete
SRF	Stress reduction factor
UPSH	Underground pumped-storage hydropower
VRE	Variable renewable energies

References

1. Menéndez, J.; Fernández-Oro, J.M.; Loredo, J. Economic feasibility of underground pumped storage hydropower plants providing ancillary services. *Appl. Sci.* **2020**, *10*, 3947. [CrossRef]
2. U.S. Energy Information Administration (EIA). International Energy Statistics. Available online: https://www.iea.org/ (accessed on 20 August 2020).
3. International Renewable Energy Agency. *Renewables and Electricity Storage: A Technology Roadmap for Remap 2030*; IREA: Abu Dhabi, UAE, 2015.
4. Wong, I.H. An underground pumped storage scheme in the bukit Timah granite of Singapore. *Tunn. Undergr. Space Technol.* **1996**, *11*, 485–489. [CrossRef]
5. Kucukali, S. Finding the most suitable existing hydropower reservoirs for the development of pumped-storage schemes: An integrated approach. *Renew. Sustain. Energy Rev.* **2014**, *37*, 502–508. [CrossRef]
6. Minakshi, M.; Mitchell, D.R.G.; Jones, R.T.; Pramanik, N.C.; Jean-Fulcrand, A.; Garnweitner, G. A hybrid electrochemical energy storage device using sustainable electrode materials. *Chem. Sel.* **2020**, *5*, 1597–1606. [CrossRef]
7. Divakaran, A.M.; Hamilton, D.; Manjunatha, K.N.; Minakshi, M. Design, development and thermal analysis of reusable Li-Ion battery module for future mobile and stationary applications. *Energies* **2020**, *13*, 1477. [CrossRef]
8. Menéndez, J.; Schmidt, F.; Konietzky, H.; Fernández-Oro, J.M.; Galdo, M.; Loredo, J.; Díaz-Aguado, M.B. Stability analysis of the underground infrastructure for pumped storage hydropower plants in closed coal mines. *Tunn. Undergr. Space Technol.* **2019**, *94*. [CrossRef]
9. Uddin, N.; Asce, M. Preliminary design of an underground reservoir for pumped storage. *Geotech. Geol. Eng.* **2003**, *21*, 331–355. [CrossRef]
10. Carneiro, J.F.; Matos, C.R.; van Gessel, S. Opportunities for large-scale energy storage in geological formations in mainland Portugal. *Renew. Sustain. Energy Rev.* **2019**, *99*, 201–211. [CrossRef]
11. Khaledi, K.; Mahmoudi, E.; Datcheva, M.; Schanz, T. Analysis of compressed air storage caverns in rock salt considering thermomechanical cyclic loading. *Environ. Earth Sci.* **2016**, *75*. [CrossRef]
12. Liu, J.; Zhao, X.D.; Zhang, S.J.; Xie, L.K. Analysis of support requirements for underground water-sealed oil storage cavern in China. *Tunn. Undergr. Space Technol.* **2018**, *71*, 36–46. [CrossRef]
13. Chen, J.; Lu, D.; Liu, W.; Fan, J.; Jiang, D.; Yi, L.; Kang, Y. Stability study and optimization design of small-spacing two-well (SSTW) salt caverns for natural gas storages. *J. Energy Storage* **2020**, *27*. [CrossRef]
14. Rutqvist, J.; Kim, H.M.; Ryu, D.W.; Synn, J.H.; Song, W.K. Modeling of coupled thermodynamic and geomechanical performance of underground compressed air energy storage in lined rock caverns. *Int. J. Rock Mech. Min. Sci.* **2012**, *52*, 71–81. [CrossRef]
15. Zhu, W.S.; Li, X.J.; Zhang, Q.B.; Zheng, W.H.; Xin, X.L.; Sun, A.H.; Li, S.C. A study on sidewall displacement prediction and stability evaluations for large underground power station caverns. *Int. J. Rock Mech. Min. Sci.* **2010**, *47*, 1055–1062. [CrossRef]

16. Harza, R.D. Hydro and pumped storage for peaking. *Power Eng.* **1960**, *64*, 79–82.
17. Isaaksson, G.; Nilsson, D.; Sjostrand, T. Pumped storage power plants with underground lower reservoirs. In Proceedings of the Seventh World Power Conference, Moscow, Russia, 20–24 August 1968; Section 2. p. 160.
18. Sorensen, K.E. Underground reservoirs: Pumped storage of the future? *Civ. Eng.* **1969**, *39*, 66–70.
19. Dames and Moore. An Assessment of Hydroelectric Pumped Storage. National Hydroelectric Power Resources Study. Under Contract No. DACW-31-80 C-0090 to US Army Corps of Engineers. 1981. Available online: https://www.iwr.usace.army.mil/Portals/70/docs/iwrreports/IWR019-000001-000517.pdf (accessed on 20 August 2020).
20. Tam, S.W.; Blomquist, C.A.; Kartsounes, G.T. Underground pumped hydro storage—An overview. *Energy Sources* **1979**, *4*, 329–351. [CrossRef]
21. Braat, K.B.; Van Lohuizen, H.P.S.; De Haan, J.F. Underground pumped hydro-storage project for the Netherlands. *Tunn. Tunn.* **1985**, *17*, 19–22.
22. Menendez, J.; Loredo, J.; Fernandez-Oro, J.M.; Galdo, M. Energy storage in underground coal mines in NW Spain: Assessment of an underground lower water reservoir and preliminary energy balance. *Renew. Energy* **2019**, *134*, 1381–1391. [CrossRef]
23. Menéndez, J.; Loredo, J.; Fernández-Oro, J.M.; Galdo, M. Underground Pumped-Storage Hydro Power Plants with Mine Water in Abandoned Coal Mines. In Proceedings of the 2017 International Mine water Association Congress 'Mine Water and Circular Economy', Lappeenranta, Finland, 25–30 June 2017; Wolkersdorfer, C., Sartz, L., Sillanpää, M., Häkkinen, A., Eds.; LUT Scientific and Expertise Publications: Lappeenranta, Finland, 2017; pp. 6–13.
24. Wessel, M.; Madlener, R.; Hilgers, C. Economic Feasibility of Semi-Underground Pumped Storage Hydropower Plants in Open-Pit Mines. *Energies* **2020**, *13*, 4178. [CrossRef]
25. Luick, H.; Niemann, A.; Perau, E.; Schreiber, U. Coalmines as Underground Pumped Storage Power Plants (UPP)—A contribution to a sustainable energy supply? *Geophys. Res. Abstr.* **2012**, *14*, 4205.
26. Madlener, R.; Specht, J.M. *An Exploratory Economic Analysis of Underground Pumped-Storage Hydro Power Plants in Abandoned Coal Mines*; FCN Working Paper No. 2/2013; FCN: Aachen, Germany, 2013.
27. Pujades, E.; Orban, P.; Archambeau, P.; Kitsikoudis, V.; Erpicum, S.; Dassargues, A. Underground pumped-storage hydropower (UPSH) at the martelange mine (Belgium): Interactions with groundwater flow. *Energies* **2020**, *13*, 2353. [CrossRef]
28. Kitsikoudis, V.; Archambeau, P.; Dewals, B.; Pujades, E.; Orban, P.; Dassargues, A.; Pirotton, M.; Erpicum, S. Underground Pumped-Storage Hydropower (UPSH) at the Martelange Mine (Belgium): Underground Reservoir Hydraulics. *Energies* **2020**, *13*, 3512. [CrossRef]
29. Menéndez, J.; Fernandez-Oro, J.M.; Galdo, M.; Loredo, J. Efficiency analysis of underground pumped storage hydropower plants. *J. Energy Storage* **2020**, *28*. [CrossRef]
30. Barton, N.; Lien, R.; Lunde, J. Engineering classification of rock masses for the design of tunnel support. *Rock Mech.* **1974**, *6*, 189–236. [CrossRef]
31. Grimstad, E.; Barton, N. Updating the Q-system for NMT. In Proceedings of the Int. Symp. Sprayed Concrete, Fagernes, Norway, 17–21 October 1993; Norwegian Concrete Association: Oslo, Norway, 1993; pp. 46–66.
32. Barton, N. Some new Q-value correlations to assist in site characterization and tunnel design. *Int. J. Rock Mech. Min. Sci.* **2002**, *39*, 185–216. [CrossRef]
33. Goel, R.K.; Jethwa, J.L.; Paithankar, A.G. Indian experiences with Q and RMR systems. *Tunn. Undergr. Space Technol.* **1995**, *10*, 97–109. [CrossRef]
34. Ulusay, R. *The ISRM Suggested Methods for Rock Characterization, Testing and Monitoring: 2007–2014*; Springer International Publishing: Cham, Switzerland, 2015.
35. Ray, A.K. Influence of cutting sequence on development of cutters and roof falls in underground coal mine. In Proceedings of the 28th International Conference on Ground Control in Mining, Morgantown, WV, USA, 28–30 July 2009; Peng, S.S., Ed.; West Virginia University: Morgantown, WV, USA, 2009.

MDPI
St. Alban-Anlage 66
4052 Basel
Switzerland
Tel. +41 61 683 77 34
Fax +41 61 302 89 18
www.mdpi.com

Applied Sciences Editorial Office
E-mail: applsci@mdpi.com
www.mdpi.com/journal/applsci